Integrated Optics

Integrated Optics

Edited by
Dietrich Marcuse
Member, Technical Staff
Bell Telephone Laboratories

A volume in the IEEE PRESS Selected Reprint Series, prepared under the sponsorship of the IEEE Microwave Theory and Techniques Group and the IEEE Electron Devices Group.

IEEE PRESS

The Institute of Electrical and Electronics Engineers, Inc. New York

International Standard Book Numbers:
Clothbound: 0-87942-021-9
Paperbound: 0-87942-022-7

Library of Congress Catalog Card Number 72-92691

PRINTED IN THE UNITED STATES OF AMERICA

Contents

Introduction 1

Part 1—Review Articles

1. A Survey of Integrated Optics, S. E. Miller 5
   IEEE Journal of Quantum Electronics, February 1972
2. Light Waves in Thin Films and Integrated Optics, P. K. Tien 12
   Applied Optics, November 1971

Part 2—Two-Dimensional Optics

1. Optical Guided-Wave Focusing and Diffraction, R. Schubert and J. H. Harris 33
   Journal of the Optical Society of America, February 1971
2. Geometrical Optics in Thin Film Light Guides, R. Ulrich and R. J. Martin 41
   Applied Optics, September 1971

Part 3—Waveguide Papers

1. A Circular-Harmonic Computer Analysis of Rectangular Dielectric Waveguides, J. E. Goell 53
   Bell System Technical Journal, September 1969
2. Dielectric Rectangular Waveguide and Directional Coupler for Integrated Optics, E. A. J. Marcatili 80
   Bell System Technical Journal, September 1969
3. Wave Propagation in Thin-Film Optical Waveguides Using Gyrotropic and Anisotropic Materials as Substrates, S. Wang, M. L. Shah, and J. D. Crow 112
   IEEE Journal of Quantum Electronics, February 1972

Part 4—Mode Launching

1. Modes of Propagating Light Waves in Thin Deposited Semiconductor Films, P. K. Tien, R. Ulrich, and R. J. Martin 119
   Applied Physics Letters, May 1, 1969
2. Evanescent Field Coupling into a Thin-Film Waveguide, J. E. Midwinter 123
   IEEE Journal of Quantum Electronics, October 1970
3. Variable Tunneling Excitation of Optical Surface Waves, J. H. Harris and R. Schubert 131
   IEEE Transactions on Microwave Theory and Techniques, March 1971
4. Holographic Thin Film Couplers, H. Kogelnik and T. P. Sosnowski 138
   Bell System Technical Journal, September 1970
5. Grating Coupler for Efficient Excitation of Optical Guided Waves in Thin Films, M. L. Dakss, L. Kuhn, P. F. Heidrich, and B. A. Scott 144
   Applied Physics Letters, June 15, 1970
6. Experiments on Light Waves in a Thin Tapered Film and a New Light-Wave Coupler, P. K. Tien and R. J. Martin 146
   Applied Physics Letters, May 1, 1971

Part 5—Radiation Losses Caused by Bends and Surface Roughness

1. Bends in Optical Dielectric Guides, E. A. J. Marcatili 151
   Bell System Technical Journal, September 1969
2. Optical Waveguide Scattering and Griffith Microcracks, J. H. Harris, D. P. GiaRusso, and R. Schubert 181
   Proceedings of the IEEE, July 1971
3. Properties of Irregular Boundary of RF Sputtered Glass Film for Light Guide, Y. Suematsu, K. Furuya, M. Hakuta, and K. Chiba 183
   Proceedings of the IEEE, June 1972
4. Mode Conversion Caused by Surface Imperfections of a Dielectric Slab Waveguide, D. Marcuse 185
   Bell System Technical Journal, December 1969
5. Radiation Losses of Dielectric Waveguides in Terms of the Power Spectrum of the Wall Distortion Function, D. Marcuse 214
   Bell System Technical Journal, December 1969

Part 6—Integrated Optics Lasers

1. Junction Lasers Which Operate Continuously at Room Temperature, I. Hayashi, M. B. Panish, P. W. Foy, and S. Sumski 227
   Applied Physics Letters, August 1, 1970
2. Surface Lasers, F. Varsanyi 230
   Applied Physics Letters, September 15, 1971
3. Stimulated Emission in a Periodic Structure, H. Kogelnik and C. V. Shank 232
   Applied Physics Letters, February 15, 1971
4. A Thin-Film Ring Laser, H. P. Weber and R. Ulrich 235
   Applied Physics Letters, July 15, 1971

Part 7—Modulators and Light Deflectors

1. Deflection of an Optical Guided Wave by a Surface Acoustic Wave, L. Kuhn, M. L. Dakss, P. F. Heidrich, and B. A. Scott 241
   Applied Physics Letters, September 15, 1970
2. Voltage-Induced Optical Waveguide, D. J. Channin 244
   Applied Physics Letters, September 1, 1971
3. Observation of Propagation Cutoff and its Control in Thin Optical Waveguides, D. Hall, A Yariv, and E. Garmire 246
   Applied Physics Letters, August 1, 1970
4. Efficient GaAs-Al_xGa_{1-x} as Double-Heterostructure Light Modulators, F. K. Reinhart and B. I. Miller 249
   Applied Physics Letters, January 1, 1972
5. Digital Electro-Optic Grating Deflector and Modulator, J. M. Hammer 252
   Applied Physics Letters, February 15, 1971

Part 8—Parametric Devices

1. Optical Second Harmonic Generation in Form of Coherent Cerenkov Radiation from a Thin-Film Waveguide, P. K. Tien, R. Ulrich, and R. J. Martin 257
   Applied Physics Letters, November 15, 1970
2. Wideband CO_2 Laser Second Harmonic Generation Phase Matched in GaAs Thin-Film Waveguides, D. B. Anderson and J. T. Boyd 261
   Applied Physics Letters, October 15, 1971
3. Low-Power Quasi-cw Raman Oscillator, E. P. Ippen 264
   Applied Physics Letters, April 15, 1970

Part 9—Film Deposition Techniques

1. Optical Waveguides Formed by Proton Irradiation of Fused Silica, E. R. Schineller, R. P. Flam, and D. W. Wilmot 269
   Journal of the Optical Society of America, September 1968
2. Properties of Ion-Bombarded Fused Quartz for Integrated Optics, R. D. Standley, W. M. Gibson, and J. W. Rodgers 275
   Applied Optics, June 1972
3. Sputtered Glass Waveguide for Integrated Optical Circuits, J. E. Goell and R. D. Standley 279
   Bell System Technical Journal, December 1969
4. Thin Organosilicon Films for Integrated Optics, P. K. Tien, G. Smolinsky, and R. J. Martin 283
   Applied Optics, March 1972
5. Embossed Optical Waveguides, R. Ulrich, H. P. Weber, E. A. Chandross, W. J. Tomlinson, and E. A. Franke 289
   Applied Physics Letters, March 15, 1972

Author Index 292
Subject Index 293
Editor's Biography 296

Integrated Optics

Introduction

The invention of the laser has given an enormous stimulus to the entire field of optics. Many traditional domains of optics, such as Raman spectroscopy, have benefited from the availability of a powerful, coherent light source. Dormant applications, like holography, have been brought to life, and many entirely new fields have been opened up. One of these new offspring of an old art is integrated optics. Like integrated electronics, integrated optics does not really provide new possibilities in principle, but it makes practical many well-known applications that without this technique would remain too cumbersome or expensive to be utilized.

Traditional optical apparatus must be aligned with extreme accuracy and is thus susceptible to the smallest amount of vibration and temperature change. Integrated optics concentrates light in thin-film waveguides that are deposited on the surface or inside a substrate. Because of the short wavelength of light, dielectric light waveguides can be extremely small in their dimensions. The reduced size of integrated optics circuits thus makes it possible to achieve a much higher density of components compared to conventional optical equipment aligned on steel rails or on heavy optical benches. An additional advantage of the small size and rugged construction of dielectric light waveguides is their insensitivity to vibrations and temperature changes in their environment.

Integrated optics circuits can be made two dimensional if the light is allowed to spread out and propagate in a thin dielectric layer on top of a glass substrate. These circuits become one dimensional if the dielectric layers are deposited as thin strips that guide light not only in the dimension of the substrate surface but also confine it to a narrow region defined by the narrow waveguide.

The field of integrated optics is still in its infancy and thus it is too early to say what functions it will perform most successfully. Some proponents hope to utilize it for two-dimensional optical data processing. Lenses, prisms, and gratings can be built by changing either the thickness or the dielectric constant of the thin two-dimensional film guiding the light. It is thus possible to construct the two-dimensional counterpart of optical analog computers that would perform such functions as Fourier transformations and taking convolutions. Other applications involve light modulation by interactions of the light beam and acoustical waves, both being confined to the region in or near the dielectric layer that is deposited on a substrate. Light modulation and deflection by means of the electrooptic effect in optically nonlinear media appear particularly promising. Either the light guiding film or the substrate (or both) can be made from electrooptic materials. Because of the small dimensions, high electric fields per unit length can be obtained with moderate voltages. This advantage of integrated optics makes it possible to build modulators and light deflectors using but a few volts and very little power of the modulating signal. The possibility of building optical filters, directional couplers, and even lasers in very small dimensions makes integrated optics a natural field for some or all terminal functions of communications apparatus.

This brief introduction to the possibilities of integrated optics makes it clear that this new field has not yet advanced to the point of commercial utilization. Many of the possible applications are only suggestions that have not yet proven practical. This raises the question of the need for a reprint book in such a new field. It is the editor's opinion that the need for reprint books is particularly pressing in this new field. Once a subject has matured to the point where many textbooks become available, the need for a reprint book declines. It appears to the editor that the purpose of a reprint book is to bridge the gap between the time when only research papers are available in technical journals and the period of maturity where textbooks have been written explaining the subject matter in detail. A reprint book collects the principal ideas and contributions of research workers between two covers and saves the student or worker the inconvenience of having to find the pertinent literature by sifting through the library. This reprint book on integrated optics is intended to provide the workers in the field with easy access to some of the important papers and to stimulate students and people near the field of integrated optics to join and help in the exploration and development of a new, exciting area of optics.

The papers collected in this reprint volume have been selected in the hope of providing a representative sample of the work presently being done in integrated optics. This collection of papers cannot be complete; the number of papers available is sufficient to fill several large volumes. Having to make a selection from the wealth of available material is difficult. Many excellent contributions must be left out. The editor has divided the field into parts and has tried to include the most representative papers, avoiding repetitions. The choice of which paper was to be included and which was to be rejected is subjective and represents the editor's prejudice.

The book begins with two review articles that provide a good introduction to the aims and aspirations of integrated optics. The two papers in Part 2 deal with the subject of two-dimensional optics. The ways of constructing lenses, prisms, and diffraction gratings are discussed and demonstrated with experimental results. Part 3 provides three theoretical papers that analyze the properties of optical waveguides for integrated optics applications utilizing isotropic and anisotropic materials.

A waveguide is useless unless it can be excited. The part on mode launching describes several promising and successful methods of mode excitation and their analytical treatment. The prism coupler is probably the most widely used method of feeding light into integrated optics waveguides. However, the grating coupler and the tapered waveguide coupler may well prove to be more suitable in practice; they incorporate the launching device directly into the integrated optics circuit without the need of positioning and aligning an external element, the prism, on the film.

Radiation losses are a serious problem of dielectric optical waveguides. Several theoretical and experimental papers deal with the problems of radiation losses caused by bent waveguides and unintentional surface roughness.

Integrated optics lasers are sure to become one of the most important areas of future research. This field is certainly only in its infancy since dye lasers as well as semiconductor lasers still suffer from the problem of short lifetimes. Four papers describe the state of the art at present.

Modulators and light deflectors have been demonstrated by various means. The five papers presented in this collection describe the most important ideas that have been tried to date.

Parametric interactions in integrated optics waveguides combine two typical advantages of this new field. Because of the high field intensities possible in dielectric waveguides the need for high pump powers is reduced. An additional advantage is the possibility of achieving phase matching between the various waves involved in the parametric interaction by utilizing the dispersion of optical waveguides. Parametric devices using Gaussian beams in essentially infinitely extended media must rely on the anisotropy of electrooptic materials or on anomalous dispersion to achieve phase matching. Integrated optics circuits provide additional flexibility by offering the possibility of using waveguide dispersion for the purpose of phase matching. Three papers of our collection deal with parametric devices which are surely in their infancy at present.

The final part presents methods of film deposition that are crucial if integrated optics is to be a success. Several promising methods exist at this time. Deposition of glass layers on glass substrates has been tried. Deposition of organic films by several techniques appears promising. Ion implantation offers the possibility of defining dielectric waveguides inside the substrate, removing them from the surface. The new technique of embossing dielectric waveguides into a plastic substrate seems to offer the possibility of a cheap method of mass producing integrated optics circuits.

If the similarity in name is any indication of a similar fate, we might expect that integrated optics may burgeon into as profitable and exciting an area as its cousin, integrated electronics. Many new and hitherto unthought of possibilities and applications are sure to emerge in the future. We thus expect that the literature on this field will grow to many times its present size.

It is hoped that the present volume will make a useful contribution in helping to direct attention to the new field of integrated optics, thereby winning it new adherents. It is also hoped that it will help to inform students and workers in the field of the achievements that have already been obtained and of the potential that still lies in the future.

Part 1
Review Articles

A Survey of Integrated Optics

STEWART E. MILLER, FELLOW, IEEE

Abstract—In order for optical transmission systems to compete successfully with other techniques, the circuits must be comparably small, comparably rugged, and comparably priced. The avenue to achieving these characteristics is via miniature guided-wave structures in contrast to beam-mode propagation via lenses and mirrors.

Passive miniature transmission lines with satisfactory losses have been formed in glass and fused silica using ion bombardment and reactive sputtering.

Certain crystalline systems seem well adapted to thin-film nonlinear interactions at low absolute-power levels. Nonlinear liquids in hollow glass fibers can be used to achieve interaction lengths of a meter or more with the field concentrated to a few wavelengths width in the transverse direction.

Photolithographic techniques based on visible or ultraviolet exposure of the photoresist yield waveguide-edge roughness greater than that needed for optical circuitry. However, electron beam exposure of the photoresist seems capable of yielding satisfactory results.

THIS paper is concerned with an exciting new field in laser applications going under the heading Integrated Optics. The activity is epitomized by miniaturization of the needed optical communication components with concomitant improvement in efficiency, ruggedness, and stability.

Like many new activities, this one has its roots in work that predates the objectives and needs that are now recognized. Specifically, reference is made to the pioneering work of Osterberg and Smith (Fig. 1) [1]. Using incoherent visible light an image was transmitted along the surface of a sheet of glass, through a gap and into another sheet, prism, and focusing apparatus. The surface effect was detectable in most glasses, but was enhanced in Pilkington float glass, which had a graded increase in index of refraction close to the surface; this produced an excellent surface waveguide.

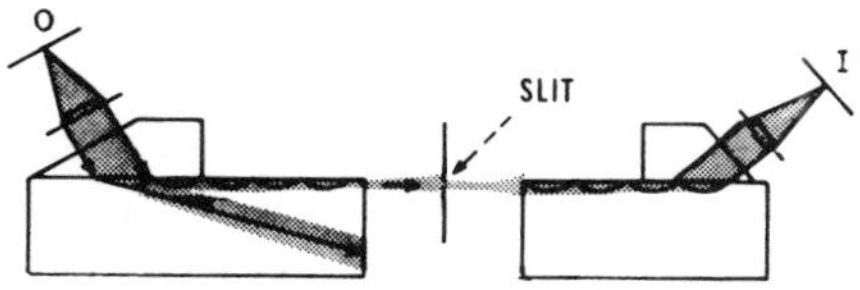

Fig. 1. Schematic of image transmission experiment [1].

Later work at the Wheeler Laboratory under contract with NASA aimed at producing two-dimensional waveguides for laser beams [2]. Fig. 2 represents the use of proton irradiation in fused silica to bury a waveguide beneath the surface. (Note the plot of index versus depth.) Index increases as large as 0.6 percent were reported. Other experiments by Schineller of the Wheeler Laboratory were done with sheets of glass immersed in a liquid having an index only a fraction of a percent lower than that of the glass, and in films of liquid between sheets of glass. Similarly, more recent work at the University of Washington by Shubert and Harris involved thin films of phenolic polyurethane epoxy 1 to 4 μm thick between sheets of glass. A He–Ne laser beam was divided in one experiment and used to excite either the lowest order or next order slab-waveguide modes, by varying the excitation components as depicted in Fig. 3. Observations confirmed the expected modal field distributions.

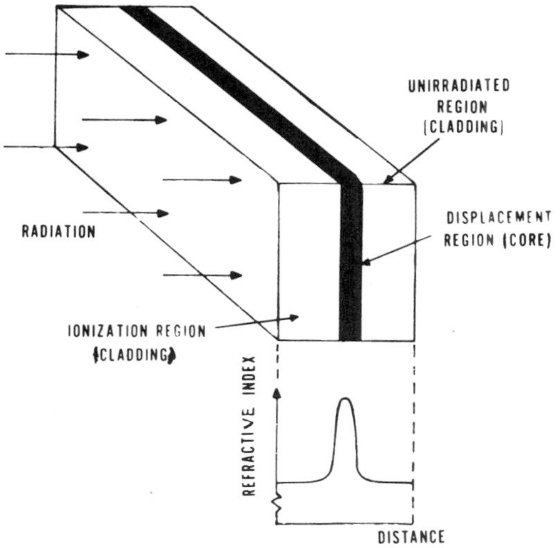

Fig. 2. Schematic of proton irradiation to create a buried optical waveguide.

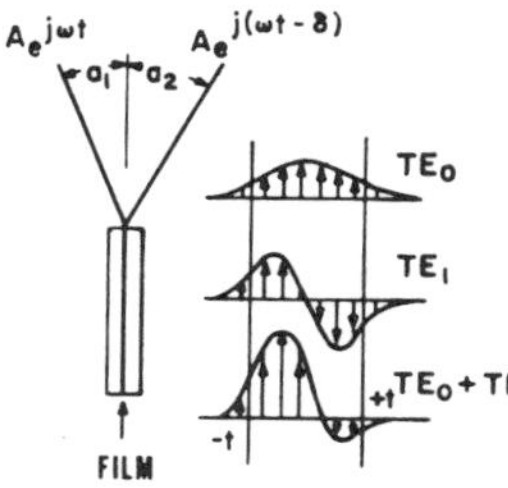

Fig. 3. Modal distributions for laser beam transmission in a film

Other thinking by the Wheeler group involved [Fig. 4(a)] implantation of two guides by using two different energies of ion bombardment, and (Fig. 4(b)) the use of irradiation through a mask to write a circuit using three-dimensional guides. To the knowledge of the writer, these ideas were never carried through experimentally.

At about this point in time the laser transmission art had grown up to a degree where the competitive relationship between prospective laser transmission systems and alternative transmission systems could be appreciated. The result was a new view [3]–[8].

Manuscript received June 14, 1971; revised September 13, 1971.
The author is with the Crawford Hill Laboratory, Bell Telephone Laboratories, Inc., Holmdel, N. J. 07733.

Reprinted from *IEEE J. Quantum Electron.*, vol. QE-8, part 2, pp. 199–205, Feb. 1972.

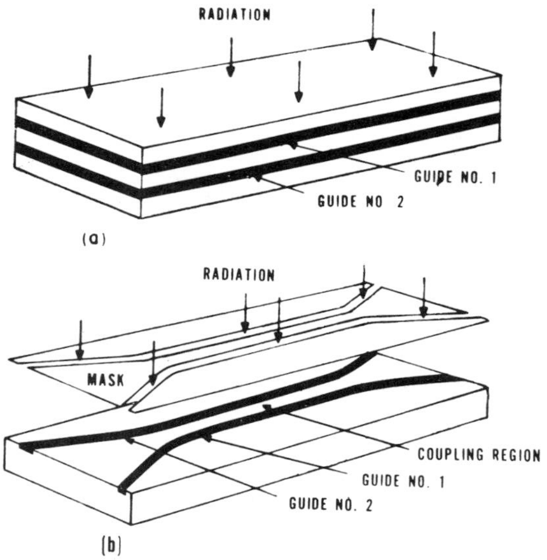

Fig. 4. Early proposals for optical waveguides. (a) Two parallel planar guides produced by irradiation with two energy levels. (b) Three-dimensional guides produced by irradiation through a mask.

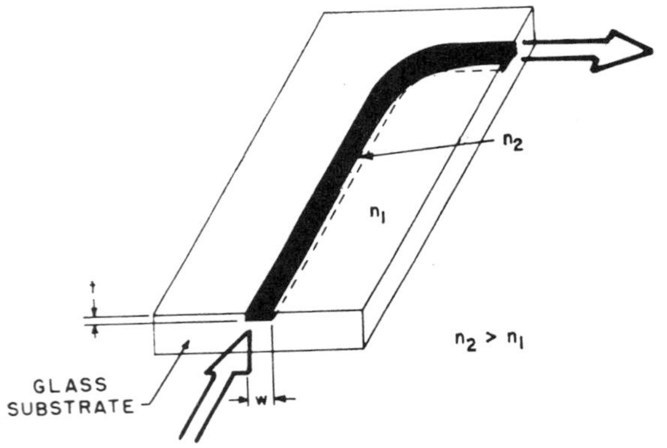

Fig. 5. Schematic of general three-dimensional dielectric waveguide for integrated optics.

Laser experiments are conveniently carried out using unguided beam modes. The laser output typically has a cross section 1 mm or so in diameter, and this beam is reflected, refocused as needed, and propagated through half-silvered mirrors, polarization separators, and nonlinear materials in beam-mode form. The physics of beam modes is elegant and the arrangement generally is well adapted to fundamental experiments on materials in the simplest form. However, beam-mode circuitry is poorly adapted to commercial application; the resulting apparatus is sensitive to ambient temperature gradients, to absolute temperature changes, to airborne acoustic effects, and to mechanical vibrations of the separately mounted parts. By going to miniature optical circuitry one solves many of these problems. Fig. 5 shows an elementary transmission line, a dielectric waveguide. The guide has an index of refraction only 1 percent (or less) higher than the surrounding region, and under these conditions a single mode of propagation can be maintained with guide width w in the 3-to-5-μm region. Mode properties and allowable bending radii will be discussed at a later point, but consider first the power density in the miniature guide. With a guided energy cross section of 3 by 5 μm an absolute power level of 150 mW yields a power density of 10^6 W/cm^2. Powers of under 100 mW will give negligible nonlinearity in typical glasses, which satisfies the need for passive circuitry. Alternatively, by using long fibers of nonlinear materials the concentration of power enables nonlinear effects to be utilized with modest absolute power levels.

TABLE I

Maximum Dimension for Single-Mode Operation in a Dielectric Rod of Index n_1 Surrounded by a Region of Index n_2

Core Aspect Ratio	Rod Dimension n_1/n_2 1.001	1.01	1.05	1.50
1	$15.3\lambda/n_1$	$4.9\lambda/n_1$	$2.25\lambda/n_1$	$0.92\lambda/n_1$
2	9.5×19	3.05×6.1	1.4×2.8	0.6×1.2
4	6.7×26.8	2.1×8.5	0.95×3.8	0.39×1.37

There is in the literature extensive theory on round dielectric waveguides, and recently this has been extended in work in England, Japan, and the United States to rectangular dielectric guides and guides of very small index difference. Fig. 6 is a typical plot [6] of normalized propagation constant (ordinate) versus normalized frequency (abscissa). The lowest order mode has no cutoff; it propagates with the phase constant of a plane wave in the outer region at low frequency and goes smoothly toward the phase constant of a plane wave in the core-type material at high frequencies. Other mode configurations are unguided below some discrete frequency, above which they too approach the phase constant of a plane wave in the core-type material. Fig. 7 shows the computed intensity distributions for several of the lowest order modes in a rectangular dielectric guide [6].

What are typical dimensions? As shown in Table I, for single-mode operation, for a square core, and for a 1-percent index difference, the core dimensions are $4.9\lambda/n_1$, which is computed to be about 3 μm for GaAs laser wavelength. For core aspect ratios up to at least 4:1 the two waveguide dimensions approximately average out to the square-core value, to maintain single-mode operation, as illustrated by the more exact values of Table I.

Also of interest is the tolerable bending radius for dielectric waveguide circuitry. There are two limitations on bending—*mode conversion* if the guide supports more than one mode, and *radiation* as found by Richtmyer [9] and Marcatili [8]. The radiation effect can be seen with reference to Fig. 8. Any dielectric waveguide will radiate if bent; this is a consequence of the fact that the curved region part of the field (which extends to infinity) is being asked to move at a velocity faster than the speed of light, which it cannot do. Consequently that portion of the field radiates. One learns, however, that the radia-

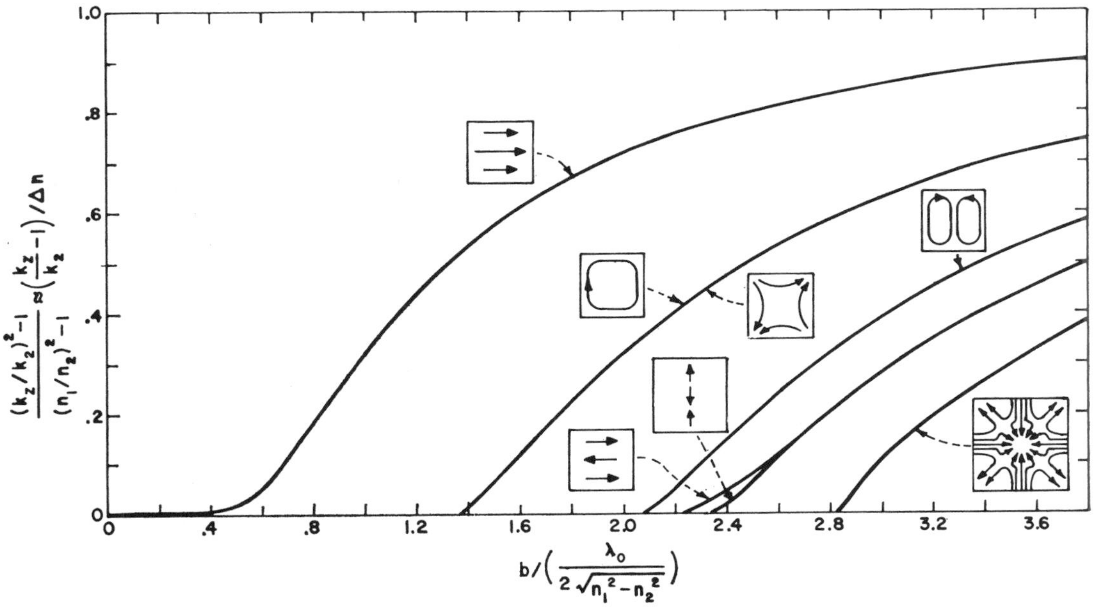

Fig. 6. Normalized propagation constant (ordinate) versus normalized frequency (abscissa) for a square dielectric rod (index n_1) immersed in a medium of index n_2; with λ_0—free-space wavelength, k_z—modal phase constant, k_2—$2\pi n_2/\lambda_0$, $\Delta n = n_1 - n_2$, b—rod width.

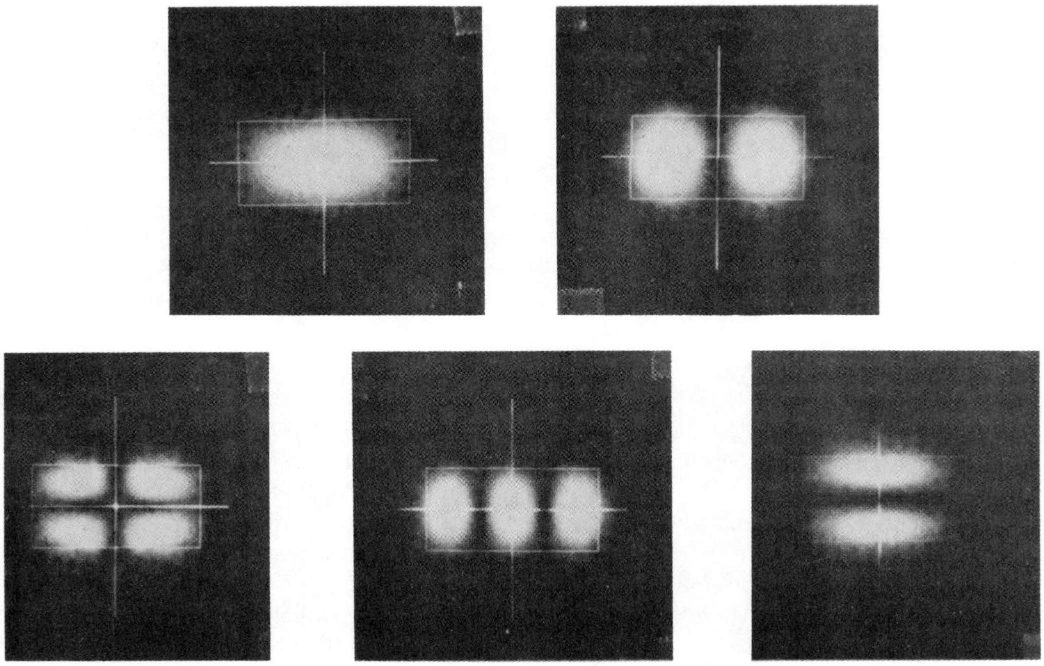

Fig. 7. Intensity distributions for guided modes in rectangular dielectric waveguide [6]. The similarity to unguided beam modes is striking.

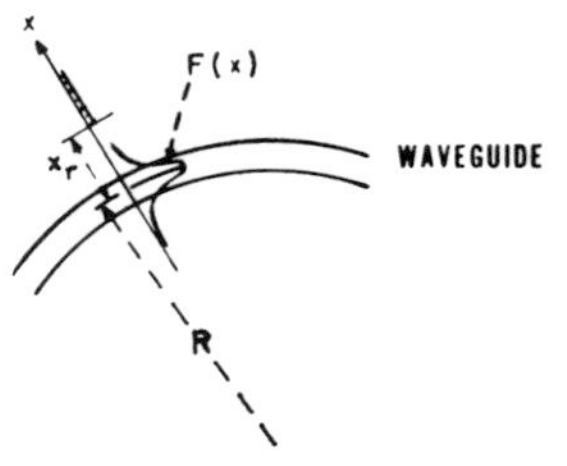

Fig. 8. Schematic of field on curved dielectric waveguide.

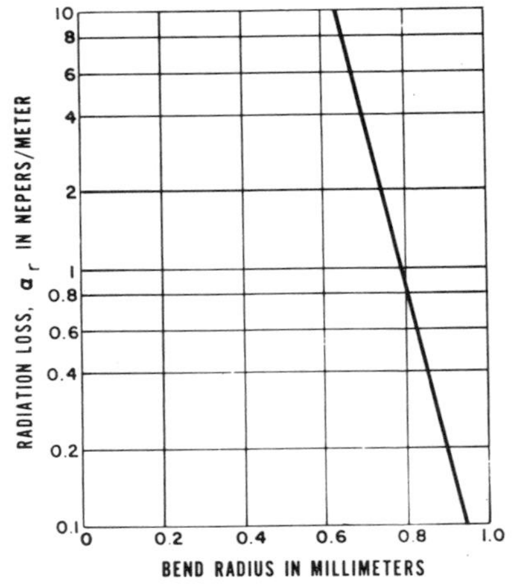

Fig. 9. Radiation loss versus bend radius for two-dimensional dielectric waveguide.

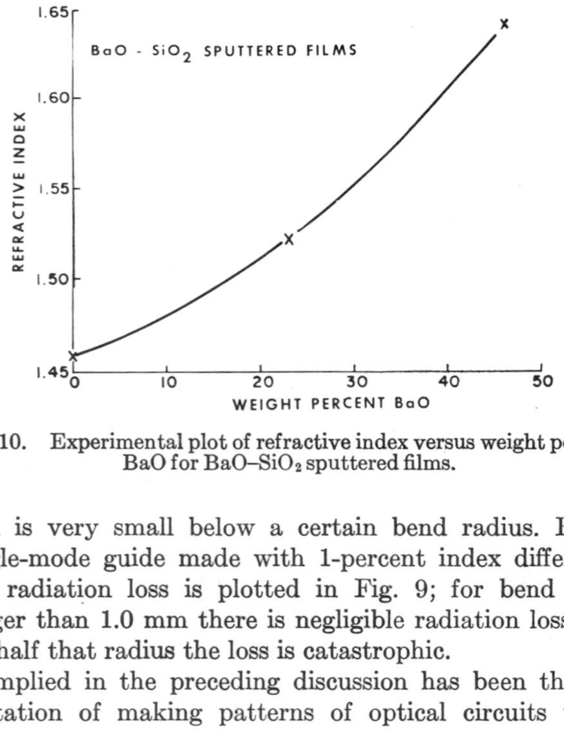

Fig. 10. Experimental plot of refractive index versus weight percent BaO for BaO–SiO_2 sputtered films.

tion is very small below a certain bend radius. For a single-mode guide made with 1-percent index difference the radiation loss is plotted in Fig. 9; for bend radii longer than 1.0 mm there is negligible radiation loss and for half that radius the loss is catastrophic.

Implied in the preceding discussion has been the expectation of making patterns of optical circuits using photolithographic techniques. One example is, a directional filter. A ring resonant at frequency f is coupled loosely to two straight guides, and transfers the energy in the frequency region near f to the auxiliary waveguide. A second example is the possibility of forming a modulator by using an electrooptic material for the guide and achieving low-voltage drive by using metal electrode spacings in the order of tens of microns. As a further example, an elementary resonant circuit could be formed by placing grating reflectors at the two ends of a section of waveguide. Other illustrations of gratings used in a diffraction mode will be discussed below.

Very low-loss films are needed as a starting point for integrated optical circuitry and one way to produce them is by sputtering glassy materials. Fig. 10 shows an experimental plot due to Goell and Standley [10] for sputtered BaO–SiO_2 films—index of refraction on the ordinate versus fractional BaO content on the abscissa. It should be emphasized that great care in maintaining cleanliness is needed; when this is done losses as low as 0.5 dB/in can be achieved for 0.63-μm transmission along the film.

Also needed are means for efficiently launching laser beams into thin-film circuits; Fig. 11 shows two widely used techniques. The prism method has been contributed to by Tien *et al.* [11], by Harris *et al.* [12], and by Midwinter [13]. The grating coupler was first published by Dakss *et al.* [14] and an improved version using Bragg effects in thick films to suppress unwanted orders of diffracted beam was described by Kogelnik and Sosnowski [15]. The prism coupler has the advantage of versatility, a single unit being adaptable to many samples of film and to a variety of film types. The grating coupler is beautifully adapted to the integrated circuit format, with superior ruggedness and stability. Both the prism and thick-film grating couplers are capable of transferring more than 70 percent of the incident light beam into the film.

For reasons of completeness and later discussion a rapid review of the use of photolithography and the sputtered films to produce a circuit is in order. First one starts with a *very* clean substrate, and sputters on a low-loss base layer. Next a slightly higher index film is sputtered to be the guide. Then using a photoresist, a pattern of the desired optical circuit, and conventional photolithography the excess material can be removed leaving the optical guides shown in Fig. 12 in cross section for a directional coupler. Finally, a covering layer (of index $n_2 < n_1$) may be sputtered on to protect the surface and provide a symmetrical guiding structure.

The most serious difficulty in the process just described is the edge definition produced by conventional photolithography (Fig. 13). When visible or ultraviolet radiation is used to expose the photoresist, there are statistical variations in the position of the edge between the exposed and the unexposed regions. These variations are usually increased by a subsequent chemical etching step. The requirements on waveguide wall smoothness have been developed in theoretical work by Marcuse [16], who showed that an edge variation of 500 Å rms would lead to a loss of 10 percent per cm. It is conclued that an edge is needed which is smooth to about 100 to 200 Å. The technique [17] developed at IBM and elsewhere for using electron beam exposure of photoresist appears capable of that precision. It is certainly anticipated that further

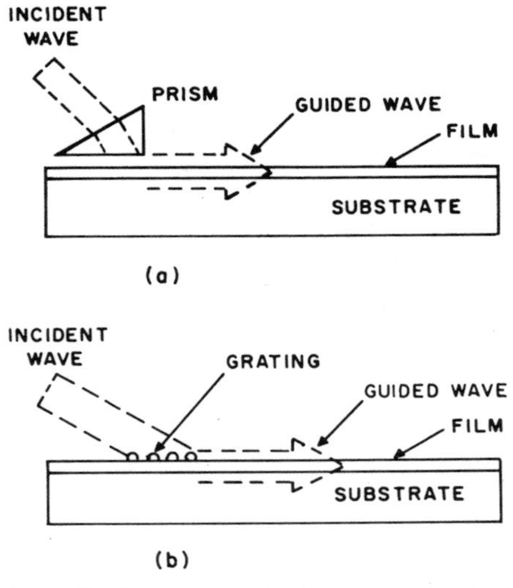

Fig. 11. Schematic of two methods of transforming freely propagating wave into guided wave in thin film. (a) Prism coupler. (b) Grating coupler.

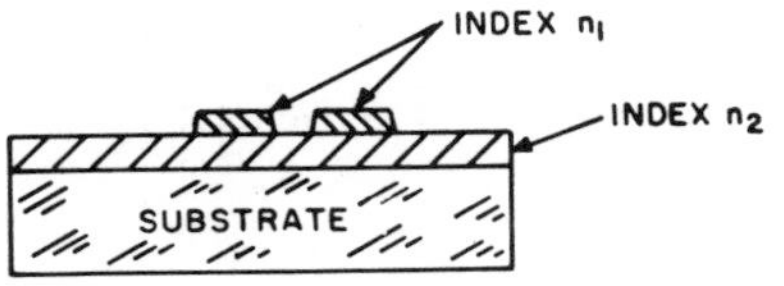

Fig. 12. Section through integrated optical circuit.

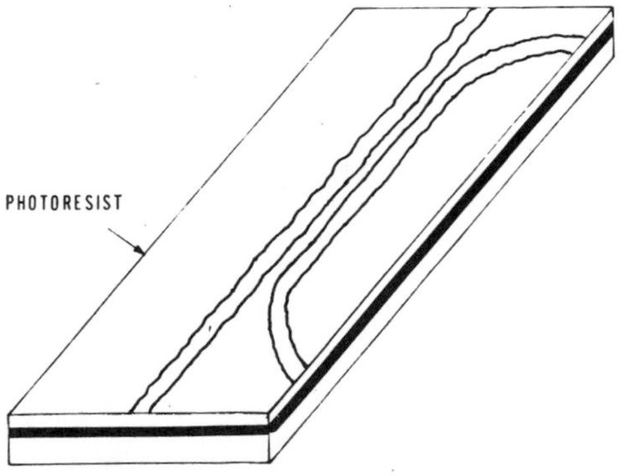

Fig. 13. Schematic illustrating edge roughness in photolithographically produced patterns.

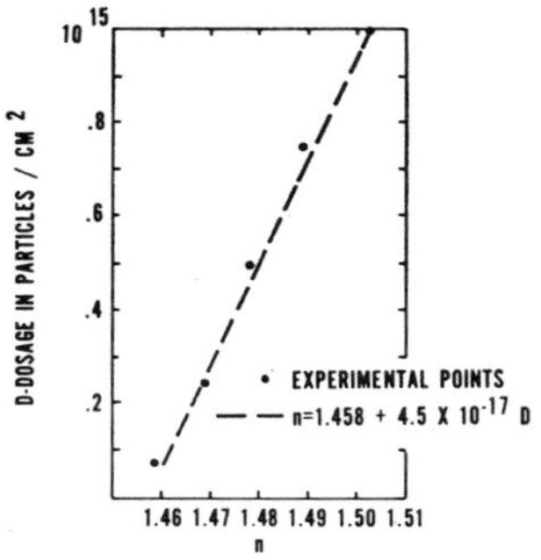

Fig. 14. Experimental data relating change of index of refraction (abscissa) to dosage (ordinate) for Li^{+7} irradiation of fused quartz.

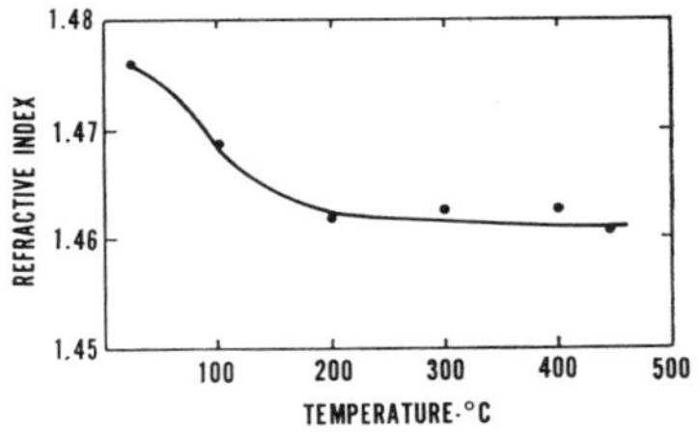

Fig. 15. Typical change in refractive index due to post anneal of Li bombarded sample.

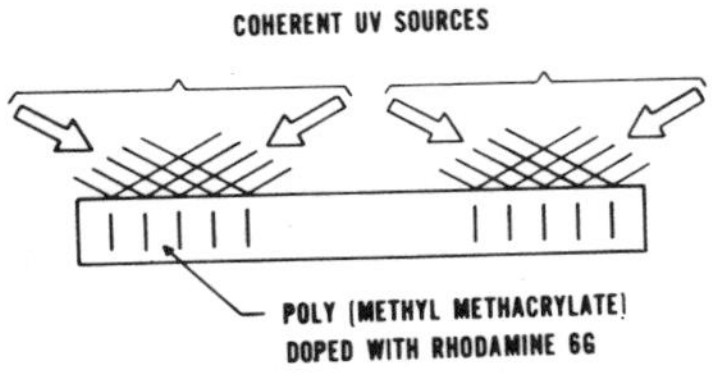

Fig. 16. Schematic of method used to create an index grating in bulk poly (methyl methacrylate) using coherent ultraviolet radiation.

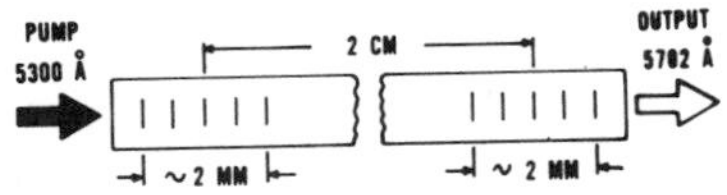

Fig. 17. Dye laser using structure fabricated as shown in Fig. 16.

work on optical circuits made using electron beam photolithography will be done.

An alternate process would be to use photolithography to produce a durable mask, and then to implant ions to raise the index and form the guide. Work has progressed on ion bombardment for optical circuitry by Standley *et al.* [18]. In Fig. 14 is shown the index increase on the abscissa versus dosage in particles per square centimeter of Li^7 into a fused quartz substrate on the ordinate. Initial index increases on the order of 2 percent can be readily created; however, the index change is found to be due to damage of the material structure rather than chemical change. Fig. 15 shows how an annealing cycle changes the Li bombarded sample. After the 500°C anneal, the index should be stable at room temperature, and an index difference of about 1/3 percent remains. This is usable for many possible applications, although smaller than desired for some.

Another technique for making miniature optical circuitry is illustrated in Fig. 16 taken from the work of Kaminow *et al.* [19]. Using a coherent ultraviolet source, a pair of standing-wave patterns was set up in properly prepared poly(methyl methacrylate), and a permanent refractive index change occurred in the bulk sample. The standing-wave pattern from the UV source caused the formation of two phase gratings, which act as frequency-selective reflectors for a wave passing between them. The substrate was also doped with rhodamine 6G dye, which was known to be a suitable gain medium when pumped. Then Fig. 17 by pumping at 5300 Å through the gratings, lasing was observed in the 2-cm-long structure.

Other uses of dyes in thin-film structures have been reported. Holographic techniques were used to produce a thin-film grating, which was impregnated with rhodamine 6G to form a distributed feedback laser [20]. In another study [21] a ring laser was produced by depositing a doped polyurethane film on a rod and by pumping with a fan beam from a nitrogen laser. Despite the poor life properties of present dyes, these are significant stepping stones to the miniature long-life structures that are sought.

Another broad area of work relevant to integrated optics employs light transmission along semiconductor diodes as schematically depicted in Fig. 18. In 1963 Yariv *et al.* showed that p–n junctions guide light quite efficiently [22], [23]. Reinhart *et al.* [24] reported a GaP phase modulator requiring only 1 mW modulator power per megahertz bandwidth at the 1-rad modulation level, and new work by Hall *et al.* showed propagation cutoff effects in GaAs diodes adaptable to the modulator function [25].

Recent new developments on the junction-diode form of laser will have profound influence on laser technology. The structure of Fig. 18 was altered to put the junction in a GaAs film about 0.5 μm thick, with epitaxially grown layers of AlGaAs placed immediately adjacent on both sides of the junction layer. The larger band gap and lower index of refraction of AlGaAs, as compared to GaAs, provided confinement of the light wave and of the electrons near the thin GaAs junction. By this new materials technology, workers [26], [27] have produced injection lasers that run CW at room temperature, an accomplishment that had eluded many workers for years. Note that the new epitaxial-layer technology, capable of a continuous range of indexes, is applicable rather broadly in integrated optics and may be expected to yield additional results of importance.

Another merging of an older art with thin-film optics is illustrated in Fig. 19; shown there is an acoustooptic deflector due to Kuhn *et al.* [28]. The acoustic carrier sets up an acoustic wave in the α-quartz substrate, which causes a grating in the glass film deposited on the quartz. The optical wave is diffracted in the presence of the acoustic carrier; digital modulation of the optical wave may therefore be achieved by gating the acoustic carrier. A less attractive feature of acoustooptic modulators has been a rather appreciable modulator power requirement, due to the transducer from electrical modulator power to acoustical power. However, transducer efficiencies have been improved, giving some relief on this feature.

An optical deflector analogous to that of Fig. 19 using the electrooptic effect has been described by Hammer [29]. The quartz and glass film of Fig. 19 was replaced by a lithium niobate crystal, and the index grating was induced by setting up a spatially periodic electric field along the desired path. Digital modulation of the optical diffraction results from gating the electric field. Low modulation powers seem feasible, particularly if the concept were applied in thin-film form. Other uses of periodic optical coupling effects have been proposed [30].

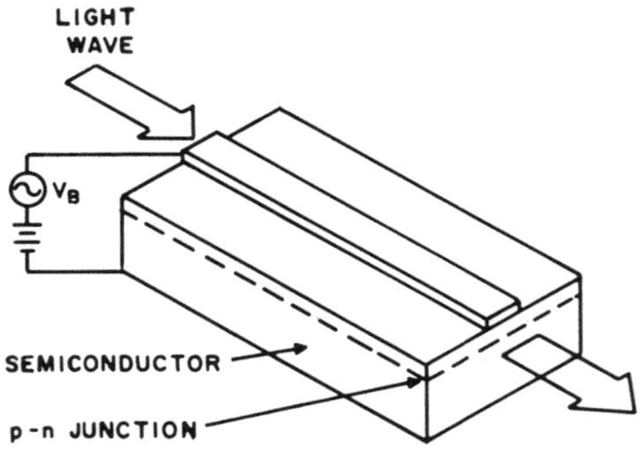

Fig. 18. Schematic of semiconductor junction device useful as laser or light modulator.

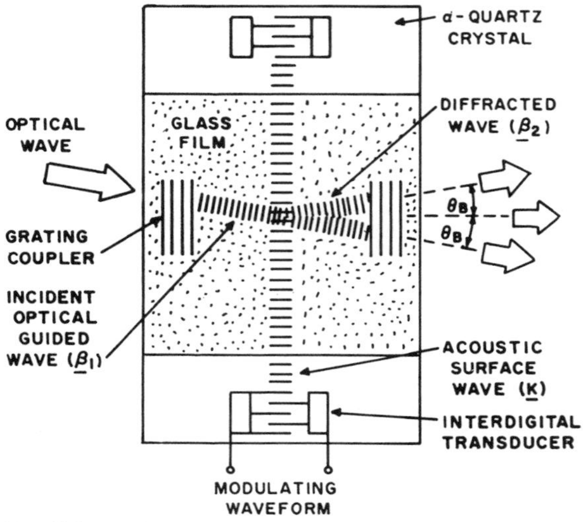

Fig. 19. Schematic of acoustooptic deflector. Light travels in glass film and is deflected by index grating due to acoustic wave in α-quartz [27].

A new class of photochromic materials has been described by Tomlinson *et al.* [31]. A change of index of refraction may be established in a material and removed or reestablished repeatedly. Using an anthracene ester in poly(methylmethacrylate), for example, dimers can be made using 365-nm radiation and cleaved using 303-nm radiation; the dimers and broken dimers are stable at room temperature and may be used in holographic information stores or in optical circuits for longer wavelength radiation.

Very successful experiments have been reported on obtaining nonlinear optical effects at reduced power levels in fibers and thin films. Ippen [32] reported a Raman oscillator in CS_2 at 5 W input power using a hollow glass fiber to confine the CS_2 and the guided beams at the pump and Stokes wavelengths to a 12-μm diameter. Tien *et al.* [33] reported second-harmonic generation using a ZnS film 0.2 μm thick on a ZnO substrate; the wave guided by the film extended into the substrate, which developed an appreciable second-harmonic beam at an input power of 2 W.

Concluding Remarks

The moment of application of integrated optics to complete receiver assemblies has not yet arrived, although possible forms of such application have been conceived [10]. The earliest application is likely to be in relatively simple devices, such as in laser-modulator integration. These might be very important economically nonetheless. Really extensive applications of integrated optics such as would be required in a frequency-division-multiplex form of large-capacity optical transmission system would appear to be at least ten years ahead.

No consistent attempt has been made to place a value judgment on the alternative techniques described in this paper. This is in part due to the fact that most of the studies are at an early stage where it would be premature to be too selective about emphasis placed on further investigations.

The interest in this field runs strong. To the research physicist thin-film or fiber optics presents an opportunity to study phenomena not accessible in larger bulk samples. To the applied physicist and engineer, integrated optics presents the solution to practical problems encountered in useful applications of laser wavelengths. To both it should be fruitful professionally and profitable economically.

References

[1] H. Osterberg and L. W. Smith, "Transmission of optical energy along surfaces: Parts I and II," *J. Opt. Soc. Amer.*, vol. 54, pp. 1073–1084, Sept. 1964.

[2] E. R. Schineller, R. P. Flam, and D. W. Wilmot, "Optical waveguides formed by proton irradiation of fused silica," *J. Opt. Soc. Amer.*, vol. 58, pp. 1171–1176, Sept. 1968.

[3] N. S. Kapany, J. J. Burke, K. L. Frame, and R. E. Wilcox, "Coherent interactions between optical waveguides and lasers," *J. Opt. Soc. Amer.*, vol. 58, pp. 1176–1183, Sept. 1968.

[4] R. Shubert and J. H. Harris, "Optical surface waves on thin films and their application to integrated data processors," *IEEE Trans. Microwave Theory Tech.*, vol. MTT-16, pp. 1048–1054, 1968.

[5] S. E. Miller, "Integrated optics: An introduction," *Bell Syst. Tech. J.*, vol. 48, pp. 2059–2069, Sept. 1969.

[6] J. E. Goell, "A circular harmonic computer analysis of rectangular dielectric waveguide," *Bell Syst. Tech. J.*, vol. 48, pp. 2133–2160, Sept. 1969.

[7] E. A. J. Marcatili, "Dielectric rectangular waveguide and directional coupler for integrated optics," *Bell Syst. Tech. J.*, vol. 48, pp. 2071–2102, Sept. 1969.

[8] E. A. J. Marcatili, "Bends in optical dielectric guides," *Bell Syst. Tech. J.*, vol. 48, pp. 2103–2132, Sept. 1969. E. A. J. Marcatili and S. E. Miller, "Improved relations describing directional control in electromagnetic wave guidance," *ibid.*, pp. 2161–2188.

[9] R. D. Richtmyer, "Dielectric resonators," *J. Appl. Phys.*, vol. 10, pp. 391–398, June 1939.

[10] J. E. Goell and R. D. Standley, "Integrated optical circuits," *Proc. IEEE*, vol. 58, pp. 1504–1512, Oct. 1970. J. E. Goell, R. D. Standley, and T. Li, "Optical waveguides bring laser communication closer," *Electronics*, pp. 60–67, Aug. 31, 1970.

[11] a. P. K. Tien, R. Ulrich, and R. J. Martin, "Modes of propagating light waves in thin deposited semiconductor films," *Appl. Phys. Lett.*, vol. 14, p. 291, 1969. b.——, "Experiments on light waves in thin tapered films and a new light-wave coupler," *ibid.*, 1971.

[12] a. J. H. Harris, R. Shubert, and J. N. Polky, "Beam coupling to films," *J. Opt. Soc. Amer.*, Aug. 1970. b. J. H. Harris and R. Shubert, "Variable tunneling excitation of optical surface waves," *IEEE Trans. Microwave Theory Tech.*, vol. MTT-19, pp. 269–276, Mar. 1971.

[13] J. E. Midwinter, "Evanescent field coupling into a thin-film waveguide," *IEEE J. Quantum Electron.*, vol. QE-6, pp. 583–590, Oct. 1970.

[14] M. L. Dakss, L. Kuhn, P. F. Heidrich, and B. A. Scott, "Grating coupler for efficient excitation of optical guided waves in thin films," *Appl. Phys. Lett.*, vol. 16, pp. 523–525, June 15, 1970.

[15] H. Kogelnik and T. P. Sosnowski, "Holographic thin film couplers," *Bell Syst. Tech. J.*, vol. 49, p. 1602, Sept. 1970.

[16] D. Marcuse, "Mode conversion caused by surface imperfections of a dielectric slab waveguide," *Bell Syst. Tech. J.*, vol. 48, pp. 3187–3215, Dec. 1969.

[17] I. Haller, M. Hatzakis, and R. Srinivasan, "High-resolution positive resists for electron-beam exposure," *IBM J. Res. Develop.*, pp. 251–256, May 1968.

[18] R. D. Standley, W. M. Gibson, and J. W. Rodgers, "Properties of ion bombarded fused quartz for integrated optics," presented at the Spring Meeting of the Optical Society of America, Tucson, Ariz., Apr. 5–8, 1971.

[19] I. P. Kaminow, H. P. Weber, and E. A. Chandross, "A poly-(methyl methacrylate) dye laser with international diffraction grating resonator," *Appl. Phys. Lett.*, vol. 18, pp. 497–499, June 1, 1971.

[20] H. Kogelnik and C. V. Shank, "Stimulated emission in a periodic structure," *Appl. Phys. Lett.*, vol. 18, pp. 152–154, Feb. 15, 1971.

[21] H. P. Weber and R. Ulrich, "A thin film ring laser," presented at the Spring Meeting of the Optical Society of America, Tucson, Ariz., Apr. 5–8, 1971.

[22] A. Yariv and R. C. C. Leite, "Dielectric-waveguide mode of light propagation in p-n junctions," *Appl. Phys. Lett.*, vol. 2, p. 55, 1963.

[23] W. L. Bond, B. G. Cohen, R. C. C. Leite, and A. Yariv, "Observation of the dielectric-waveguide mode of light propagation in p–n junctions," *Appl. Phys. Lett.*, vol. 2, p. 57, 1963.

[24] F. K. Reinhart, D. F. Nelson, and J. McKenna, "Electro-optic and waveguide properties of reverse-biased gallium phosphide p–n junctions," *Phys. Rev.*, vol. 177, pp. 1208–1221, Jan. 15, 1969.

[25] D. Hall, A. Yariv, and E. Garmire, "Observation of propagation cut-off and its control in thin optical waveguides," *Appl. Phys. Lett.*, vol. 17, p. 127, 1970.

[26] Zh. I. Alferov, V. M. Andreev, D. Z. Garbuzov, Ju. V. Zhilgoev, E. P. Morozov, E. L. Portnoy, and V. G. Trofim, *Fiz. Tekh. Poluprov.*, vol. 4, p. 1826, 1970 (transl.: *Sov. Phys.—Semicond.*, vol. 4, p. 1573, 1971).

[27] I. Hayashi, M. B. Panish, P. W. Foy, and S. Sumski, "Junction lasers which operate continuously at room temperature," *Appl. Phys. Lett.*, vol. 17, pp. 109–111, Aug. 1, 1970.

[28] L. Kuhn, M. L. Dakss, P. F. Heidrich, and B. A. Scott, "Deflection of an optical guided wave by a surface acoustic wave," *Appl. Phys. Lett.*, vol. 17, pp. 265–267, Sept. 15, 1970.

[29] J. M. Hammer, "Digital electro-optic grating deflector and modulator," *Appl. Phys. Lett.*, vol. 18, pp. 147–149, Feb. 13, 1971.

[30] S. E. Miller, "Some theory and applications of periodically coupled waves," *Bell Syst. Tech. J.*, vol. 48, pp. 2189–2219, Sept. 1969.

[31] W. J. Tomlinson, E. A. Chandross, R. L. Fork, A. A. Lamola, and C. A. Pryde, "A new class of photochromic materials," presented at the Spring Meeting of the Optical Society of America, Tucson, Ariz., Apr. 5–8, 1971. W. J. Tomlinson, E. A. Chandross, R. L. Fork, C. A. Pryde, and A. A. Lamola, "Reversible photodimerization: A new type of photochromism," *Appl. Opt.*, Mar. 1972.

[32] E. P. Ippen, "Low-power quasi-cw Raman oscillator," *Appl. Phys. Lett.*, vol. 16, pp. 303–305, Apr. 15, 1970.

[33] P. K. Tien, R. Ulrich, and R. J. Martin, "Optical second harmonic generation in form of coherent radiation from thin film waveguide," *Appl. Phys. Lett.*, vol. 17, pp. 447–450, Nov. 15, 1970.

Light Waves in Thin Films and Integrated Optics

P. K. Tien

Integrated optics is a far-reaching attempt to apply thin-film techno ogy to optical circuits and devices, and, by using methods of integrated circuitry, to achieve a better and more economical optical system. The specific topics discussed here are physics of light waves in thin films, materials and losses involved, methods of coupling a light beam into and out of a thin film, and nonlinear interactions in waveguide structures. The purpose of this paper is to review in some detail the important development of this new and fascinating field, and to caution the reader that the technology involved is difficult because of the smallness and perfection demanded by thin-film optical devices.

I. Introduction

Since the invention of the solid and gas lasers a decade ago, there have been immense advances in optoelectronics. For example, laser transitions now cover the spectral region from ultraviolet light to millimeter waves. Optical modulation, frequency mixing, and parametric oscillations have been extensively studied. Almost all the experiments in the past were performed invariably in bulk materials and with a light beam of nearly gaussian intensity distribution. Recently, however, the introduction of the concept of integrated optics[1,2] and the development of the prism–film[3] and grating[4,5] couplers have raised important questions concerning the future needs of optical systems: Can we conform to the idea of the integrated optics and develop optical modulators, frequency converters, and parametric oscillators in a planar thin-film form which are more efficient than their bulk counterparts? Is there any advantage of using a waveguide structure for nonlinear, electrooptical, or light-wave scattering experiments? While many are devoted to the development of ever better bulk crystals for optical devices, could it be that what we really need in the long run are thin crystal films and epitaxial layers?

One cannot answer the above questions until the complex technology involved in the fabrication of thin-film structures is solved. On the other hand, as far as optical systems are considered, certain features of the thin-film devices appear to be definitely advantageous. First, all the elements of a thin-film device are exposed on the surface and are easily accessible for probing, measurement, or modification. Second, compared to microwaves, the optical wavelength is a factor of 10^4 smaller. The thin-film optical devices can be made very small and they can be placed one next to the other on a single substrate, forming an optical system which is naturally more compact, less vulnerable to the environmental changes, and more economical. Third, since the film has a thickness comparable to the optical wavelength and since most of the light energy is confined within the film, the light intensity inside the film can be very large even at a moderate laser power level. For example, 1 W of laser power can easily result in a power density of 22 MW/cm^2 in a ZnS film 0.46 μm thick. This large power density is important in nonlinear interactions. Finally, the phase velocity of a light wave in a thin-film waveguide depends on the thickness of the film and the mode of propagation. This provides new possibilities in the design of experiments and devices.

Although this field of integrated optics or thin-film optoelectronics is still in a very elementary stage, in the past two years there have been a number of very important advances which generated substantial excitement. In what follows we will briefly enumerate some of these advances.

The first advance is the great improvement in the method of coupling a light wave propagating in free space into a well-defined mode of the thin-film guide. This was brought about through the invention of the prism–film coupler as described by Tien *et al.*[3] Other theories and experiments of the prism–film coupler have been given by Tien and Ulrich,[6] Ulrich,[7] Midwinter,[8] Harris *et al.*,[9] and Harris and Shubert.[10] Recently, a coupling efficiency of 88% was reported by Ulrich.[11] Another type of coupler, which uses a grating in place of the prism, was reported by Dakss *et al.*[4] and by Kogelnik and Sosnowski.[5] Recently an extremely

The author is with Bell Telephone Laboratories, Inc., Holmdel, New Jersey 07733.

Received 3 May 1971.

Reprinted with permission from *Appl. Opt.*, vol. 10, pp. 2395–2413, Nov. 1971.

simple technique of coupling which involves a tapered film edge was described by Tien and Martin.[12] A second advance which has occurred involves the materials of the films used for light-wave propagation. Initial experiments of Tien *et al.*[3] used sputtered ZnO and vacuum-evaporated ZnS films, which have large scattering losses. Then, excellent sputtered glass films were developed by Goell and Standley.[13] Other useful materials include polyurethane and polyester epoxy films developed by Harris *et al.*,[9] Ta_2O_5 films by Hensler *et al.*,[14] and polymerized organosilicon films by Tien *et al.*[15] The organosilicon films made from vinyltrimethylsilane and hexamethyldisiloxane monomers have a loss of the order of 0.04 dB/cm as opposed to a number which was several hundred times larger in some of the earlier experiments. Other advances involve nonlinear and electrooptic effects observed in thin films and laser oscillation in iterated film structures. Tien *et al.*[16] have observed optical second harmonic generation using a ZnS film on a single-crystal ZnO substrate. The nonlinear interaction in this experiment was enhanced by the large concentration of light energy in the vicinity of the film. Kuhn *et al.*[17] succeeded in deflection of an optical guided wave in a thin film by its interaction with a surface acoustic wave. Hall *et al.*[18] have shown that it is possible to control the intensity of the light wave which propagates through a thin film such as a GaAs depletion layer by applying an appropriate bias. The latter two experiments indicate the possibility of efficient modulation of light in thin-film waveguides. Finally, laser oscillation was reported by Kogelnik and Shank[19] in a thin-film grating structure which is doped with a dye of Rhodamine 6G.

The basic problem considered here is simply a dielectric or semiconductor film that is deposited on a substrate. When a light beam in space is fed into a film, the light beam adapts itself so that it is confined within the thickness of the film. However, the transverse dimension of the light beam is not restricted. The film can thus be considered as a slab waveguide. Mathematical solutions of the waveguide show many possible modes of light-wave propagation. Within the plane of the film, the light wave in any of these waveguide modes is allowed to propagate in any direction and can be reflected or refracted at any given boundary. The problem can be thought of as two-dimensional optics. Moreover, the light wave can be made to interact with the material of the film or of the substrate, or with externally applied electric or microwave fields so that certain functions such as modulation of light, parametric interaction, etc., can be achieved. Integrated optics is therefore an interdisciplinary science; it involves materials, film fabrication, electronics, and physical optics. In this paper, we select a few topics that are essential to the understanding of this new and complex field and discuss them in sufficient detail, hoping that the paper can serve as an introductory review for newcomers as well as a useful reference for those already in the field. The specific topics chosen are waveguide and radiation modes, the light-wave couplers, materials and losses of thin-film waveguides, and nonlinear interactions among optical guided waves. We will not discuss passive optical circuits and their fabrication, which have been reviewed in several excellent papers.[1,2,20]

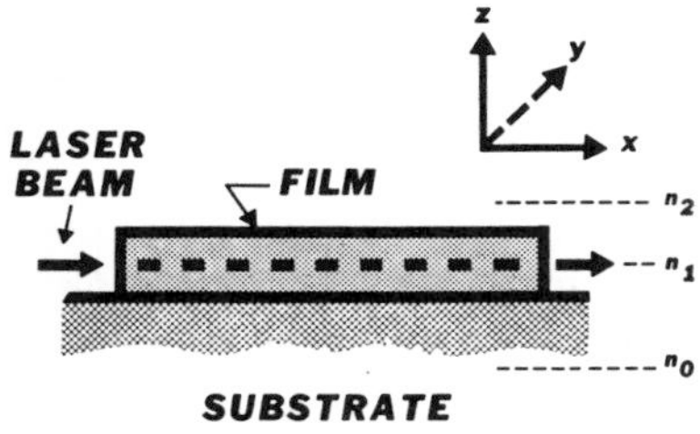

Fig. 1. Coordinate system that will be used throughout the paper. The light wave propagates in the film parallel to the x axis. The surface of the film is in the xy plane and its thickness in the z direction.

II. Waveguide and Radiation Modes

The film considered here has a thickness on the order of 1 μ or less; it is so thin that it has to be supported by a substrate. We thus consider three media: a film, an air space above, and a substrate below. As shown in Figure. 1, the thickness of the film is in the X–Y plane. For a thin film to support propagating modes and to act as a dielectric waveguide for the light waves, the refractive index of the film n_1 must be larger than that of the substrate n_0 and naturally that of the air space above n_2. A typical experiment is shown in Fig. 2. Here the entire surface of a 7.6-cm by 2.5-cm microscope glass slide is coated with a layer of an organic film made from a vinyltrimethylsilane monomer by gas discharge. At a wavelength of 6328 Å of the

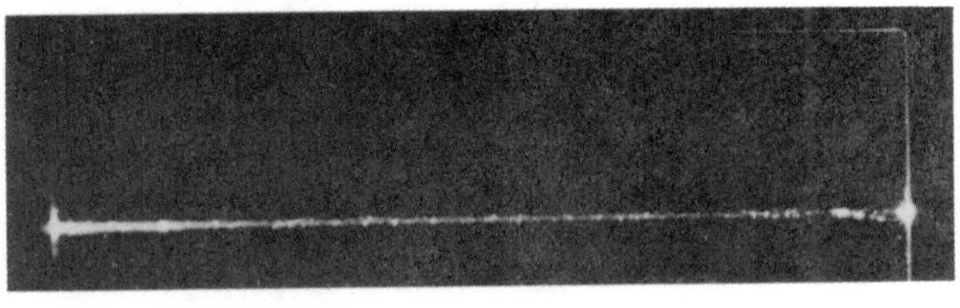

Fig. 2. Light beam in an organic film. The film is coated on a 2.5-cm by 7.6-cm microscope glass slide and it serves as a dielectric waveguide for the light wave.

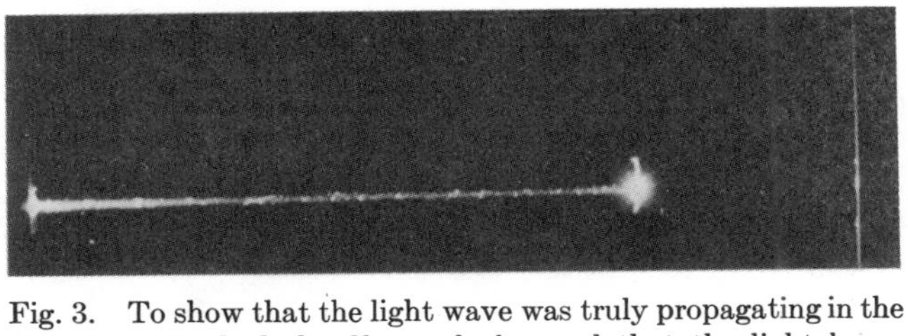

Fig. 3. To show that the light wave was truly propagating in the film, we scratched the film and observed that the light beam stopped immediately at the scratched point, which then radiated brightly.

helium–neon laser, the refractive index of the film is 1.5301, which is larger than that of the glass (1.5125) and also that of the air (1.00). A light beam was fed into the film at the left side of the figure. It propagated through the entire length of the film and then radiated into the free space at the right side of the film. To show that the light wave was truly propagating inside the film, we scratched the film as shown in Fig. 3. The light beam then stopped at the scratched point, which radiated brightly as an antenna. Mathematically, the problem involves a solution of the Maxwell equations that matches the boundary conditions at the film–substrate and film–air interfaces. The solutions indicate three possible modes of propagation. The light wave can be bound and guided by the film as the *waveguide modes.* It can radiate from the film into both of the air and substrate spaces as the *air modes,* or it can radiate into the substrate only as the *substrate modes.* The air and substrate modes are the radiation modes discussed by Marcuse.[21] The modes described above can be explained simply and elegantly by the Snell law of refraction and the related total internal reflection phenomenon in optics.

Let (Fig. 4a) n_0, n_1, and n_2 be the refractive indices and θ_0, θ_1, θ_2 be the angles measured between the light paths and the normals of the interfaces in the substrate, film, and air, respectively. Here $n_1 > n_0 > n_2$. We have then from the Snell law

$$\sin\theta_2/\sin\theta_1 = n_1/n_2, \tag{1}$$

and

$$\sin\theta_0/\sin\theta_1 = n_1/n_0. \tag{2}$$

Let us increase θ_1 gradually from 0. When θ_1 is small, a light wave, for example, starts from the air space above the film, can be refracted into the film, and is then refracted again into the substrate (Fig. 4a). In this case, the waves propagate freely in all the three media—air, film, and substrate—and they are the radiation fields that fill all the three spaces (air modes). Next, as θ_1 is increased to a value larger than the critical angle $\sin^{-1}(n_2/n_1)$ of the film–air interface as shown in Fig. 4(b), the impossible condition incurred in Eq. (1), $\sin\theta_2 > 1$, indicates that the light wave is totally reflected at the film–air boundary. Now the wave can no longer propagate freely in the air space. We thus describe a solution that the light energy in the film radiates into the substrate only (substrate modes). Finally, when θ_1 is larger than the critical angle $\sin^{-1}(n_0/n_1)$ of the film–substrate interface, the light wave as shown in Fig. 4(c) is totally reflected at both the upper and lower surfaces of the film. The energy flow is then confined within the film; that is to be expected in the waveguide modes.

It is interesting to note that in the waveguide modes, the light wave in the film follows a zigzag path (Fig. 4c). The light energy is trapped in the film as the wave is totally reflected back and forth between the two film surfaces. We can represent this zigzag wave motion by two wave vectors A_1 and B_1 in Fig. 5(a). We then divide the wave vectors into the vertical and horizontal components in Fig. 5(b). The horizontal

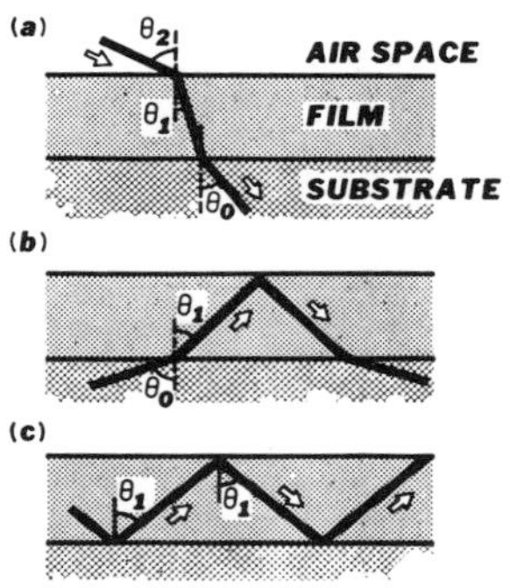

Fig. 4. (a) When $\theta_1 < \sin^{-1}(n_2/n_1)$, the light wave shown represents the air mode. According to ray optics, the light wave originated in the film is refracted into both the substrate and air space. (b) As θ_1 increases so that $\sin^{-1}(n_2/n_1) < \theta_1 < \sin^{-1}(n_0/n_1)$, the light wave shown now represents the substrate mode. It is refracted into the substrate but is totally reflected at the film–air interface. (c) When θ_1 increases further so that $\theta_1 > \sin^{-1}(n_0/n_1)$, the light wave shown is totally reflected at both the film–air and film–substrate interfaces. It is confined in the film as is to be expected in the waveguide mode.

components of wave vectors A_1 and B_1 are equal, indicating that the waves propagate with a constant speed in a direction parallel to the film. The vertical component of the wave vector A_1 represents an upward-traveling wave; that of the wave vector B_1, a downward-traveling wave. When the upward- and downward-traveling waves are superposed, they form a standing wave field pattern across the thickness of the film. By changing θ_1, we change the direction of the wave vectors A_1 and B_1 and thus their horizontal and vertical components. Consequently, we change the wave velocity parallel to the film as well as the standing wave field pattern across the film.

Since we discuss here a planar geometry, the waves described above are plane waves. They are TE waves if they contain the field components E_y, H_z, and H_x; they are TM waves if they contain the field components H_y, E_z, and E_x. Here x is the direction of the wave propagation parallel to the film. The wave vectors A_1 and B_1 discussed above have thus a magnitude kn_1, where $k = \omega/c$ and ω and c are, respectively, the angular frequency of the light wave and the speed of

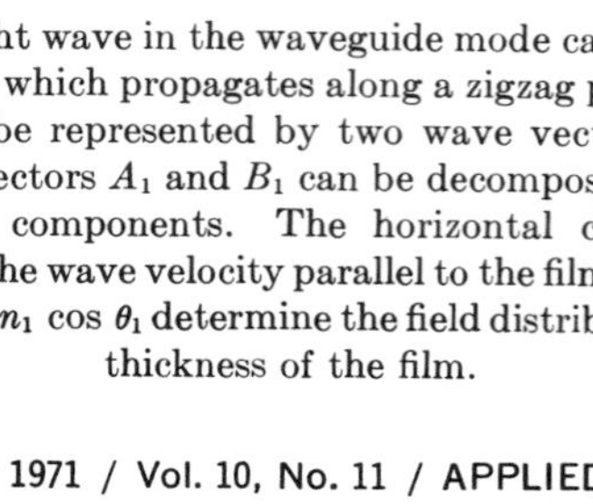

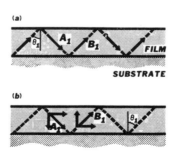

Fig. 5. (a) Light wave in the waveguide mode can be considered as a plane wave which propagates along a zigzag path in the film. The wave can be represented by two wave vectors A_1 and B_1. (b) The wave vectors A_1 and B_1 can be decomposed into vertical and horizontal components. The horizontal components $kn_1 \sin\theta_1$ determine the wave velocity parallel to the film. The vertical components $\pm kn_1 \cos\theta_1$ determine the field distribution across the thickness of the film.

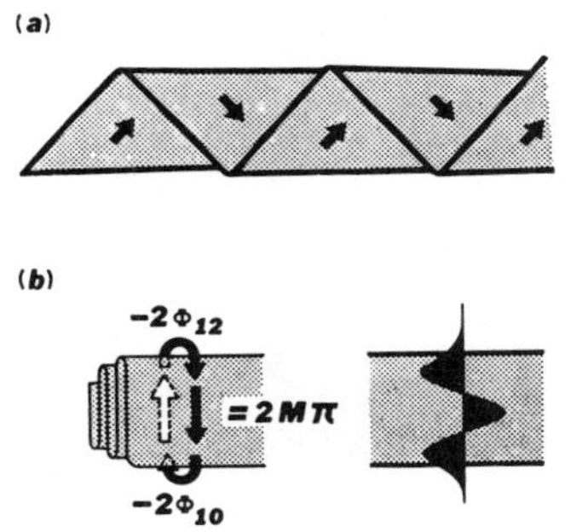

Fig. 6. (a) In wave optics, a light wave in the waveguide mode is an infinitely wide sheet of plane wave which folds back and forth in a zigzag manner between the top and the bottom surface of the film. (b) A light wave propagating inside the film is totally reflected at the two film surfaces. The figure shows that in order for the wave and its reflections to add in phase, the total phase change for the light wave to travel across the thickness of the film, up and down in one round trip, must be equal to $2m\pi$. The figure also shows that the light wave suffers a phase change of $-2\Phi_{12}$ and $-2\Phi_{10}$ at the upper and lower film surfaces, respectively. These phase changes determine the field distribution across the thickness of the film, which is shown at the right of the figure for the $m = 3$ waveguide mode.

light in vacuum. In the picture of the wave optics, the vectors A_1 and B_1 are the normals of the wavefronts, when an infinitely wide sheet of plane wave folds back and forth in a zigzag manner between the two film surfaces (Fig. 6a). Now consider an observer who moves with the wave in the direction parallel to the film. He does not see the horizontal components of the wave vectors. What he observes is a plane wave that folds upward and downward, one directly on top of the other as shown in Fig. 6(b). The condition, then, for all those multiple reflected waves to add in phase, as seen by this observer, is that the total phase change experienced by the plane wave for it to travel one round trip, up and down across the film, should be equal to $2m\pi$, where m is an integer. Otherwise, if after the first reflections from the upper and lower film surfaces, the phase of the reflected wave differs from the original wave by a small phase δ, the phase differences after the second, third,..., reflections would be $2\delta, 3\delta, \ldots$, and then the waves of progressively larger phase differences would add finally to zero. As shown in Fig. 5(b), the vertical components of the wave vectors A_1 and B_1 have a magnitude $kn_1 \cos\theta_1$. The phase change for the plane wave to cross the thickness W of the film twice (up and down) is then $2kn_1W \cos\theta_1$. In addition, the wave suffers a phase change of $-2\Phi_{12}$ due to the total reflection at the upper film boundary and, similarly, a phase change of $-2\Phi_{10}$ at the lower film boundary. Here, the phases $-2\Phi_{12}$ and $-2\Phi_{10}$ represent, in fact, the Goos-Haenchen shifts.[22] Consequently, in order for the waves in the film to interfere constructively, we have

$$2kn_1W \cos\theta_1 - 2\Phi_{10} - 2\Phi_{12} = 2m\pi, \qquad (3)$$

which is the condition for the waveguide modes. Here $m = 0, 1, 2, 3, \ldots$, is the order of the mode. According to Born and Wolf[23] on the theory of total reflection,

$$\tan\Phi_{12} = (n_1^2 \sin^2\theta_1 - n_2^2)^{\frac{1}{2}}/(n_1 \cos\theta_1);$$
$$\tan\Phi_{10} = (n_1^2 \sin^2\theta_1 - n_0^2)^{\frac{1}{2}}/(n_1 \cos\theta_1) \qquad (4)$$

for the TE waves, and

$$\tan\Phi_{12} = n_1^2(n_1^2 \sin^2\theta_1 - n_2^2)^{\frac{1}{2}}/(n_2^2n_1 \cos\theta_1);$$
$$\tan\Phi_{10} = n_1^2(n_1^2 \sin^2\theta_1 - n_0^2)^{\frac{1}{2}}/(n_0^2n_1 \cos\theta_1) \qquad (5)$$

for the TM waves.

It is clear that in spite of the zigzag wave motion described above, the wave in a waveguide mode appears to propagate in the horizontal direction only; the vertical part of the wave motion simply forms a standing wave between the two film surfaces. To avoid confusion, it is desirable to use β and v exclusively for the phase constant and the wave velocity parallel to the film. Thus,

$$\beta = kn_1 \sin\theta_1, \qquad v = c(k/\beta). \qquad (6)$$

Another quantity which will also be used frequently is the ratio β/k. As shown in Eqs. (6), it is the ratio of the speed of light in vacuum to the speed of wave propagation in the waveguide.

After substituting Eqs. (4) or (5) into Eq. (3), we find that both Eqs. (3) and (6) are transcendental equations. Fortunately, the transcendental functions involve θ_1 only. For a given n_0, n_1, n_2, and m we may easily compute both β/k and W for a common θ_1, and then tabulate β/k and W by assigning different values for θ_1. The curves showing W vs β/k using m as the parameter are the mode characteristics of the waveguide. They will be shown later, for example, in Fig. 23.

To summarize, any radius of the quarter-circle shown in Fig. 7 represents a possible direction for the

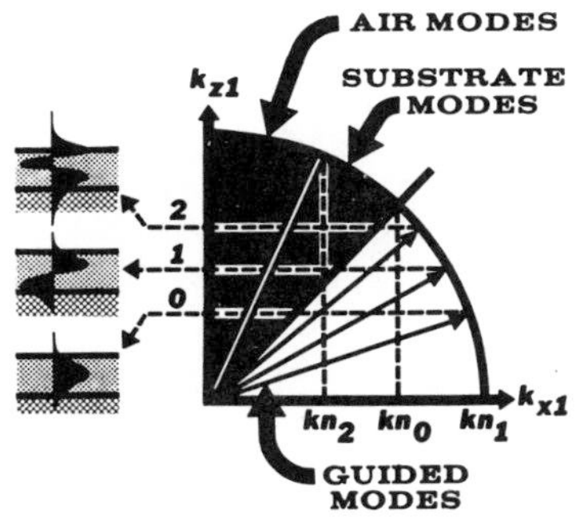

Fig. 7. Any radius of the quarter-circle at the right side of the figure represents a possible direction for the wave vector B_1. In the black region of the circle, the wave vector represents the substrate or air mode. In the white region of the circle, the wave vector represents the waveguide mode, but only a discrete set of the directions in this region satisfies the equation of the waveguide modes. Each direction of this discrete set represents one waveguide mode and each waveguide mode has its own field distribution as shown in the left side of the figure.

wave vector B_1 described above, and θ_1 is the incident angle measured between the wave vector and the vertical axis. The waveguide modes occur in the range $\sin^{-1}(n_0/n_1) < \theta_1 < \pi/2$. Within this range of θ_1, there is a discrete set of the directions which satisfies the equation of the modes (3). Each direction corresponds to one waveguide mode of the film. The horizontal component of the wave vector, $kn_1 \sin\theta_1$, determines the wave motion parallel to the film, while its vertical component, $kn_1 \cos\theta_1$, determines the standing wave field pattern across the film. As shown in the left side of Fig. 7, when $m = 0$, the standing wave pattern has a form similar to a half-sine wave. When $m = 1$, it has a form similar to a full sine wave, and so on. The air and substrate modes occur in the range $0 < \theta_1 < \sin^{-1}(n_0/n_1)$; they occupy the black region of the quarter-circle. As we vary θ_1 continuously from 0 to $\sin^{-1}(n_2/n_1)$ for the air modes and $\sin^{-1}(n_2/n_1)$ to $\sin^{-1}(n_0/n_1)$ for the substrate modes, the corresponding θ_0 and θ_2 sweep through the entire space of the substrate and the air space. It is thus possible to express any radiation field by superposing waves of the air and substrate modes. What we have discussed here is therefore simply an expansion of the solution of the Maxwell equation into plane waves of all possible directions.

III. Wave Equation and the Field Distribution

Having described the modes of light-wave propagation purely on an intuitive basis, we may now derive them mathematically. For simplification, assume the light wave in the film to be infinitely wide in the Y direction so that $\partial/\partial y = 0$ (Fig. 1). Let X be the direction of the wave propagation parallel to the film. The Maxwell equations in E_y for TE waves (or H_y for TM waves) can be reduced to the wave equation below.

$$\partial^2 E/\partial x^2 + \partial^2 E/\partial z^2 = -(kn_j)^2 E, \quad j = 0, 1, \text{ or } 2, \tag{6}$$

where n_j is the refractive index of the medium j. The subscripts $j = 0,1$, and 2 denote the substrate, the film, and the air space, respectively. A time dependence $\exp(-i\omega t)$ is used in Eq. (6), where $i = \sqrt{-1}$. The solution of the wave equation is in the form of $\exp(ik_{xj}x) \exp(\pm ik_{zj}z)$, which may be substituted into Eq. (6) to obtain

$$k_{xj}^2 + k_{zj}^2 = (kn)_j^2. \tag{7}$$

The boundary conditions at the film–air and film–substrate interfaces demand a same wave motion parallel to the film in all the three media considered; we may thus put

$$k_{x0} = k_{x1} = k_{x2} = \beta. \tag{8}$$

All the fields thus vary in time and x according to the factor $\exp(-i\omega t + i\beta x)$. This common factor will be omitted in all the later expressions for simplification. Combining Eqs. (7) and (8), we obtain an important relation,

$$k_{zj} = (k^2 n_j^2 - \beta^2)^{\frac{1}{2}}. \tag{9}$$

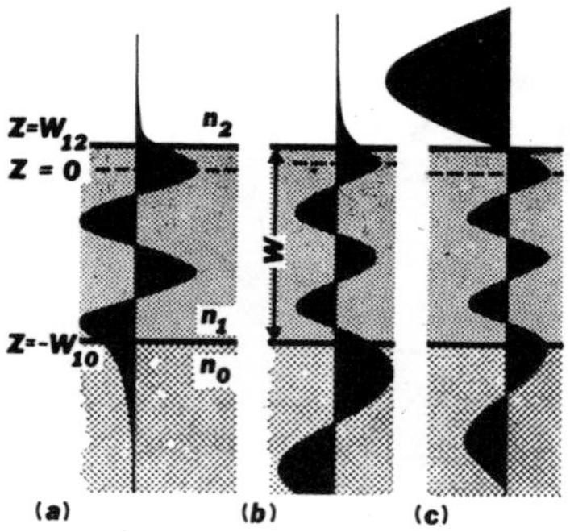

Fig. 8. The electric field distribution of (a) a TE waveguide mode; (b) a TE substrate mode; (c) a TE (even) air mode.

In the film, k_{x1} and k_{z1} are the horizontal and vertical components of the wave vector A_1 or B_1 discussed before. They are, respectively, $k_{x1} = \beta = kn_1 \sin\theta_1$ and $k_{z1} = kn_1 \cos\theta_1$. In the waveguide modes, we find from Eq. (9) and from the condition $\sin^{-1}(n_0/n_1) < \theta_1 < \pi/2$ that $kn_0 < \beta < kn_1$, k_{z1} is real, and k_{z0} and k_{z2} are imaginary. The field distribution in Fig. 8(a) is thus a standing wave in the film and exponential in the substrate and in the air space. Next, for the substrate modes, we find from Eq. (9) and from the condition $\sin^{-1}(n_2/n_1) < \theta_1 < \sin^{-1}(n_0/n_1)$ that k_{z1} and k_{z0} are real, but k_{z2} is imaginary. The fields in this case are standing waves in the film and in the substrate, but exponential in the air space (Fig. 8b). Finally, for the air modes, we find that $0 < \theta_1 < \sin^{-1}(n_2/n_1)$, and k_{z0}, k_{z1}, and k_{z2} are all real. The fields in all the three media are now standing waves (Fig. 8c). It is convenient to denote k_{zj} by b_j when it is real and by ip_j when it is imaginary. For $n_0 \neq n_2$, such as the case that is considered throughout this paper, the waveguide is asymmetric. We choose $z = W_{12}$ and $z = -W_{10}$ as the upper and lower film surfaces. The thickness of the film is then $W = W_{10} + W_{12}$.

The field distributions are derived by choosing $z = 0$ at the position where E_y is maximum for any waveguide, substrate, or even air mode, and $E_y = 0$ for any odd air mode. It is important to note that these positions of $z = 0$ are different for different modes in an asymmetric waveguide. These choices are necessary in order to simplify mathematics so that we can visualize the field distributions of various modes easily. To avoid confusion, we consider below E_y of a TE wave only.

For the waveguide modes, as mentioned earlier, the wave suffers a phase change of $-2\Phi_{12}$ at the upper film surface, and a phase change of $-2\Phi_{10}$ at the lower film surface because of the internal total reflections. The fields at the two film surfaces must therefore be $\pm A \cos\Phi_{12}$ and $\pm A \cos\Phi_{10}$, respectively, where A is a constant. Let the field at $z = 0$ be a maximum value, A. Then, we choose $k_{z1}W_{12}$ (or b_1W_{12}) $= \Phi_{12}$ so that the field at the upper film surface, $z = W_{12}$, can be $A \cos\Phi_{12}$. Similarly we choose $b_1W_{10} = \Phi_{10} + m\pi$ so that the field at the lower film surface, $z = -W_{10}$, can be $A \cos\Phi_{10}$ if $m =$ even and $-A \cos\Phi_{10}$ if $m =$ odd as shown in Fig. 8(a). These choices give $b_1W =$

Table I. Electric Field Distribution in (a) a Waveguide Mode, (b) a Substrate Mode, and (c) the Even and Odd Air Modes[a]

Waveguide Mode $\sin^{-1}(n_0/n_1) < \theta_1 < \pi/2;\ kn_0 < \beta < kn_1$			
Medium	k_{xj}	k_{zj}	E_y (TE wave)
Film	$=\beta$	$=b_1$	$A\cos b_1 z$
Substrate	$=\beta$	$=ip_0$	$A\cos(\Phi_{10}+m\pi)\exp[-p_0(\lvert z\rvert - W_{10})]$
Air-space	$=\beta$	$=ip_2$	$A\cos\Phi_{12}\exp[-p_2(\lvert z\rvert - W_{12})]$

Substrate Mode $\sin^{-1}(n_2/n_1) < \theta_1 < \sin^{-1}(n_0/n_1);\ \sin^{-1}(n_2/n_0) < \theta_0 < \pi/2;\ kn_2 < \beta < kn_0$			
Medium	k_{xj}	k_{zj}	E_y (TE wave)
Film	$=\beta$	$=b_1$	$A\cos b_1 z$
Substrate	$=\beta$	$=b_0$	$\frac{1}{2}A\,[\cos(b_1W_{10}) - i(b_1/b_0)\sin(b_1W_{10})]\cdot\exp[-ib_0(\lvert z\rvert - W_{10})] + \text{c.c.}$
Air space	$=\beta$	$=ip_2$	$A\cos\Phi_{12}\exp[-p_2(\lvert z\rvert - W_{12})]$

Even and Odd Air Modes $0 < \theta_1 < \sin^{-1}(n_2/n_1);\ 0 < \theta_0 < \sin^{-1}(n_2/n_0);\ 0 < \theta_2 < \pi/2;\ 0 < \beta < kn_2$				
Medium	k_{xj}	k_{zj}		
Film	$=\beta$	$=b_1$	Even	$A\cos b_1 z$
			Odd	$A\sin b_1 z$
Substrate	$=\beta$	$=b_0$	Even	$\frac{1}{2}A[\cos(b_1W_{10}) - i(b_1/b_0)\sin(b_1W_{10})]\exp[-ib_0(\lvert z\rvert - W_{10})] + \text{c.c.}$
			Odd	$-\frac{1}{2}A[\sin(b_1W_{10}) + i(b_1/b_0)\cos(b_1W_{10})]\exp[-ib_0(\lvert z\rvert - W_{10})] + \text{c.c.}$
Air space	$=\beta$	$=b_2$	Even	$\frac{1}{2}A[\cos(b_1W_{12}) - i(b_1/b_2)\sin(b_1W_{12})]\exp[-ib_2(\lvert z\rvert - W_{12})] + \text{c.c.}$
			Odd	$\frac{1}{2}A[\sin(b_1W_{12}) + i(b_1/b_2)\cos(b_1W_{12})]\exp[-ib_2(\lvert z\rvert - W_{12})] + \text{c.c.}$

[a] In deriving these expressions, we have chosen $z = 0$ at the position where E_y is either zero or maximum. These positions of $z = 0$ are therefore different for different modes.

$b_1W_{10} + b_1W_{12} = \Phi_{12} + \Phi_{10} + m\pi$, which satisfies Eq. (3). The boundary conditions require E_y and $\partial E_y/\partial_z$ to be continuous at the two interfaces. We have, therefore,

$$E_y = A\cos\Phi_{12}\exp[-p_2(\lvert z\rvert - W_{12})]$$

in the air space and

$$E_y = A\cos(\Phi_{10} + m\pi)\exp[-p_0(\lvert z\rvert - W_{10})]$$

in the substrate.

For the substrate modes, we again assume a maximum field A at $z = 0$ and choose $b_1W_{12} = \Phi_{12}$ (Fig. 8b). The field at $z = W_{12}$ is still $A\cos\Phi_{12}$ and that in the air space is still $A\cos\Phi_{12}\exp[-p_2(z - W_{12})]$. The field at the lower film surface is then $A\cos(b_1W_{10})$ and that in the substrate is

$$\tfrac{1}{2}A[\cos(b_1W_{10}) - i(b_1/b_0)\sin(b_1W_{10})]\exp[-ib_0(\lvert z\rvert - W_{10})] + \text{the complex conjugate.}$$

For the air modes, the even and odd modes must be treated separately. For an asymmetric waveguide, we can choose the $z = 0$ plane anywhere between $z = W_{12}$ and $z = -W_{10}$. However, once it is chosen, the same $z = 0$ plane should be used for all the air modes. For the even modes, the field is a maximum at $z = 0$ and the fields at the two film surfaces are $A\cos b_1W_{12}$ and $A\cos b_1W_{10}$, respectively (Fig. 8c).The boundary conditions require the fields in the substrate and in the air space in the form

$$\tfrac{1}{2}A[\cos(b_1W_{1j}) - i(b_1/b_j)\sin(b_1W_{1j})]\exp[-ib_j(\lvert z\rvert - W_{1j})] + \text{the complex conjugate,}$$

where $j = 0$ and 2. For the odd modes the field is zero at $z = 0$ and is $A\sin(b_1W_{12})$ and $-A\sin(b_1W_{10})$ at the film surfaces. The fields in the substrate and air space are then

$$\pm\tfrac{1}{2}A[\sin(b_1W_{1j}) + i(b_1/b_j)\cos(b_1W_{1j})]\exp[-ib_j(\lvert z\rvert - W_{1j})] + \text{the complex conjugate,}$$

where the plus sign is for $j = 2$ and the minus sign is for $j = 0$. The results discussed above are summarized in Table I.

Mathematically, the field distributions described above are identical to those of the problem of a square potential well in quantum mechanics. Here the air space and the substrate are the potential barriers. We divide the wave energy here into the horizontal and vertical components, keeping the total energy constant. It is the vertical component of the wave energy that negotiates the potential barriers mentioned above. The wave vector represents the momentum and its square, the wave energy. Within the interval $\beta = kn_1$ and $\beta = kn_0$, because of the large horizontal component of the wave vector β, the vertical component of the energy is small enough so that the wave, or the particle, is trapped in the potential well. The mode spectrum or the energy level is thus discrete (waveguide modes). As the horizontal component of the momentum is reduced to a value $\beta < kn_0$, the vertical component of the wave energy is large enough to overcome the lower potential barrier. The wave function spills over the entire substrate space and we enter into the region of the substrate modes. The mode spectrum or the energy level is now continuous. As we increase further the vertical component of the wave energy by reducing β below kn_2, the wave can spill over the upper and the lower barriers. The mode spectrum remains continuous and it belongs to the air modes.

IV. Light-Wave Couplers

The development of the light-wave couplers in the past two years is an important step forward in thin-film optoelectronics. We can now couple a laser beam efficiently into and out of any thin-film structure and can excite there any single mode of light-wave propagation. In both the prism–film and grating couplers, we feed a light beam into a film through a broad surface of the film and thus avoid the difficult problem of focusing a light beam through a rough film edge. Since the film and the prism (or the grating) are coupled over a length of many optical wavelengths, we can imagine energy transfer taking place continuously between them as waves propagate over the coupled region. It is possible to discuss this type of distributed couplers by a unified theory. We can further show that these couplers have the same optimum coupling efficiency of about 81%, provided that both the coupling strength and the intensity of the incoming laser beam are uniformly distributed over the entire coupling length. An even better efficiency can be achieved by varying the coupling strength along the coupling length in a prescribed manner. By simplifying an earlier theory[6] and by using illustrative figures, we will describe below the principles of the couplers and derive their coupling efficiency in very simple terms.

Figure 9 shows a prism–film coupler. In order to excite all possible waveguide modes in the film, the refractive index of the prism n_3 should be larger than that of the film n_1. An incoming laser beam enters the prism and is totally reflected at the base of the prism. Because of the total reflection, the field in the prism is a standing wave that continues into an exponentially decreasing function below the base of the prism. The part of the field that extends below the prism base is called the evanescent field, since it decreases rapidly away from the prism and does not represent a free radiation. If we represent the incoming wave in the prism by a wave vector A_3 (Fig. 10), it has a magnitude kn_3 and can be decomposed into a horizontal component $kn_3 \sin\theta_3$ and a vertical component $kn_3 \cos\theta_3$. The boundary conditions of the electromagnetic fields at the prism base require that the fields below and above the prism base have the same horizontal wave motion. The evanescent field varies therefore as $\exp(ikn_3x \sin\theta_3)$ in x. Now we place the prism on top of a thin film, maintaining a small but uniform air gap between the base of the prism and the top surface of the film. For effective coupling, the spacing of the air gap is on the order of one-eighth to one-fourth of the vacuum optical wavelength. The evanescent field below the prism then penetrates into the film and excites a light wave into the film. We call this coupling process the optical tunneling. As discussed in Sec. I, the film has many waveguide modes. If the horizontal component of the wave vector A_1 or B_1 of one of the waveguide modes happens to be equal to that of the incoming light wave in the prism $kn_3 \sin\theta_3$, the light wave in the prism is coupled exclusively to this waveguide mode and the laser beam is said to be in a synchronous direction. It is therefore possible to couple the light wave to any waveguide mode by simply choosing a proper direction θ_3 for the incoming laser beam.

When the laser beam is in a synchronous direction, the waves in the prism and in the film have the same horizontal wave motion. The fields at the two opposite sides of the air gap are in phase at every point along x. As shown in Fig. 9, the field in the waveguide mode has an exponential tail extending upward above the film. The evanescent field of the prism is an exponential extending downward below the prism. These two exponential tails overlap in the air gap. The parts of the fields that overlap are common to the prism and the film and constitute the coupling between them.

Let a_3 and b_3 be the field amplitudes of the incoming and reflected waves in the prism and let a_1 and b_1 be the field amplitudes of the zigzag waves in the waveguide mode of the film (Fig. 10). The a_1 and b_1 waves are represented by the wave vectors A_1 and B_1 in the

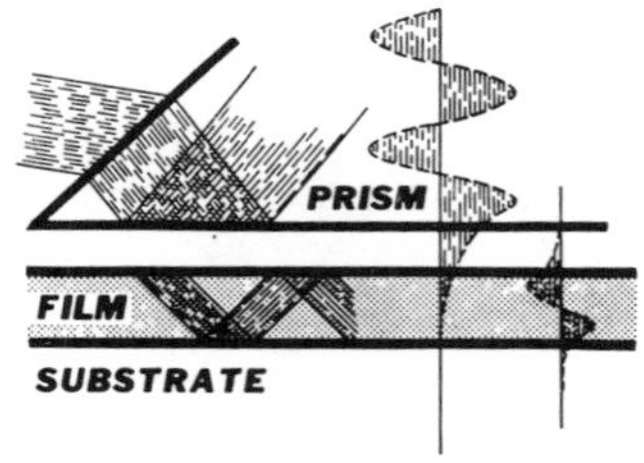

Fig. 9. In a prism–film coupler, the light wave from a laser is totally reflected at the prism base. The field distributions in the prism and in the film show that their evanescent fields overlap each other in the gap region.

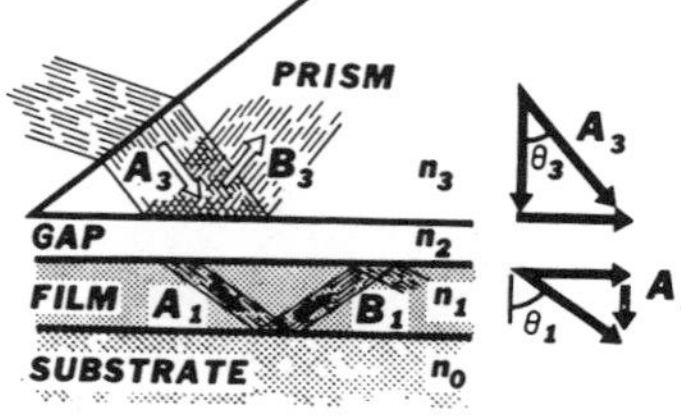

Fig. 10. In order that a waveguide mode be excited in a film by a prism–film coupler, the horizontal component of the wave vector for the wave in the prism must be equal to that of the wave in the film. The prism–film coupler, therefore, excites a single waveguide mode only, and by changing the direction of the incoming light wave, any waveguide mode can be excited.

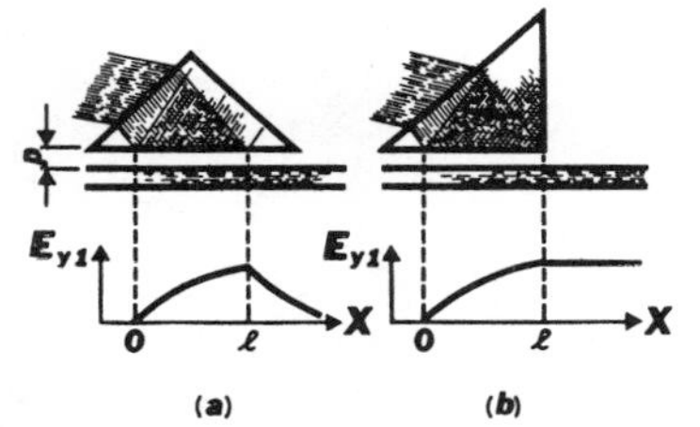

Fig. 11. (a) The light energy transferred from the prism to the film in the region $0 < X < l$ is returned to the prism in the region $X > l$; the net energy retained in the film is therefore zero. (b) By using a right-angle prism, the coupling between the prism and the film discontinues in the region $X > l$. The light wave coupled into the film in the region $0 < X < l$ is therefore retained in the film and continues to propagate in the film.

earlier discussion, and a_1 may be considered as the reflection of the b_1 wave, or vice versa, so that $|a_1| = |b_1|$ at any x. Let all the wave amplitudes be normalized such that $a_j a_j^*$ or $b_j b_j^*$ is the Poynting vector in the direction normal to the film, where $j = 1$ or 3. Because of the coupling described above, the energy is continuously transferred from the prism to the film along the coupling length which starts from $x = 0$. Since the Maxwell equations are linear in field amplitudes, we expect that a_1 (or b_1) increases in x according to a_3, or that da_1/dx should be linearly proportional to a_3. On the other hand, as soon as the wave energy in the film builds up, it continuously leaks into the prism, since the energy transfer is possible in both ways between the prism and the film. We ought to expect, then, that da_1/dx is also proportional $(-a_1)$. We have thus,

$$da_1/dx = Ta_3 - Sa_1, \tag{10}$$

where T and S are the coupling constants that depend on the geometrical configuration and the refractive indices of the media. Near $x = 0$, a_1 is small and so is the term Sa_1 in Eq. (10); a_1 increases linearly from $x = 0$ according to Ta_3x. At a large x, a_1 grows to an amplitude so that Sa_1 approaches a value that nearly cancels the term Ta_3 in Eq. (10); $da_1/dx = 0$ and a_1 reaches a saturation. The wave amplitude in the film cannot therefore increase indefinitely by simply increasing the coupling length.

In Fig. 11(a), we assume that a_3 is uniformly distributed between $x = 0$ and $x = l$, which are the left and right edges of the laser beam incident on the prism base. The amplitude a_1 increases in x until the point $x = l$. Beyond that point, $a_3 = 0$ in Eq. (10) and the equation indicates that a_1 should decrease exponentially to zero according to $\exp[-Sx]$. All the energy fed into the film between $x = 0$ and l is returned to the prism at $x > l$ and therefore the net energy transfer from the prism to the film is zero. If the film is not perfect and scatters the light, a more complex phenomenon occurs. Since the incident laser beam is in the synchronous direction, the light energy is coupled into one of the waveguide modes of the film. However, the energy in the original waveguide mode can be rapidly scattered into other waveguide modes before it is coupled back to the prism. The returned light wave in the prism therefore consists of many waveguide modes; each of them appears in its own synchronous direction. We thus observe a series of bright lines at the right side of the prism. They are called the m lines.[3] For good film, the m lines are thin and weak.

In Fig. 11(b), we use a rectangular prism. Here the rectangular corner of the prism is placed at $x = l$. Contrary to the earlier case, here the coupling between the prism and the film no longer exists beyond $x = l$. The wave energy which is fed into the film between $x = 0$ and l is retained in the film as the wave continues to propagate beyond $x = l$. Therefore, for coupling light energy into or out of a film, we always use a rectangular prism and place the right edge of the laser beam as close as possible to the rectangular corner of the prism.

We notice that because of their directions, the a_3 wave in the prism is coupled only with the a_1 wave in the film and, similarly, the b_1 wave is coupled only to the b_3 wave. Of course, a_1 and b_1 must increase and decrease together in x, since each of them is the reflection of the other. To calculate the coupling efficiency, we consider first the prism-film coupler in Fig. 12, which is used as an output coupler, to couple light energy out of the film. There is a light wave propagating in the film, which is represented by the a_1 and b_1 waves in the film. There is no input laser beam and thus $a_3 = 0$ everywhere. Beyond $x = 0$, the film is coupled to the prism; consequently, the amplitude of the b_1 (or a_1) wave decreases as light energy in the film leaks into the prism. By replacing a_1 by b_1 and putting $a_3 = 0$ in Eq. (10), we have

$$\begin{aligned} b_1(x) &= b_1(0)\exp[-Sx], \quad x > 0, \\ &= b_1(0) \qquad\qquad\quad , \quad x < 0. \end{aligned} \tag{11}$$

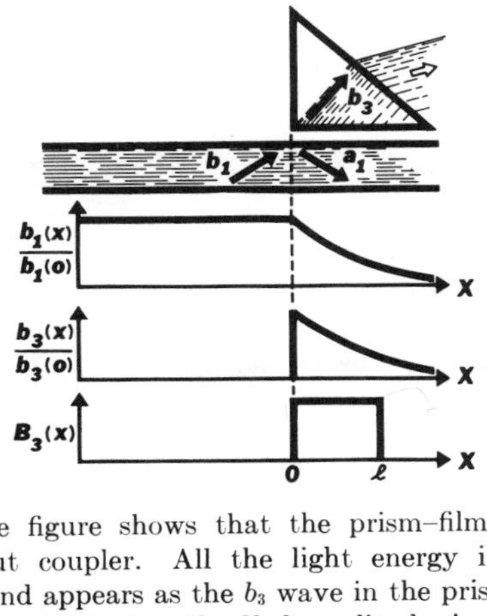

Fig. 12. The figure shows that the prism–film coupler is a perfect output coupler. All the light energy in the film is coupled out and appears as the b_3 wave in the prism. $b_1(x)$ and $b_3(x)$ show respectively how the filed amplitudes in the film and in the prism vary along the coupling length. $B_3(x)$ is the distribution of an incoming laser beam, which is used to demonstrate how to calculate the coupling efficiency.

It is easy to see that the total power flow in the film, which is proportional to $b_1(x)b_1{}^*(x)$, must decrease as $\exp[-2Sx]$. The power lost in the film in a small distance dx is then proportional to $\partial[b_1(x)b_1{}^*(x)]/\partial x$ and it must reappear in the prism as the b_3 wave. We have thus immediately

$$b_3(x) = b_3(0)\exp[-Sx], \quad x > 0$$
$$= 0 \quad , \quad x < 0. \tag{12}$$

Both $b_1(x)$ and $b_3(x)$ are plotted in Fig. 12. We see from Eq. (11) that eventually all the power in the film will be transferred into the prism. An output coupler is always at perfect output coupler, provided that the coupling length is sufficiently long. Now the reciprocity of the linear optics indicates that we can reverse the process. Consequently, if we apply a laser beam in the prism in the direction opposite to that of the b_3 wave discussed above, and if it has an amplitude distribution exactly as $b_3(x)$ in Fig. 12, all the applied laser energy should then enter into the film as to be expected in a perfect input coupler. On the other hand, if we apply a laser beam which is uniform over the cross section as shown in $B_3(x)$ in Fig. 12, we expect that a part of $B_3(x)$ which matches $b_3(x)$ is accepted into the film and the rest is reflected at the prism base. We can therefore define an overlap integral[11]

$$\eta = \frac{\left[\int_{-\infty}^{+\infty} B_3(x)b_3{}^*(x)dx\right]^2}{\int_{-\infty}^{+\infty} B_3(x)B_3{}^*(x)dx \int_{-\infty}^{+\infty} b_3(x)b_3{}^*(x)dx}, \tag{13}$$

which specifies the correlation between the amplitude distribution of the input laser beam and that required for a perfect coupler. In fact, η in Eq. (13) is the coupling efficiency. Since here $B_3(x)$ is constant between $x = 0$ and l, and $b_3(x)$ is exponential beyond $x = 0$, the integral in Eq. (13) can easily be performed. We have

$$\eta = 2/Sl[1 - e^{-Sl}]^2, \tag{14}$$

which is identical to the formula given in the earlier papers.[6,7] By maximizing the expression (14) with respect to Sl, we find the optimum coupling length $Sl = 1.25$, and the optimum coupling efficiency $\eta = 81\%$. We notice that all the properties of the coupler depend on the parameter S which can be computed from the geometrical configuration and the refractive indices of the coupler, or it can be determined experimentally. A detailed calculation[6] shows

$$S = e^{-2p_2 d}\ \sin 2\Phi_{12} \sin 2\Phi_{32}/(W_{\text{eff}} \tan\theta_1), \tag{15}$$

where $p_2 = [\beta^2 - (kn_2)^2]^{\frac{1}{2}}$; d is the spacing of the air gap between the prism and the film, W_{eff} is the effective thickness of the film, and Φ_{32} may be obtained from Eq. (4) or (5) by replacing the subscript 1 by 3. The effective thickness W_{eff} will be discussed later in Sec. VII. The use of W_{eff} instead of the actual thickness of the film W is due to the Goos-Haenchen shifts.[24] The coupling efficiency does not differ significantly if a gaussian input field distribution is used.

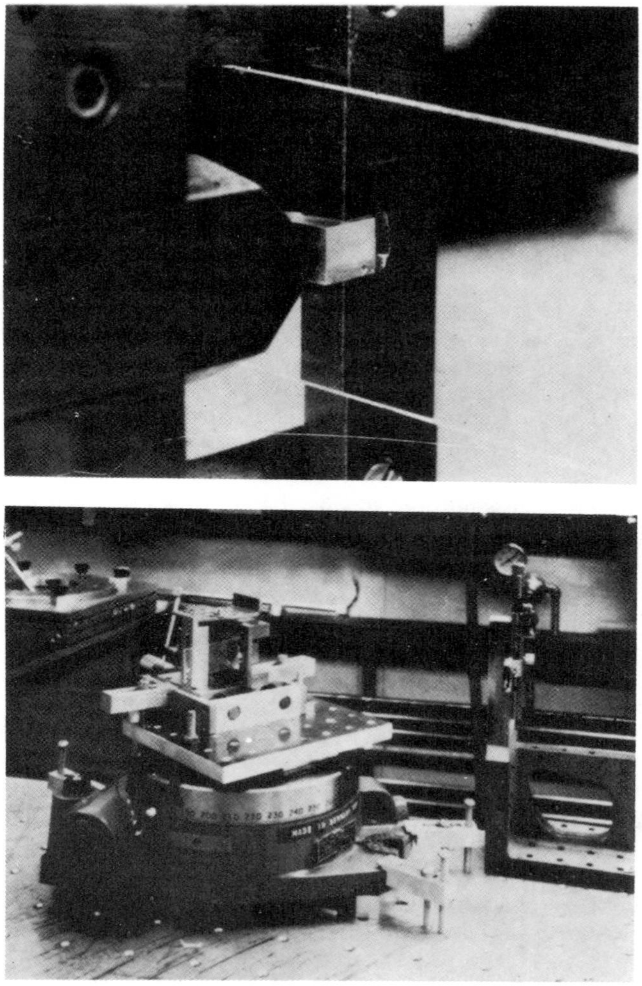

Fig. 13. The top photograph shows how a thin film coated on the glass substrate is pressed against the base of the prism in a prism–film coupler. The lower photograph shows that the entire prism–film assemblage is mounted on a turntable so that the incident laser beam can enter into the prism at any angle.

From the above discussion, we realize that when the prism–film coupler is used as an output coupler, it can easily be made 100% efficient. The light which is not coupled out, say up to the point $x = x_a$, remains in the film. It thus can always be coupled out at $x > x_a$, provided that the coupling length is sufficiently long. In contrast, an input coupler is 100% efficient only when the input light is properly distributed along the coupling gap, since the uncoupled light is immediately lost upon being reflected at the prism base.

The field distribution of a perfect coupler, $b_3(x)$, in Fig. 12 tends to be more uniform and thus matches better the uniform input field distribution, if we can decrease the coupling strength at $x = 0$ and increase it at $x = l$ (refer to Fig. 12) by properly varying the gap spacing between film and prism. We accomplish this in the following way.

Figure 13 shows a prism–film coupler which is mounted on a turntable so that a laser beam can enter

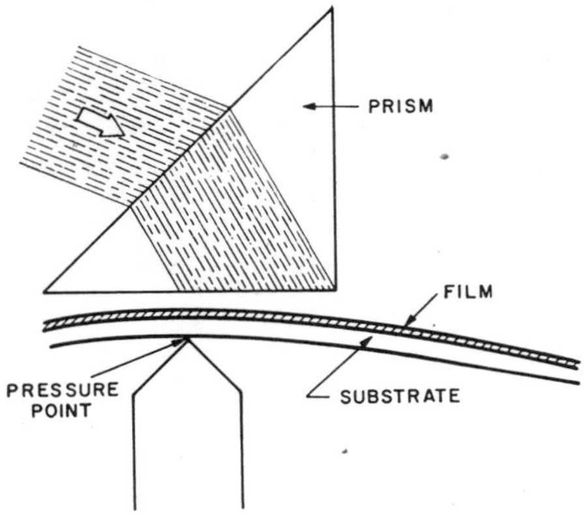

Fig. 14. When the prism–film coupler is used as an input coupler, the pressure is applied at a point about a few tenths of a millimeter from the rectangular corner of the prism. Because of this pressure, the glass substrate bends slightly so that the gap between the prism base and the film is the smallest at the pressure point. To achieve a high coupling efficiency, the incident laser beam should fill the region between the pressure point and the rectangular corner of the prism.

the prism at any angle. The film is coated on a glass slide and pressed against the base of the prism by a knife edge, while the dust particles between the prism and the film act as the spacers. The pressure point is about 1 mm or less away from the rectangular corner of the prism. The pressure applied to the back side of the glass slide actually bends the slide slightly so that the gap between the prism base and the film is smaller at the pressure point and larger at the corner of the prism, as shown in Fig. 14. This provides a stronger coupling at $x = l$ and weaker coupling at $x = 0$ (Fig. 12) and between these two points the input laser beam is located. We thus approach the ideal condition mentioned earlier. In practice, it is easy to obtain a coupling efficiency of about 60%. Beyond that percentage, one needs a prism having a sharp rectangular corner, since any imperfection in the corner would radiate light energy and thus limit the attainable coupling efficiency. Figure 15 shows a laser beam that enters an organic film through the prism–film coupler at the right. The light beam propagated through the film and then was taken out of the film by another prism–film coupler at the left.

The prism–film coupler serves many useful functions in an experiment. By correlating the measured values of the synchronous directions with a theoretical calculation on the waveguide modes, one can independently determine the refractive index and the thickness of the film.[3] The prism–film coupler has been extensively used to determine the refractive index of organosilicon films.[15] The accuracy obtained is within 1 part in 1000 for the refractive index and 1% for the thickness. We often use a rutile prism for semiconductor films and a glass prism for organic and glass films.

Figure 16 shows a grating coupler.[4,5] A phase grating made of photoresist or dichromated gelatin is fabricated by holographic technique on a thin film. A laser beam incident on the phase grating at an angle θ has a phase variation in the x direction according to $\exp[i(2\pi/\lambda_0)(\sin\theta)x]$, where λ_0 is the vacuum laser wavelength. As the beam passes through the grating, it obtains an additional spatial phase modulation $\Delta\Phi \sin(2\pi x/d)$, where $\Delta\Phi$ is the amplitude of the spatial phase modulation caused by the grating and is sometimes called the phase depth of the grating; d is the periodicity of the grating. The light wave reaching to the top surface of the film contains many fourier components, $\exp\{i(2\pi/\lambda_0)[\sin\theta + m(2\pi/d)]x\}$, where m is an integer. If one of these components matches the wave motion of one of the waveguide modes of the film, the light beam is exclusively coupled to this mode and the light energy is fed into the film. We can analyze the grating coupler by the same method used previously for the prism–film coupler. The only difference is the calculation of the parameter S which, in practice, is determined experimentally anyway.

The tapered-film coupler[12] is operated on an entirely different principle. It utilizes the cutoff property of an asymmetric waveguide. We remember in the discussion of the waveguide modes that the thickness of the film W can be divided into two parts W_{12} and W_{10}

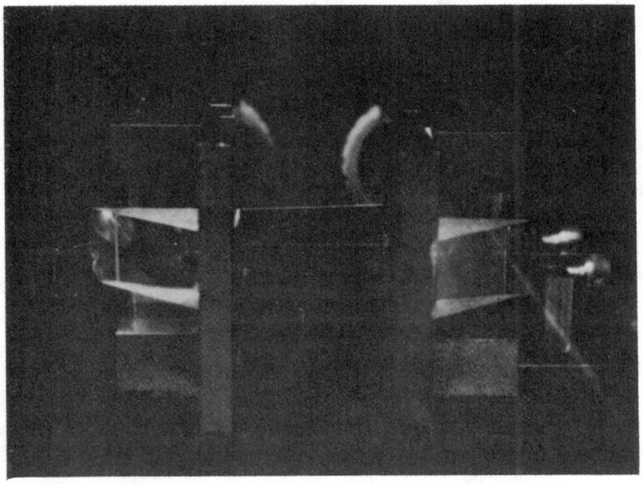

Fig. 15. The photograph shows a light beam which is fed into an organic film by the prism–film coupler at the right. The light wave propagates inside the film and is then coupled out of the film at the left by another prism-film coupler. (This figure is reproduced in color on the cover of the November 1971 *Applied Optics* issue.)

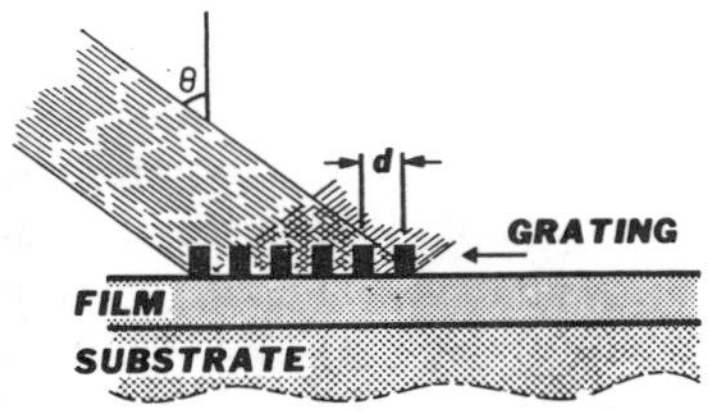

Fig. 16. A grating light-wave coupler.

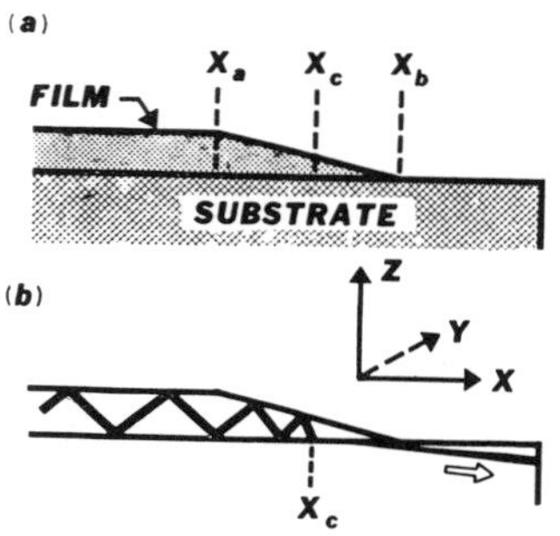

Fig. 17. (a) A tapered film light-wave coupler is simply a tapered film edge deposited on a substrate. The figure shows that the film is tapered to nothing between X_a and X_c. (b) As a light wave originally propagating inside the film enters into the tapered region of the film, the angle between the zigzag light path and the vertical Z axis becomes smaller and smaller. At the cut-off point, $X = X_c$, the angle is smaller than the critical angle of the film–substrate interface and the light wave is refracted into the substrate. The tapered film edge then serves as an output light-wave coupler.

and that they are proportional to Φ_{12} and $(\Phi_{10} + m\pi)$, respectively. As β varies from kn_1 to kn_0, both Φ_{12} and Φ_{10} decrease. At the cutoff point of the waveguide modes, $\beta = kn_0$, $\Phi_{10} = 0$, and W is minimum. Therefore, there is a minimum thickness of the film, less than which a waveguide mode of the order m cannot propagate. Now consider, in Fig. 17(a), a film that is deposited on a substrate and is tapered to nothing in a distance between $x = x_a$ and $x = x_b$, typically in the order of 10 to 100 vacuum wavelengths. The film is in the X–Y plane and the tapered edge is parallel to the Y axis. Consider a light wave in a waveguide mode of the film propagating toward the tapered edge in the direction normal to it. Since the thickness of the film decreases continuously in the tapered region, the waveguide mode is cut off at $x = x_c$. A detailed calculation shows that within a distance of about eight vacuum wavelengths in the vicinity of x_c, the waveguide mode is gradually converted into the substrate modes and the light wave reappears in the substrate as the radiation field. The far-field pattern of the radiation field shows that more than 80% of its energy is concentrated within an angle of 15° below the film–substrate interface. We can understand the problem better by considering again the ray optics. Figure 17(b) shows a light wave that propagates in a zigzag path from the left side of the film toward the tapered film edge. As it enters into the tapered region, the angle between the light path and the Z axis becomes smaller and smaller, and eventually, near $x = x_c$, the angle becomes smaller than the critical angle of the film–substrate interface. The light beam is then refracted into the substrate. In the experiments,[12] it is very easy to couple all the light energy out of a film through the tapered film edge. By reversing the process, we can also feed a light beam into the film by focusing it on the tapered edge through the substrate. The tapered-film coupler is simply the film itself and is particularly useful to the study of the semiconductor epitax layers. Here the refractive index of the film such as the GaAs layer is so large that it becomes difficult to find a prism of a higher refractive index that is also transparent to the radiation used in the experiment.

V. Materials and Losses

Sputtered ZnO films were first used in the light-guiding experiments.[3] They were deposited on a heated glass substrate in an argon–oxygen atmosphere. The method of deposition was developed previously by Foster *et al.*[25,26] for ultrasonic transducers. The films discussed below were grown at a substrate temperature of 400°C. They have very high resistivities. After heat treatment in vacuum, nitrogen, or hydrogen, their resistivities can be reduced from 10^6 to 10^{-2} Ω-cm. The largest mobility measured is 40 cm²/V-sec as compared with that of bulk 200 cm²/V-sec. The films have the hexagonal or Wurzite crystal structure with the c axis normal to the surface of the film and the (0002) planes parallel to it. As determined from x-ray diffraction by using a Debye-Scherrer camera, the c axis of the crystallites is oriented within 5° from the normal of the film. The refractive index of the film is 1.973 ± 0.001 at 6328 Å of the helium–neon laser wavelength as compared with the value 1.988 of bulk ZnO.

It was indeed a surprise when we found that those seemingly perfect films had a loss of more than 60 dB/cm in the light propagation experiment.[3] The excessive loss in the film was not understood until we took electron micrographs of the films. Figure 18(a) shows

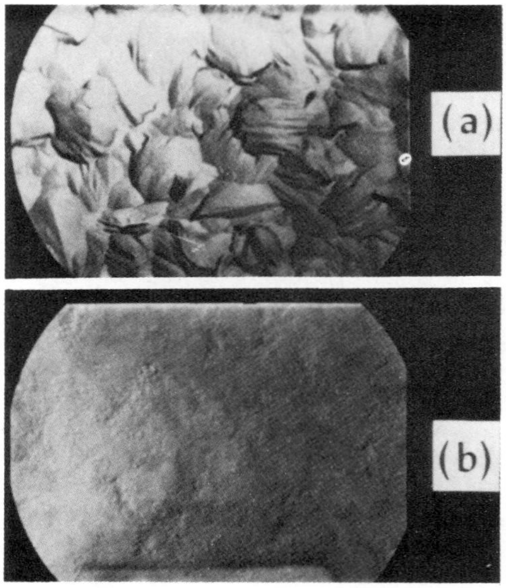

Fig. 18. (a) Electron micrograph of an oriented sputtered ZnO film showing that the sizes of the crystal sites are on the order of 0.5 μm. The entire width of the micrograph is 5 μm. (b) Electron-micrograph showing the surface of the same film after polishing.

the surface profile of a ZnO film 1.5 μ thick. This electron micrograph was taken from a platinum-shadowed carbon replica. It can be seen that the average grain size is on the order of 0.5 μ. This grain size is comparable to the optical wavelength used in the experiment and thus causes excessive scattering to the propagating light wave in the film. The film surface (Fig. 18a) has an irregular profile; the peak-to-peak surface roughness is estimated on the order of 1000 Å. It is interesting to note that it was possible to polish the surface of the film by lapping it with chromium oxide dispersed in water. Figure 18(b) shows the surface of the ZnO film after being lapped. We estimated that the residual roughness was reduced to less than 100 Å. After being polished, the films still have a loss of more than 20 dB/cm.

To avoid scattering caused by large crystals, we started to evaporate ZnS films on glass substrates which were held at room temperature. They are polycrystal films with very small crystal sites. Although we have significantly reduced the scattering loss, unfortunately additional loss was found due to the absorption arising from the long tail of the fundamental band gap. The refractive indices of the film are 2.404, 2.342, and 2.289 at the wavelengths 5322.5 Å, 6328 Å, and 1.0645 Å, respectively. The loss measured at 6328-Å wavelength is on the order of 5 dB/cm.

Now, trying to avoid both the scattering and absorption losses, we chose Ta_2O_5, which has a larger energy gap, 4.6 eV. The Ta_2O_5 films[14] were prepared by first sputtering high purity tantalum in an argon atmosphere. The β-tantalum thus deposited was then heated in pure oxygen at 500°C until completely converted to Ta_2O_5. Debye-Scherrer x-ray diffraction patterns show the films to be amorphous. In spite of many visible defects, the same conclusion can be drawn from the electron micrograph in Fig. 19. The refractive indices of the film are 2.2136, 2.2423, and 2.2767 at the wavelengths 6328 Å, 5145 Å, and 4880 Å, respectively. At the above wavelengths, the losses measured are 0.9 dB/cm for the red, 2.5 dB/cm for the green, and 4.1 dB/cm for the blue.

Fig. 19. Electron micrograph of a Ta_2O_5 film showing that the film is generally amorphous, though a number of small crystals visible in the micrograph still exist.

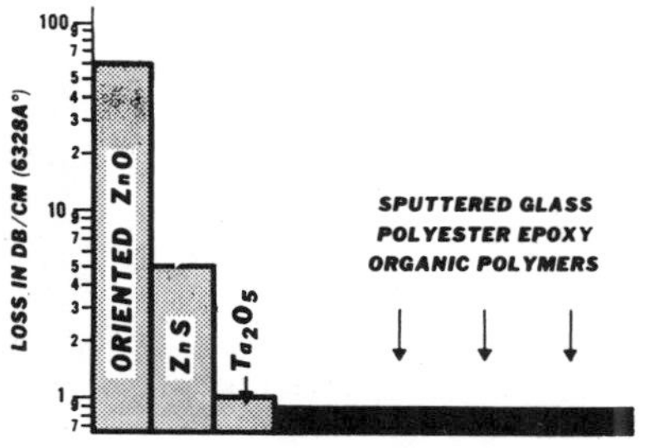

Fig. 20. Losses in decibels per centimeter measured at 6328-Å light wavelength for several semiconductor and organic films.

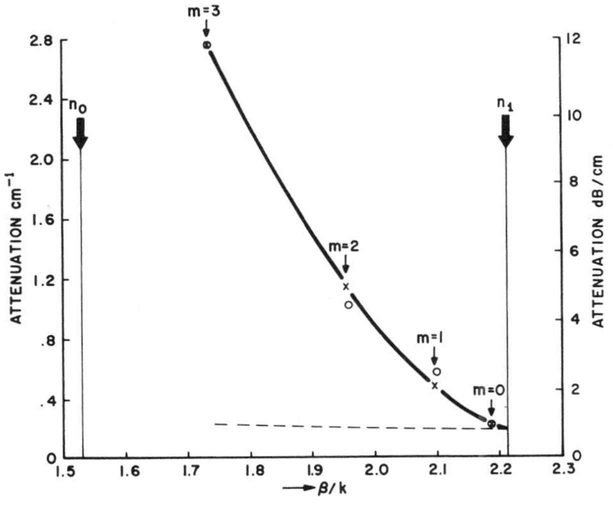

Fig. 21. Measurement (*circles*) and calculation (*crosses*) of the losses in a Ta_2O_5 film at 6328-Å light wavelength. The figure shows that the loss in db/cm or attenuation in cm^{-1} for the $m = 3$ waveguide mode is as much as 14 times that of the $m = 0$ waveguide mode. The dashed line is the volume loss in the film and the vertical distance between the dashed line and the solid curve is the surface scattering.

As shown in Fig. 20, all the low-loss films presently used in the light-guide experiments are amorphous. This includes sputtered glass,[13] polyurethane and polyester epoxy,[9] and organic polymer[15] films. All of these have a loss less than 1 dB/cm. To the author's knowledge, large transparent single-crystal semiconductor films other than epitax layers are not available at the present, although they are essential to the development of thin-film devices.

VI. Simple Theory of the Surface Scattering

The losses of the films quoted above are the losses of the $m = 0$ waveguide mode. In fact, the loss increases rapidly in the higher order modes. The circles in Fig. 21 are the measurement made in a Ta_2O_5 film by Hensler *et al.*[14] for a TE wave at 6328-Å light wavelength. The data show that the loss expressed in decibels per cm (or intensity attenuation in cm^{-1}) for the $m = 3$ mode is as much as fourteen times that of the $m = 0$ mode. In this figure, the losses of different modes are plotted vs β/k, where β is the horizontal component of the wave vector A_1 or B_1 as discussed at the end of Sec. II. It was pointed out earlier that the waveguide modes occur between $\beta/k = n_0$ and n_1, and that the value of β/k associated with each mode depends on the thickness of the film and the refractive indices of the film and substrate. It is evident in Fig. 21 that the loss for the $m = 0$ mode depends on the value of β/k; it represents neither volume loss of the material nor scattering loss caused by surface roughness. Therefore, it seems that there must be a better way to define the loss in a film other than that of a particular waveguide mode.

The losses in Fig. 21 were measured by a system of lenses, filter, slit, and detector which collected the light scattered from a small area of the film into a detector through an adjustable slit. By moving the detector system away from the input coupler but keeping it along a path parallel to the light beam in the film, we measured a decrease of scattered light along the light path, which should represent the intensity attenuation of the light wave propagating inside the film. The losses measured therefore included the volume absorption and scattering as well as the surface scattering. The same method of measurement was used by Goell and Standley[13] for their glass films.

Another method which was used extensively to measure losses in organosilicon films is the transmission measurement.[15] It involved two prism–film couplers in an arrangement similar to that shown in Fig. 15. One prism–film coupler excited a light streak in the film and a second coupler, several centimeters from the first, coupled the light wave out of the film. The efficiency of the input coupler was not measured, but it remained intact during the entire experiment. We accomplished this by monitoring the light scattered from a small section of the streak near the input end so that any change of the input conditions could be detected and corrected. The efficiency of the output coupler was always adjusted to be 100%. We mentioned earlier in Sec. IV that the output coupler can easily be made 100% efficient. In our experiment, the output coupler was applied at different points along the light streak. At each point the coupling was adjusted until the streak disappeared completely beyond the coupling point. The light emerging from the output prism was then detected. The measurements thus obtained at different points along the streak were used to evaluate the loss of the film.

It is possible to estimate the loss of a film based on the sensitivity of the eye. The sensitivity of the eye covers a range of about 27 dB. Thus if the length of the light streak as observed by the naked eye is x cm, the loss should be $27/x$ dB/cm.

The surface scattering of a symmetric slab waveguide has been calculated by Marcuse[21] based on the radiation modes. His results after certain approximations and evaluated to the limit of long correlation length agree with the simple theory that we will develop below. For an asymmetric waveguide ($n_2 \neq n_0$) which is considered in this paper, the two surfaces of the film scatter differently. Since it is practically impossible to measure the scattering losses of the two surfaces separately, we hope to develop a crude theory in which we can lump all the surface properties of the film into a single parameter. We wish further to use this parameter to calculate the losses of different waveguide modes and compare them with the measurement. Of course, the theory that we will discuss is very crude. However, it establishes a guideline by which we can gain insight into this complex problem.

Returning to Fig. 21, first, we must separate the volume loss from surface scattering. As β/k approaches n_1, the fields at two film surfaces vanish. We should then expect surface scattering to vanish and the residue loss at $\beta/k = n_1$ to be volume loss only. At the other values of β/k, the volume loss should be proportional to the length of the zigzag path. We showed earlier that the light wave, in the waveguide mode, is a plane wave which propagates along a zigzag path and a zigzag path is considerably longer than the actual length of the film. We have ignored here, for simplification, the fields extending outside the film. According to the above argument, the volume loss is proportional to $(\sin\theta_1)^{-1}$ or $(\beta/kn_1)^{-1}$. It is plotted as the dashed line in Fig. 21. We find that the volume loss of the $m = 3$ mode is merely 30% larger than that of the $m = 0$ mode as compared with a factor of 14 in the total losses. Consequently, almost all the losses in the higher order modes and the large variation of the losses among different waveguide modes are due to surface scattering only.

To develop a crude theory for surface scattering, we resort to the hundred-year-old Rayleigh criterion.[27] Figure 22 shows a plane wave incident on the upper surface of the film. To cover a unit length of the film in the x direction, the plane wave has a width $\cos\theta_1$ in the direction parallel to the wavefront. Considering a TE wave, the power carried by the incident beam is $(c/8\pi)n_1E_y^2 \cos\theta_1$ in gaussian units, where E_y is the field amplitude. Again, we are taking $\partial/\partial y = 0$ and considering a space of unit length in the y direction. According to the Rayleigh criterion, the specularly reflected beam from the upper film surface has a power

$$\frac{c}{8\pi} n_1 E_y^2 \cos\theta_1 \exp\left[-\left(\frac{4\pi\sigma_{12}}{\lambda_1}\cos\theta_1\right)^2\right]. \tag{16}$$

We use the double subscripts 12 and 10 to denote the film–air and film–substance interfaces, respectively. Note that λ_1 is the wavelength in the film. The surface

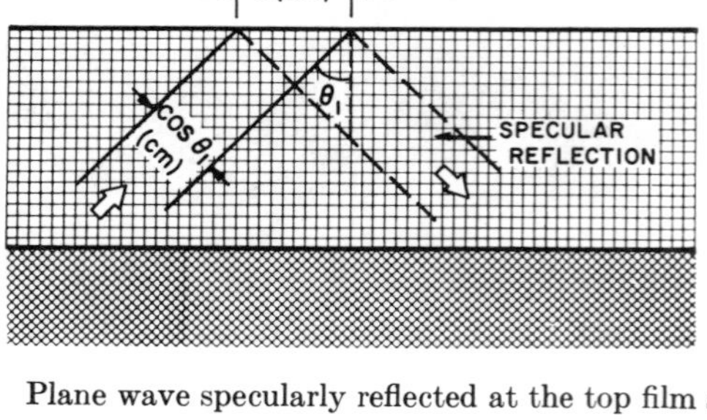

Fig. 22. Plane wave specularly reflected at the top film surface.

scattering is usually characterized by two statistical quantities: the statistical variation of the surface about the mean and the correlation length of the surface variation. Here σ_{12} in Eq. (16) is the variance of the surface roughness. A recent calculation by Marcuse[21] shows that the Rayleigh criterion applies only to the case of long correlation length. Since scattering observed in our loss measurements was always dominated by forward scattering, the assumption of the long correlation length may be correct. The limitation of the expression $\exp[-(4\pi\sigma_{12}\cos\theta_1/\lambda_1)^2]$ is also discussed by Beckmann and Spizzichino.[27] In spite of its shortcomings, the expression is widely used because of its simplicity. The power lost by surface scattering at two surfaces of the film is therefore

$$(c/8\pi)n_1E_y\cos\theta_1\{1-\exp[-K^2(\cos\theta_1)^2]\} \cong (c/8\pi)n_1E_y{}^2K^2\cos^3\theta_1, \quad (17)$$

where we have assumed that loss per unit length of the film is small, and

$$K = (4\pi/\lambda_1)(\sigma_{12}^2 + \sigma_{10}^2)^{\frac{1}{2}}. \quad (18)$$

It can be easily shown from the field distribution discussed in Sec. III that the total power flow in the film for any waveguide mode is

$$(c/4\pi)n_1E_y{}^2\sin\theta_1[W + (1/p_{10}) + (1/p_{12})], \quad (19)$$

where W is the thickness of the film, and p_{10} and p_{12} have been defined earlier. Dividing Eq. (17) by Eq. (19), we obtain the power attenuation per unit length of the film

$$\text{Attenuation} = K^2\left(\frac{1}{2}\frac{\cos^3\theta_1}{\sin\theta_1}\right)\left\{\frac{1}{[W+(1/p_{10})+(1/p_{12})]}\right\}. \quad (20)$$

We can also express the attenuation in decibels per unit length after multiplying Eq. (20) by 4.343. It is the loss caused by surface scattering only. We have thus expressed the loss as a product of three independent factors. The first factor, K, depends solely on the surface properties of the film and is a dimensionless quantity. The second factor involves θ_1 only and thus depends on the waveguide mode considered. The third factor is the reciprocal of the effective film thickness and shows explicitly that the loss is inversely proportional to the thickness, as is to be expected. From this theory, it appears that the single parameter K defines all the surface properties of the film. It is a dimensionless parameter and quantitatively it compares surface roughness with the optical wavelength.

The crosses in Fig. 21 are the results calculated from Eq. (20). The parameter $K = 1.27 \times 10^{-2}$ is evaluated so that the loss of the $m = 3$ mode computed by the theory and that obtained by the measurement coincide. For this Ta_2O_5 film, the agreement between the theory and measurement is excellent. For some other films, the agreement is only moderate. The details will be discussed elsewhere.

The large scattering loss observed in the experiment, particularly that of the higher order waveguide modes, is to be expected. Again, we return to the picture of the zigzag waves. For the Ta_2O_5 film considered in Fig. 21, the light wave is reflected back and forth between the two film surfaces about 2000 times in a length of 1 cm of the film for the $m = 0$ mode and about 10,000 times for the $m = 3$ mode. Thus, for this film to have a scattering loss of 1 dB/cm in the $m = 0$ mode, the loss per reflection should be 1.5×10^{-4}, which should be compared with the loss per reflection as large as 1×10^{-3} for the better dielectric mirrors used in lasers.

VII. Field Concentration and Mode Characteristics

Before discussing nonlinear interactions in thin films, it may be necessary to learn more about mode characteristics. Also included in this section are the discussion on the cutoff of the waveguide mode, field concentration in a film, and finally the concept of the effective film thickness. Throughout this section, we will use a ZnS film on a glass substrate as the example. The light wavelength is 1.06 μm of the YAG:Nd laser. At this wavelength, $n_1 = 2.2899$ and $n_0 = 1.5040$.

We have shown at the end of Sec. II how to calculate W and β/k for a given n_0, n_1 n_2, and m. The result of the calculation for both TE and TM waves is shown in Fig. 23. These W vs β/k curves are the mode characteristics of the waveguide. The ratio β/k is called the effective refractive index, since it measures the ratio of the speed of light in vacuum to that in the waveguide in the same way as the ordinary refractive index measures the ratio of the speed of light in vacuum to that in a dense medium. First, we notice β/k ranging from n_0 to n_1 for the waveguide modes. Let us concentrate on the $m = 0$ TE mode in Fig. 23. When W is large, the effective index (β/k) approaches the refractive index of the film n_1. To this limit, the film acts as a bulk medium and all the light energy is contained within the film. The fields therefore vanish at the two film surfaces. When β/k varies from n_1 to n_0, W decreases continuously as the fields extend more

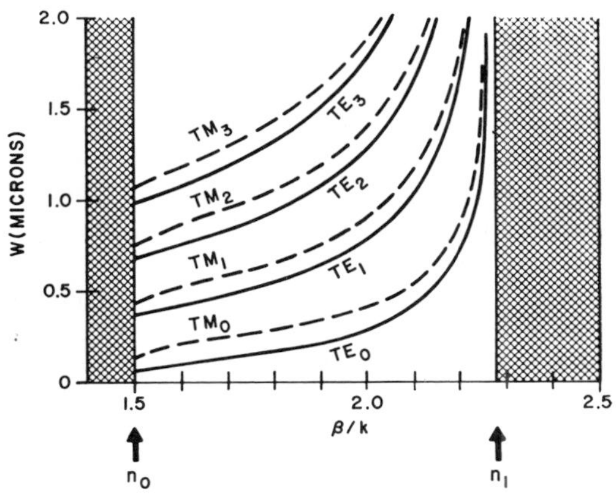

Fig. 23. Thickness W of a ZnS film deposited on a glass substrate is plotted vs the ratio of (β/k) for both TE and TM waveguide modes. Here, the ratio β/k can be considered as the effective refractive index. The left shaded region is for $\beta/k < n_0$, the right shaded region is for $\beta/k > n_1$, and in the space between the two shaded regions the waveguide modes are possible.

and more outside the film. At $\beta/k \to n_0$, the mode becomes cut off and W is the minimum thickness that can support this waveguide mode. At the cutoff, the waveguide mode is turned into a substrate mode as the fields extend infinitely into the substrate.

Next, we notice that for the same β/k, W's of different TE or TM modes are equally spaced. That is, the difference between W(TE, $m=1$) and W(TE, $m=0$) is equal to the difference between W(TE, $m=2$) and W(TE, $m=1$), and so on. We also notice that for a given m and β/k, W(TM) is always larger than W(TE) simply because the TM wave has larger Φ's [see Eqs. (4) and (5)]. The Φ's of the TM wave increase with the ratio n_1/n_0.

In designing an experiment particularly for nonlinear interactions, we often want to calculate the wave velocity accurately. This is difficult in practice. The film may not be homogeneous, and the materials may change with the environment. We do not really know the rafractive indices exactly. Moreover, the film may not be uniform and there are difficulties in measuring the film thickness exactly. It is then important to know the wave velocity, or β/k, varies with small increments in W, n_0, n_1 and n_2. For this purpose, we have calculated $[dW/d(\beta/k)]$, $[d(\beta/k)/dn_1]$, $[d(\beta/k)/dn_0]$, and $[d(\beta/k)/dn_2]$. They are given below for the TE wave only.

$$\frac{dW}{d(\beta/k)} = \frac{\beta k}{b_1{}^2}\left(W + \frac{1}{p_0} + \frac{1}{p_2}\right); \tag{21}$$

$$\frac{d(\beta/k)}{dn_1} = \frac{kn_1}{\beta}\,\frac{W + [p_{10}/(p_{10}^2 + b_1{}^2)] + [p_{12}/(p_{12}^2 + b_1{}^2)]}{W + \dfrac{1}{p_{10}} + \dfrac{1}{p_{12}}}; \tag{22}$$

$$\frac{d(\beta/k)}{dn_j} = \frac{kn_j}{\beta}\cdot\frac{b_1{}^2/[p_j(p_j{}^2 + b_1{}^2)]}{W + \dfrac{1}{p_0} + \dfrac{1}{p_2}}, \tag{23}$$

where $j = 0$ or 2. We have calculated the above expressions for the TE $m = 0$ waveguide mode. The results are shown in Fig. 24. In Fig. 24(a), we notice that $dW/d(\beta/k)$ is large for β/k near n_1 and also n_0. A large $dW/d(\beta/k)$ means that the wave velocity is not sensitive to small variations in W. In Fig. 24(b), we find that n_1 does not affect β/k very much for $\beta/k \to n_0$, and similarly n_0 does not affect β/k very much for $\beta/k \to n_1$, as is to be expected. The influence of n_2 on β/k is generally 10 times smaller than that exercised by n_1 or n_0. In spite of its smallness, it is relatively easy to vary n_2 in order to obtain a fine adjustment in the wave velocity, for example, by using a liquid of index of refraction to replace the air space on top of the film.

One can easily calculate the power carried by a TE waveguide mode based on field distribution discussed in Sec. III. We find

$$P = \frac{c}{4\pi} n_1 \sin\theta_1 E_y{}^2\left(W + \frac{1}{p_0} + \frac{1}{p_2}\right) d,$$

where E_y is the field amplitude of the A_1 or B_1 wave in the film and d is the width of the light wave in the y direction. The equation is written in gaussian units. Because the field extends outside the film, $[W + (1/p_0) + (1/p_2)]$ is the effective thickness of the film, W_{eff}. For a given P, the light intensity $E_y{}^2$ is inversely proportional to W_{eff} instead of W. Both W and W_{eff} are plotted in Fig. 25 for the $m = 0$ and m $= 1$ modes. For large concentration of light intensity inside the film, we prefer the $m = 0$ mode. We see that even for the $m = 0$ mode, we cannot increase light intensity indefinitely by simply reducing thickness of the film. When the film becomes too thin, the fields penetrate deep into the substrate and W_{eff} no longer decreases

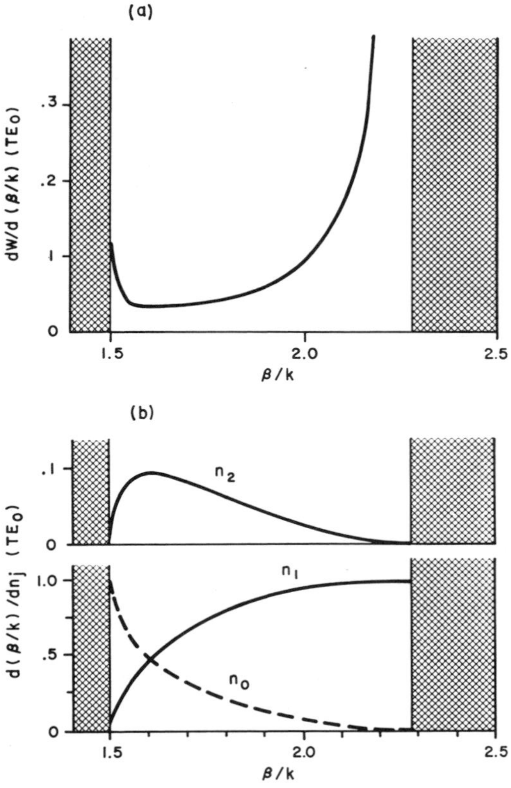

Fig. 24. (a) $dW/d(\beta/k)$ and (b) $d(\beta/k)/dn_j$ vs (β/k) curves for the $m = 0$ TE mode.

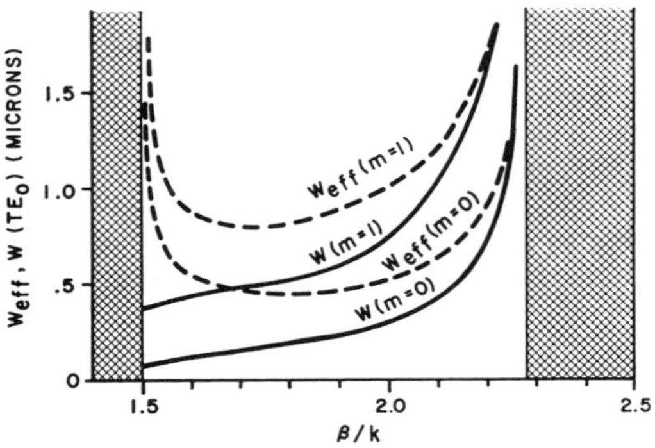

Fig. 25. W_{eff} and W vs (β/k) curves for $m = 0$ and $m = 1$ TE modes.

with W. It is interesting to note that for the $m = 0$ mode, W_{eff} approaches $\lambda_1/2$ but is never smaller than it, where λ_1 is the optical wavelength in the film medium. The minimum W_{eff} of 0.458 μm occurs at $\beta/k \rightarrow 1.82$, where the average power density and the maximum field amplitude in the film are, respectively, 21.8 MW/cm^2 and 6.78×10^4 V/cm. Here we assume that $d = 10$ μm and that 1 W of the laser power is being fed into the $m = 0$ waveguide mode of the film.

VIII. Phase-Match and Nonlinear Interactions Between Guided Waves

The advantages of performing nonlinear optics in a thin film are many. A thin film can concentrate laser energy for a long distance, whereas a focused gaussian beam diffracts rapidly away from the focused point. The phase velocity of a light wave in a waveguide mode depends on the thickness of the film and the mode of propagation. Thus, for example, by using different waveguide modes for the signal, idler, and pump waves in a parametric oscillator, we can obtain a phase match condition without relying upon the birefringence of the crystal. The crystals such as GaAs, GaP, ZnS, ZnTe, etc., which have large nonlinear coefficients but little birefringences can then be used for nonlinear experiments. The film and the substrate can be immersed in a liquid and the phase-match condition can be varied by varying the refractive index of the liquid. Finally, the nonlinear interaction can take place in the film, in the substrate, or both. All these advantages provide many alternatives to the design of the experiment. Of course, the primary purpose of developing nonlinear devices in thin-film form is that they can be used in integrated optical circuitry. In spite of these advantages, the development of thin-film parametric devices is handicapped by the lack of single-crystal films and by the difficulties in obtaining a long coherence length. We shall illustrate this fully by considering the problem of optical second harmonic generation (SHG).

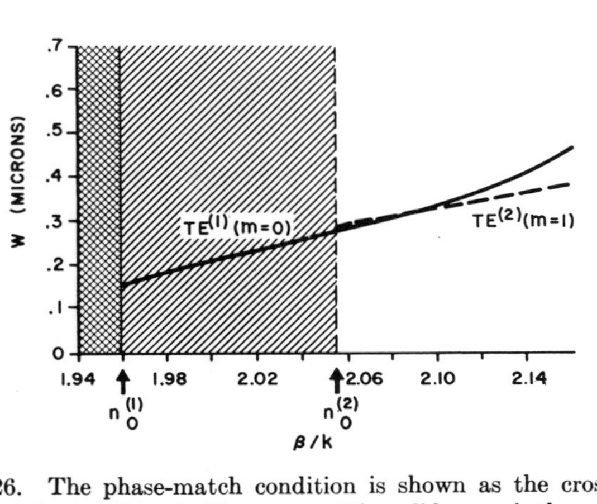

Fig. 26. The phase-match condition is shown as the crossing point of the two W vs (β/k) curves. The solid curve is the $m = 0$ TE mode of the fundamental and the dashed curve is the $m = 1$ TE mode of the harmonic.

In a parametric oscillator, the frequency of the oscillation adjusts itself so that the phase velocities of the signal, idler, and pump are matched. In the SHG, however, we do not have this kind of flexibility, and the film must have exactly the thickness required for the fundamental and the harmonic waves to propagate at the same wave velocity. Any nonhomogeneity in refractive indices and nonuniformity in thickness would reduce the efficiency of the nonlinear interaction. The phase-match condition in nonlinear optics involves only the wave velocities parallel to the film. The phase-match condition of SHG is $(\beta/k)^{(1)} = (\beta/k)^{(2)}$, where the superscripts (1) and (2) denote the fundamental and the harmonic, respectively. Basically, if we plot W vs (β/k) for the fundamental and for the harmonic, the crossing point of the curves is the phase-match condition. The problem becomes simpler if we follow the following simple rules:

1. $dW/d(\beta/k)$ for the fundamental and the harmonic are always positive and the W vs (β/k) curves of the fundamental can cross that of the harmonic only once or not at all.
2. In general, when the W vs (β/k) curves show that $W^{(1)} > W^{(2)}$ near $\beta/k = n_0$, a phase-match condition can be obtained only when $n_1^{(2)} < n_1^{(1)}$. Similarly, when $W^{(1)} < W^{(2)}$ near $\beta/k = n_0$, the phase-match can be obtained only when $n_1^{(2)} > n_1^{(1)}$.

We shall illustrate these rules in the following examples.

As the first example, we consider nonlinear interaction in the substrate. In this case, we use the electric fields that extend into the substrate. The efficiency of the interaction depends on the amount of the fundamental and the harmonic electric fields in the substrate as compared with their distributions over the film, substrate, and air space. In order to obtain a large efficiency, it is obvious that we must operate at β/k near n_0. Since the field distribution in the substrate of any waveguide mode is always exponential, it does not really matter if the fundamental and the harmonic are in the same or different waveguide modes. A good choice is $m = 0$ mode for the fundamental and $m = 1$ mode for the harmonic. In this case, birefringence of the crystal is not required. Both the fundamental and the harmonic can be TE or TM, or one of them TE and the other TM. For example, in Fig. 26, we consider a ZnS film on a single-crystal ZnO substrate. The refractive indices are $n_1^{(1)} = 2.2899$; $n_1^{(2)} = 2.4038$; $n_0^{(1)} = 1.9562$; $n_0^{(2)} = 2.0521$ for the wavelengths 1.064 μm and 0.532 μm, respectively. Here the c axis of ZnO is oriented in the y direction, and the refractive indices quoted are those of the extraordinary ray. To use nonlinear coefficient d_{33} of ZnO, both the fundamental and the harmonic must be TE waves. Phase-match condition is obtained at $(\beta/k) = 2.0877$ where the thickness of the film is 0.314 μm.

A detailed calculation shows that at $(\beta/k) = 2.0877$, $d[(\beta/k)^{(2)} - (\beta/k)^{(1)}]/dW = 0.444$. For a coherence length of 1000 μm, $d[(\beta/k)^{(2)} - (\beta/k)^{(1)}]$ should be equal or less than 2.65×10^{-3}. We find immediately that the average thickness of the film should be kept within the limits

$$W = 0.314 \pm \frac{2.65 \times 10^{-3}}{0.44} = 0.314 \pm 0.006\ \mu\text{m}.$$

It is not impossible to evaporate a ZnS film within a thickness tolerance of 0.006 μm. Because of the small birefringence, the coherence length in a bulk ZnO is less than 2 μm. Now we have obtained a coherence length of 1000 μm by using the waveguide modes. We have thus improved the coherence length by a factor of 500, which is remarkable. On the other hand, a 1.064-μm light wave can propagate at least a distance of 2.5 cm in ZnS film without suffering appreciable loss. Because of the phase-match problem we have used only a section of 1000 μm of this 2.5-cm-long beam. The advantage of the thin-film waveguide is thus not fully utilized. We have used here the problem of SHG to illustrate the difficulties involved in the integrated optics. We can gain some advantages by using the waveguide principle, but in this and other problems, the full potential of the integrated optics cannot be realized without developing techniques to control the homogeneity and uniformity of the film.

Next, as the second example, we consider nonlinear interaction in film. Although one can use different waveguide modes for the fundamental and the harmonic mismatch between the distribution of nonlinear polarization and that of the fields would reduce the efficiency of interaction sharply. We thus prefer the use of TE and TM waves of the same mode order m for the fundamental and harmonic. A study of the mode characteristics shows that W of the harmonic is always smaller than that of the fundamental at a same value of β/k near n_0. According to our second rule given earlier, we cannot obtain a phase-match condition *unless the refractive index of the film at the harmonic frequency is less than that at the fundamental frequency.* Since most of nonlinear materials have a normal dispersion in the visible spectrum, to satisfy the above condition, we must choose a material of sufficient birefringence for the film. As an example, we consider in Fig. 27 a single-crystal $LiNbO_3$ film on a quartz substrate. The c axis of $LiNbO_3$ is normal to the film. A TE ($m = 0$) wave is used for the fundamental and a TM ($m = 0$) wave for the harmonic. The nonlinear coefficient used is the d_{31} of $LiNbO_3$. The refractive indices involved are than $n_1^{(1)} = 2.365$; $n_0^{(1)} = 1.4614$; $n_1^{(2)} = 2.300$; $n_0^{(2)} = 1.4745$. Again we consider SHG from 1.064 μm to 0.532 μm. We find from Fig. 27 that a phase-match condition is obtained at $(\beta/k) = 2.2274$ where $W = 2.47$ μm. Unfortunately, the single-crystal $LiNbO_3$ film is not available at present. The discussion is therefore academic. We could use the d_{31} nonlinear coefficient of oriented ZnO or CdS films that have been developed for ultrasonic transducers, but they are too lossy, as discussed in Sec. V.

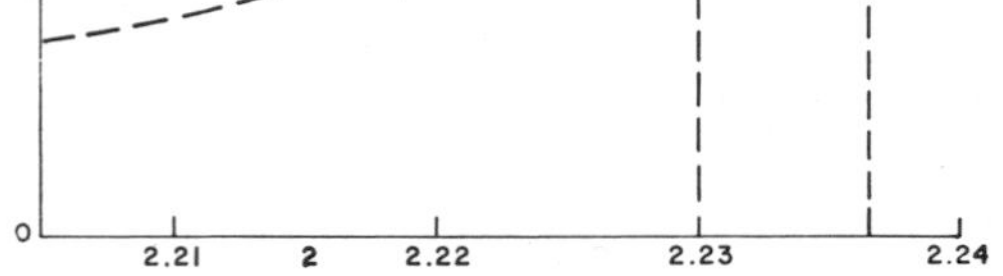

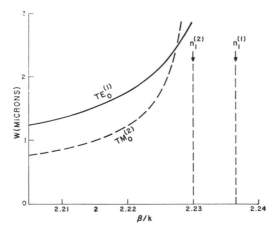

Fig. 27. The phase-match condition for SHG in a $LiNbO_3$ film on a quartz substrate is indicated as the crossing point of the two W vs (β/k) curves. The fundamental at 1.064 μm uses $m = 0$ TE mode and the harmonic at 0.532 μm uses $m = 0$ TM mode.

From our earlier discussion on nonlinear interaction in the substrate, we realize that the requirement set by the phase-match condition for the thickness and uniformity of the film is very stringent. To circumvent this problem, a novel method has been used by Tien *et al.*[16] by generating second harmonic in the form of Cerenkov radiation. The nonlinear interaction by their method is not as efficient as that under the phase-match condition, but the interaction extends to the full length of the fundamental wave. As in the case discussed before, a polycrystal ZnS film on a single-crystal ZnO substrated was used and nonlinear interaction took place in the substrate. The c axis of the ZnO was oriented in the y direction, and both the fundamental and the harmonic waves were polarized parallel to the c axis.

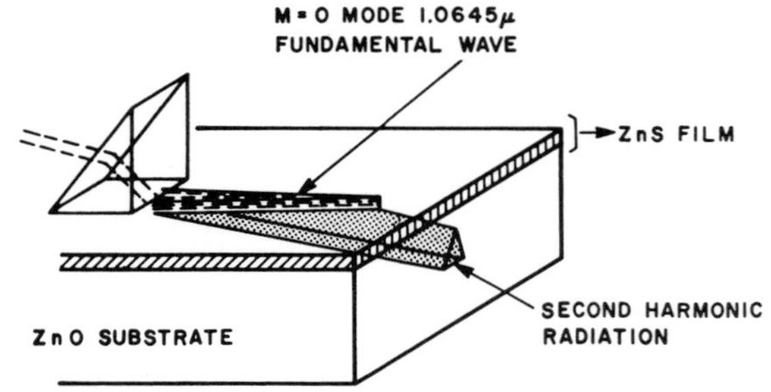

Fig. 28. Experimental arrangement of second harmonic generation in the form of Cerenkov radiation. A fundamental light wave at 1.064 μm is fed into a ZnS polycrystal film which is deposited on a single-crystal ZnO substrate. The evanescent field of the fundamental wave generates a second harmonic Cerenkov radiation in the substrate.

The experimental arrangement is shown in Fig. 28. A light beam of a YAG:Nd laser at 1.06 μm was fed into the $m = 0$ waveguide mode of the ZnS film as the fundamental wave. It propagated as $\exp[-i\omega^{(1)}t + i\beta x]$, where $n_0^{(1)} < (\beta/k) < n_1^{(1)}$. The fundamental wave excited a wave of second harmonic nonlinear polarization in the substrate via the d_{33} nonlinear coefficient of ZnO. Here the nonlinear polarization wave was a forced wave and thus it varied as $\exp[-i\omega^{(2)}t + 2i\beta x]$. Because of the normal dispersion of ZnO,

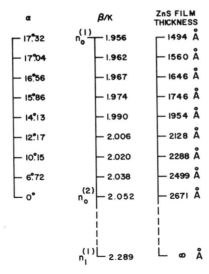

α	β/K	ZnS FILM THICKNESS
17.32°	$n_0^{(1)}$ 1.956	1494 Å
17.04°	1.962	1560 Å
16.56°	1.967	1646 Å
15.86°	1.974	1746 Å
14.13°	1.990	1954 Å
12.17°	2.006	2128 Å
10.15°	2.020	2288 Å
6.72°	2.038	2499 Å
0°	$n_0^{(2)}$ 2.052	2671 Å
	$n_1^{(1)}$ 2.289	∞ Å

Fig. 29. Any horizontal line on this table indicates the corresponding values for the thickness of the ZnS film, β/k of the fundamental wave, and the Cerenkov angle α.

$n_0^{(2)} > n_0^{(1)}$. Therefore, by using a proper thickness W for the film, it was possible to obtain

$$n_0^{(1)} < (\beta/k) < n_0^{(2)}. \quad (24)$$

Under this condition, the phase velocity of the nonlinear polarization $\omega^{(2)}/2\beta$ exceeds the phase velocity $c/n_0^{(2)}$ of the free second harmonic radiation in the substrate. Consequently, Cerenkov radiation at the second harmonic frequency is emitted and it is emitted at a Cerenkov angle α where.

$$\cos\alpha = \beta/kn_0^{(2)}. \quad (25)$$

To review, the fundamental wave propagating in the ZnS film generates a sheet of nonlinear polarization wave in the substrate immediately below the film–substrate interface. The nonlinear polarization wave then generates a second harmonic radiation in the form of Cerenkov radiation that may be considered simply as a plane wave propagating in the substrate at an angle α below the interface. If the wave vector C represents this plane wave, its horizontal component is equal to the wave vector of the nonlinear polarization wave and is also twice the fundamental wave vector β. The process described above is illustrated in Fig. 29, in which any horizontal line drawn from a value of (β/k) at the middle column gives the corresponding Cerenkov angle α in the left column and the thickness of the film W in the right column. We notice that at the upper limit of the inequality (24), $\beta/k \rightarrow n_0^{(2)}$, the Cerenkov angle vanishes, and the phase velocity of the nonlinear polarization wave is equal to that of the free wave in the substrate. At the lower limit of (24), $\beta/k \rightarrow n_0^{(1)}$, the waveguide mode of the fundamental wave becomes cut off. Figure 30 is a photograph of the experiment. The bright star in the photograph is the Cerenkov radiation. Figure 31 shows the radiation as it emerged from the side surface of the ZnO crystal. The photograph was taken through a microscope by focusing it on the side surface of the crystal.

IX. Conclusions

Most of the material presented in this paper is drawn from unpublished notes accumulated during the past two years. These notes were prepared for talks and lectures given on various occasions. Much of the time and effort has been spent in developing a method whereby one can visualize easily the waveguide and radiation modes without having to derive the Maxwell equations. A theory that consideres a zigzag plane wave has been developed for that purpose. We have used this theory in Sec. II to derive the mode equation. It was used again in Sec. IV for a unified theory of the prism and grating couplers, and in Sec. VI for a simple theory of surface scattering. This theory turned out to be most useful in the analysis of complex optical devices. For example, the prism–film coupler involves four coupled media: the prism, the air gap, the film, and the substrate. A direct solution of the Maxwell equations for four simultaneously coupled media is not simple. Here, using the theory of the zigzag wave,

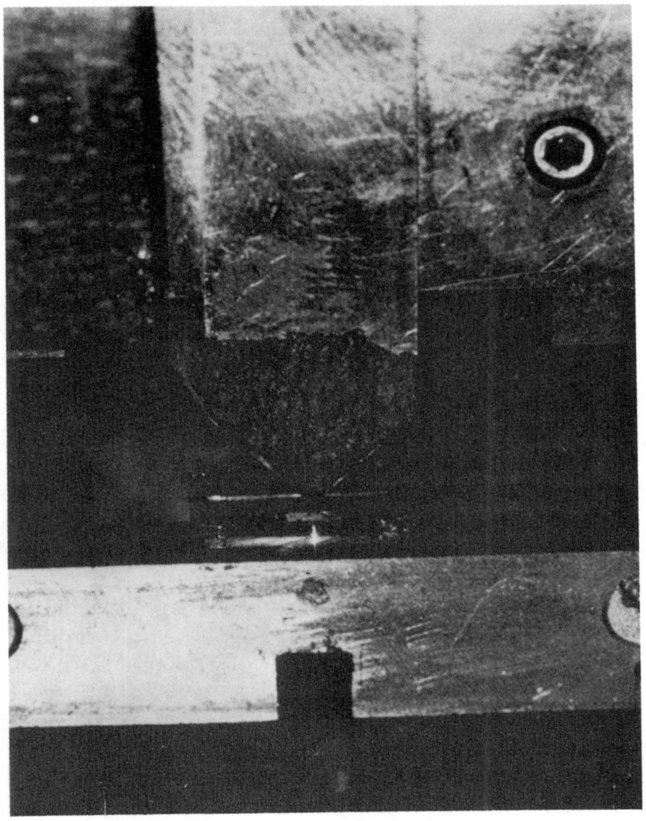

Fig. 30. Experiment of the second harmonic generation. The bright star in the figure is the second harmonic Cerenkov radiation emerging from the side surface of the ZnO substrate. The second harmonic beam thus generated is a coherent light beam of a very small aperture; the fundamental light wave at 1.064 μm is invisible.

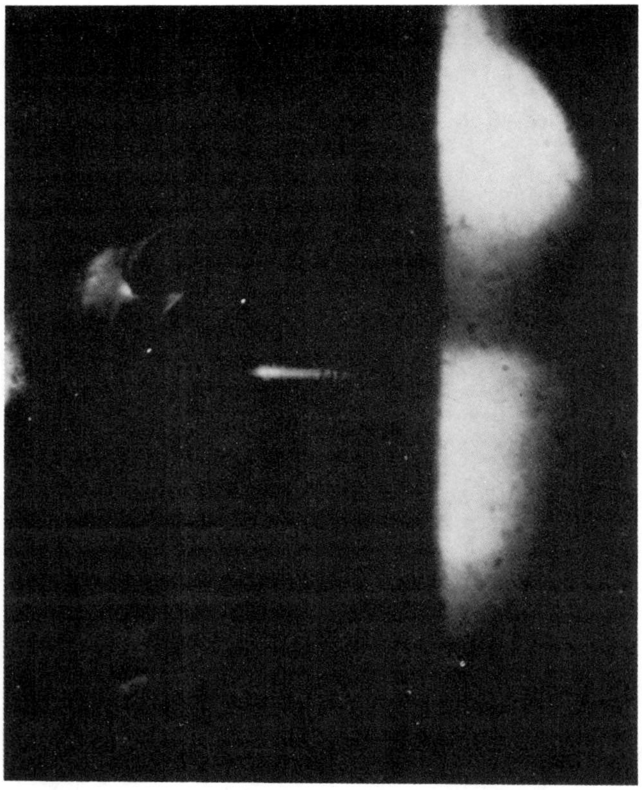

Fig. 31. Photograph of the second harmonic Cerenkov radiation taken through a microscope focused on the side surface of the ZnO substrate where the radiation emerges.

we only have to consider two sets of the waves, and each set consists of only two coupled waves.

We have not discussed in this paper the Goos-Haenchen shift, but it is included in the derivation of the mode equation by introducing the phase changes $-2\Phi_{10}$ and $-2\Phi_{12}$. It is also implied in the expression for the power flow and in the coupling constant of the prism–film coupler by introducing the effective film thickness. A detailed study of the Goos-Haenchen shift should include the discussion of energy flow and is beyond the scope of this paper.

The experiments described in this paper are difficult; one has to learn how to control the small dimensions involved in these experiments. For example, the spacing of the air gap between the prism and the film can be determined by observing Newton's rings near the pressure point. Usually, one should accurately measure the refractive indices of the prism and the substrate, tabulate the incident angles against the values of β/k, and compute the mode characteristics for each film–substrate combination.

The technology involved in integrated optics may be more difficult than that we can visualize today. Large single-crystal films are needed for the development of electrooptical and nonlinear thin-film devices. These single-crystal films, with the exception of epitax layers, simply do not exist today. Methods to control the uniformity and the thickness of the film within the accuracy of one or two atomic layers and techniques for the fabrication of structures 10^4 times smaller than their microwave counterparts are also needed. We are still observing the m lines in our best films; this indicates that considerable energy stored in the film and in the substrate is not in the main mode of propagation, and it would eventually distort the signal carried by the integrated optical circuitry. These problems described above will continue to challange us for some time to come. Looking into the future, we expect the field of integrated optics to grow rapidly, simply because there are needs for optical systems in the electronics and communication industries and there are needs for integrated optics in optical systems.

References

1. S. E. Miller, Bell Syst. Tech. J. **48,** 2059 (1969).
2. R. Shubert and J. H. Harris, IEEE Trans. MMT **16,** 1048 (1968).
3. P. K. Tien, R. Ulrich, and R. J. Martin, Appl. Phys. Lett. **14,** 291 (1969).
4. M. L. Dakss, L. Kuhn, P. F. Heidrich, and B. A. Scott, Appl. Phys. Lett. **16,** 523 (1970).
5. H. Kogelnik and T. Sosnowski, Bell Syst. Tech. J. **49,** 1602 (1970).
6. P. K. Tien and R. Ulrich, J. Opt. Soc. Am. **60,** 1325 (1970).
7. R. Ulrich, J. Opt. Soc. Am. **60,** 1337 (1970).
8. J. E. Midwinter, IEEE J. Quant. Electron. **QE-6,** 583 (1970).
9. J. H. Harris, R. Shubert, and J. N. Polky, J. Opt. Soc. Amer. **60,** 1007 (1970).
10. J. H. Harris and R. Shubert, IEEE Trans. MTT **19,** 269 (1971).
11. R. Ulrich, to be published in J. Opt. Soc. Am.
12. P. K. Tien and R. J. Martin, Appl. Phys. Lett. **18,** 398 (1971).
13. J. E. Goell and R. D. Standley, Bell Syst. Tech. J. **48,** 3445 (1969).
14. D. H. Hensler, J. D. Cuthbert, R. J. Martin, and P. K. Tien, Appl. Opt. **10,** 1037 (1971).
15. P. K. Tien, G. Smolinsky and R. J. Martin, "Thin Organosilicon Films for Integrated Optics," to be published in Appl. Opt.
16. P. K. Tien, R. Ulrich, and R. J. Martin, Appl. Phys. Lett. **17,** 447 (1970).
17. L. Kuhn, M. L. Dakss, P. F. Heindrich, and B. A. Scott, Appl. Phys. Lett. **17,** 265 (1970).
18. D. Hall, A. Yariv and E. Garmine, Appl. Phys. Lett. **17,** 127 (1970).
19. H. Kogelnik and C. V. Shank, Appl. Phys. Lett. **18,** 152 (1971).
20. J. E. Goell and R. D. Standley, Proc. IEEE **58,** 1504 (1970).
21. D. Marcuse, Bell Syst. Tech. J. **48,** 3187 (1969); **48,** 3233 (1969); **49,** 273 (1970); "Dependence of Reflection Loss on the Correlation Function" (private communication).
22. H. K. V. Lotsch, J. Opt. Soc. Am. **58,** 551 (1968).
23. M. Born and E. Wolf, *Principles of Optics* (Pergamon, New York, 1970), p. 49, Eq. (60).
24. J. J. Burke, J. Opt. Soc. Am. **61,** 676A (1971).
25. N. F. Foster, G. A. Coquin, S. A. Rozgonyi, and F. A. Vannatta, IEEE Trans. **S4-15,** 28 (1968).
26. S. A. Rozgonyi and W. J. Polito, Appl. Phys. Lett. **8,** 220 (1966).
27. P. Beckmann and A. Spizzichino, *International Series of Monographs on Electromagnetic Waves* (Oxford, New York, 1963), Chap. 5.

Part 2
Two-Dimensional Optics

Optical Guided-Wave Focusing and Diffraction*

R. Shubert and J. H. Harris
University of Washington, Department of Electrical Engineering, Seattle, Washington 98105
(Received 18 May 1970)

One-dimensional focusing and diffraction techniques for thin-film optical waveguides are examined in terms of their effectiveness and fabrication requirements. Impetus for the study is derived from potential application of optical planar guided waves to analog signal-processing systems. Methods of altering modal-phase velocities to achieve ray deflection are investigated and include the use of substrate and film-embedded refractive-index discontinuities, multilayer films, thickness-profiled layers, and absorptive-material deposition on the guiding films. Thin-film lenses having f numbers as low as 2 to 3 may be fabricated by vacuum deposition of high-index materials, such as CeO_2. Experimental results obtained with glass-sputtered films illustrating basic operation of thin-film lenses, gratings, and prisms are presented.
Index Headings: Films; Refractive index; Waveguides; Gratings.

The potential significance of integrated-optics techniques in the development of optical systems has been recognized for some time. Efforts to date in realizing this potential have been largely directed toward temporal devices, i.e., toward optical equivalents of such microwave elements as couplers, resonators, and filters, in addition to such active components as modulators, lasers, and parametric amplifiers.[1–3] Recently developed techniques that may, in principle, yield nearly complete coupling of a laser beam into a thin film and provide 10 to 10^3 power-density increases have added impetus to integrated-optics developments.[4]

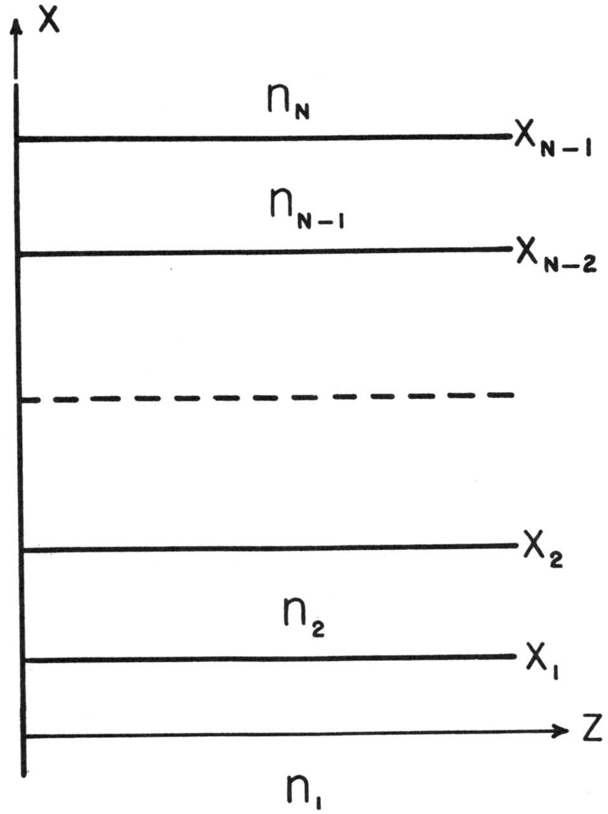

Fig. 1. General multifilm structure having N layers.

In addition to temporal devices, integrated optical systems are capable of spatial processing in a manner equivalent to traditional optical-imaging systems but with a reduction of dimensionality by 1.[5] Passive components that are useful include prisms, reflectors, beam splitters, gratings, transmission masks, and spectral filters. The basic design of passive spatial devices in optical-waveguide systems is the subject of this paper. Emphasis is placed on techniques for fabricating lenses that can provide one-dimensional imaging in thin films. Design techniques are approached by establishing achievable variations of surface-wave eigenvalues and are discussed in terms of their effectiveness for lens design (f numbers), material requirements, and fabrication considerations.

PROPAGATION IN MULTILAYER FILMS

Modal propagation of optical surface waves along single-layer dielectric films and structures with monotonic permittivity variations has been extensively investigated.[5–8] Here we consider the propagation characteristics of various structures including multilayers that offer advantages in the fabrication of practical integrated devices. Wave propagation in the N-layered (including two substrates) geometry illustrated in Fig. 1 in which refractive indices are denoted $n_p(1\leq p\leq N)$ may be represented by a superposition of plane surface waves. Propagation constants of the waves are isotropic in the plane of the layers and may be found from the zeros of the system function for the structure that is denoted $D_{1,N}$.[4] The system function is the determinant obtained by matching tangential-field components of the homogeneous wave equation at the interfaces and is conveniently represented in the partition form

$$D_{1,N}=D_{1,p}D_{p,N}-D'_{1,p}{}'D_{p,N}, \quad (1)$$

where the relations

$$D'_{1,p}=D_{1,p}(-u_p)'D_{p,N}=D_{p,N}(-u_p)$$
$$D_{p,p+1}=(u_p+u_{p+1})\exp[-i(u_p-u_{p+1}x_p)] \quad (2)$$

Reprinted with permission from *J. Opt. Soc. Am.*, vol. 61, pp. 154–161, Feb. 1971.

permit evaluation of Eq. (1) by recursion (subscript p is arbitrary). Equation (2) holds for TE waves, to which we restrict our attention. For TM waves, (u_p+u_{p+1}) is replaced by $(u_p/n_p^2+u_{p+1}/n_{p+1}^2)$. The symbol $u_p=k_0(n_p^2-n^2)^{\frac{1}{2}}$ represents the transverse wavenumber where $k_0=2\pi/\lambda_0$ and $n=\beta/k_0$ is the equivalent-mode refractive index in terms of the longitudinal propagation constant β.

$D_{1,N}$ is a function of the mode index and has zeros corresponding to the refractive indices of the layers and to the allowed eigenvalues obtained from the transcendental equation

$$D_{1,N}=0. \tag{3}$$

We consider numerical eigenvalue solutions to Eq. (3) for some typical one- and two-film guides ($N=3$, 4); these results are presented in a later section. For the two-film waveguide, the eigenvalue equation obtained from Eqs. (1)–(3) may be written

$$\tan^{-1}u_1/iu_2+\tan^{-1}[u_3/u_2 \tan(\tan^{-1}u_4/iu_3-u_3t_3)] -u_2t_2+m\pi=0, \tag{4}$$

where t_2 and t_3 are the film thicknesses, $m=0, 1, 2\cdots$ denotes the mode number, and only the principal values of the tan and $\tan^{-1}$ functions are used. Equation (4) is valid when n_2 is greater than the refractive indices of the other film and substrates. In the limit, as $t_3\rightarrow 0$, Eq. (4) reduces to the familiar eigenvalue equation for the single-film case, given by

$$\tan^{-1}u_1/iu_2+\tan^{-1}u_3/iu_2-u_2t_2+m\pi=0. \tag{5}$$

To illustrate the nature of Eq. (5), numerical solutions are plotted in Fig. 2 for refractive indices corresponding to a polyurethane film on a glass substrate in air. The curves show the number of modes in a film of thickness t_2 and give the index n of each mode. The mode index n for the asymmetrical structure must be confined between the refractive indices of the film and the most dense bounding layer; thus, if we assume $n_2>n_1>n_3$, then

$$n_2<n<n_1. \tag{6}$$

In contrast to the symmetrical case ($n_1=n_3$) where the lowest-order mode has no cutoff, each propagating mode of the asymmetrical structure has a cutoff film thickness below which it cannot propagate. The cutoff relation follows from Eq. (5) when $n\rightarrow n_1$, so that, in this limit, the first term of Eq. (5) vanishes and we have

$$(t_2/\lambda_0)^{(m)}=(1/2\pi\delta)\tan^{-1}\xi/\delta+m/2\delta, \quad m=0, 1, 2\cdots, \tag{7}$$

where

$$\delta=(n_2^2-n_1^2)^{\frac{1}{2}} \quad \xi=(n_1^2-n_3^2)^{\frac{1}{2}}. \tag{8}$$

Equation (7) gives the film thickness for which the mth TE mode is just at cutoff, and the corresponding TM modes have cutoffs close to these values.

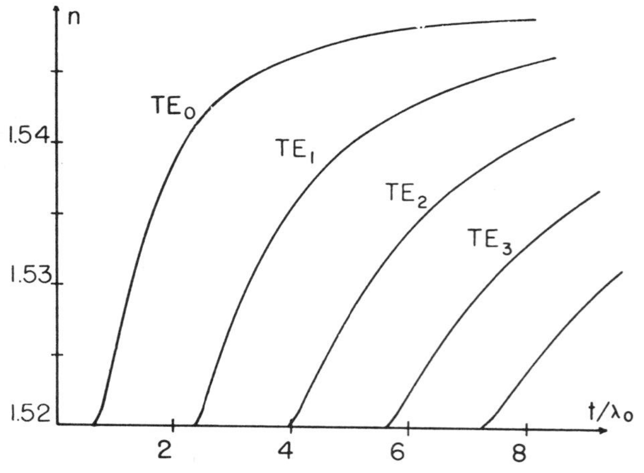

FIG. 2. Mode refractive index n vs film thickness t_2 for single-film asymmetric structure with $n_1=1.52$, $n_2=1.55$, $n_3=1.0$.

IMAGING IN THE WKB PARAXIAL APPROXIMATION

If physical properties of the optical waveguide are varied, mode indices may be controlled so as to produce lens action. In examining various methods of altering modal-propagation constants, we limit discussion to single-mode guides, in order to avoid consideration of the spatially dispersive character of the modes. Thin-film lens operation is based on the one-dimensional paraxial approximation for surface waves that is equivalent to the form for two-dimensional lenses, which produce a phase shift varying quadratically from the optic axis. In particular, if the surface wave is assumed to be incident on a region of varying mode index of the form $\hat{n}(y,z)z_1\leq z\leq z_2$, the transfer function of the region, to the thin-lens approximation, is

$$g_2(y,z_2)=\exp\left\{i\left[k_0\int_{z1}^{z2}\hat{n}(y,z)dz\right]\right\}g_1(y,z_1), \tag{9}$$

where g_1 and g_2 represent the surface wave incident on and transmitted through the region, respectively. The integral along the optic axis in the exponent is used in Eq. (9) to account for gradual changes into the region of a lens, and $\hat{n}$ is used to designate the possibly complex mode index. The reason for using smoothly varying lens regions in dielectric waveguides is discussed later in this section.

The focal-length properties of the lens specified by $\hat{n}(y,z)$ may be determined by first assuming the region of mode-index variation to lie between planes at z_1 and z_2. Under the approximation of Eq. (9), the phase at the focus a distance $(f^2+y^2)^{\frac{1}{2}}$ from the point (y,z_1) owing to an incident plane wave has the form

$$\phi=k_0\left[\int_{z1}^{z2}\hat{n}(y,z)dz+n_0(y^2+f^2)^{\frac{1}{2}}\right], \tag{10}$$

where f is the focal length. Expansion of $\hat{n}$ and $(f^2+y^2)^{\frac{1}{2}}$ in a Taylor series to second order yields a linear term that is zero if n is symmetric about $y=0$ and a quadratic term that also vanishes if

$$\frac{1}{f}=\frac{1}{n_0}\left[\int_{z_1}^{z_2}\frac{\partial^2}{\partial y^2}\hat{n}(y,z)dz\right]_{y=0}, \qquad (11)$$

where n_0 is the mode index outside the lens region. Terms higher than second order in the expansion of $\hat{n}$, which lead to lens aberrations, are assumed negligible in this discussion. Lens regions considered here are either longitudinally shaped films, resembling one-dimensional equivalents of spherical lenses or profiled-thickness films, analogous to planar lenses having radially varying refractive index. Focal length and f-number parameters for these types of structures may be obtained from Eq. (11) once $\hat{n}(y,z)$ is specified. Gradual changes of $\hat{n}(y,z)$ with z, in general, reduce the integral in Eq. (11) and thereby increase the focal length. As an example of a typical thin-film lens having smooth transitions in the direction of the optic axis, the real mode-index distribution

$$n(y,z)=n_0+\Delta n_{\max}(1-y^2/a^2)\exp[-2z^2/(d/3)^2], \qquad (12)$$

which is a gaussian of width $d/3$ along the propagation direction and parabolic in y, provides a focal length

$$f=(\pi/2)^{\frac{1}{2}}(n_0a^2/6d\Delta n_{\max}). \qquad (13)$$

This value may be compared with $n_0a^2/2d\Delta n_{\max}$ for a lens with constant index in the z direction between the planes $z_2-z_1=d$.

There are substantial differences between surface-wave propagation and propagation in standard optical systems. Of considerable importance is the fact that mode-index changes in guiding structures are accompanied by coupling to the continuous spectrum of unguided, radiative modes and result in power loss by the surface wave. These losses arise because the WKB-type wave solution expressed in Eq. (9) does not satisfy the wave equation exactly and differences must be accounted for by loss mechanisms. The extent of the losses has not as yet been established in quantitative fashion, but relevant work of Marcuse on guides of varying thickness indicates that scattering remains small if transition regions extend over many wavelengths.[9] Definitive evaluation of losses in lens regions nevertheless remains a problem of importance for device applications.

THICKNESS-PROFILED WAVEGUIDE

If the thickness of a thin-film waveguide is changed along the direction of propagation, as illustrated in Fig. 3(b), a change of the mode-propagation constant occurs. Restricting attention to the lowest-order mode ($m=0$) and using Eq. (5), we see that a decrease of the film thickness t_2 is accompanied by a corresponding decrease of the mode index n. Similarly, increasing t_2 increases n and lowers the phase velocity of the surface wave. Lens and prism structures can thus be fabricated by an appropriate shaping or profiling of a portion of the guiding film, as shown in Fig. 3(a). In practice, the micron-thickness changes required would be achieved with smooth transitions. Figure 3(b) also illustrates how a typical field component of the guided wave is altered, in passing, by a gradually sloped thickness gradient. For a film-thickness decrease, the mode moves closer to cutoff, resulting deeper penetration of the field into the substrate.

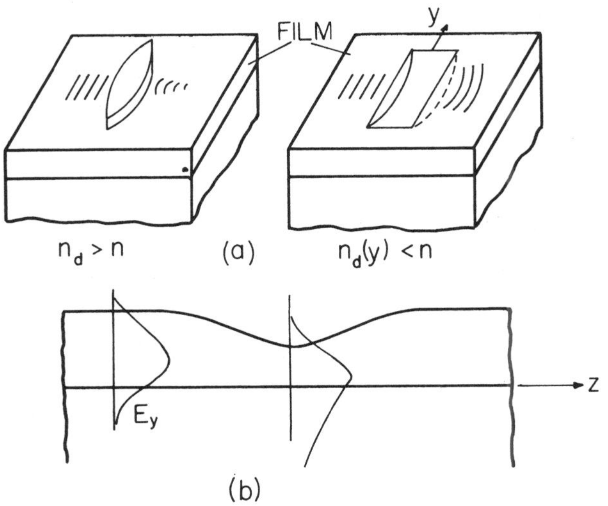

FIG. 3. (a) Schematic representation of typical thin-film lenses made by thickness profiling of guiding films; (b) illustration of the shift of transverse-electric-field distribution along a film-thickness contour, whose length is great compared to the thickness change.

To determine what magnitudes of changes of mode index are possible for films of various materials, we solved Eq. (5) numerically for a range of indices from $n_2=1.55$ to 1.75 of films on glass substrates in air; the results are shown in Fig. 4. The curves show mode-index difference $n-n_1$ as a function of film thickness t_2. As an example of use of the curves, profiling a film of high-index glass ($n_2=1.75$) from 0.2–$0.8\lambda_0$ can produce a mode-index change of approximately 0.17. The region to the left of the dashed line represents the range of thickness for single-mode operation. The index changes shown can be compared with those of physiological and glass lenses that lie in the 0.3–0.5 range. Thin films of high-refractive-index materials such as CeO_2, ZnS, and ZnO having refractive indices from 2.0 to 2.5 have been used successfully for waveguides[10] and would provide mode-index changes substantially greater than 0.5. Another significant feature of the curves is the existence of a region where the mode index varies almost linearly with the thickness of the film, a useful property for lenses fabricated by controlling deposition times.

The maximum change of mode index that can be achieved in a single-mode film may be determined as a function of layer indices. Substituting $m=0$ and $m=1$ in Eq. (7), we conclude that a single-mode guide can have film thicknesses satisfying the relation

$$\tau_{co}^{(0)} < \tau \leq \tau_{co}^{(1)}, \tag{14}$$

where $\tau = t_2/\lambda_0$ and superscripts indicate the mode order. In a single-mode guide, the maximum mode-index change Δn_{max} is achieved by changing the film thickness between the cutoffs of the zeroth- and first-order modes. This thickness change, $\Delta\tau$, may be expressed as a fraction of the largest film thickness, $\tau_{co}^{(1)}$, as

$$\Delta\tau/\tau_{co}^{(1)} = (\tau_{co}^{(1)} - \tau_{co}^{(0)}/\tau_{co}^{(1)} = 1/1 + (1/\pi)\tan^{-1}\xi/\delta. \tag{15}$$

The largest difference of propagation constant of a mode achievable with thickness variation is given by Eq. (6) as

$$\Delta n_{max} \approx n_2 - n_1 = \delta^2/(n_1+n_2) \approx \delta^2/2n_1 \quad n_1 \sim n_2. \tag{16}$$

For a practical lens, Δn_{max} should be large, to achieve greatest phase-velocity change, and $\Delta\tau/\tau_{co}^{(1)}$ small, to give minimum power loss by radiation, which dictates the choice of parameters ξ and δ.

A closer approximation to Δn_{max} in Eq. (16) is derived in the Appendix and is

$$\Delta n_{max} \cong (\delta^2/n_1)[(1+\tan^{-1}\xi/\delta)/(4+\tan^{-1}\xi/\delta)]. \tag{17}$$

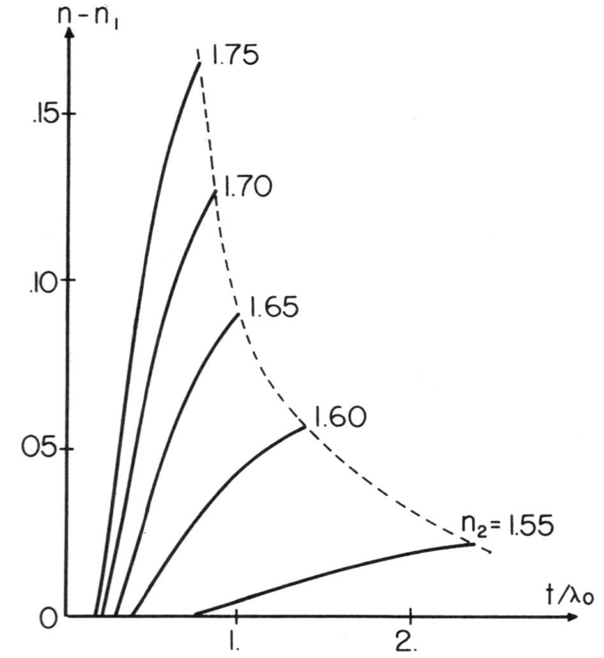

FIG. 4. Single-layer mode-index change n-n_1 vs film thickness t_2 for a range of film refractive indices with $n_1=1.52$ and $n_3=1.0$. Dashed line represents boundary of single-mode region.

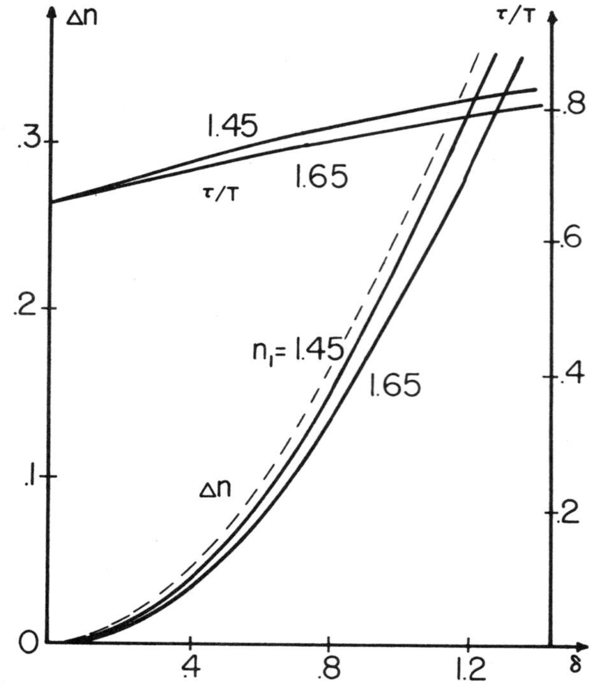

FIG. 5. Curves showing the maximum change of mode index Δn achievable on a single-mode guide by varying film thickness, as a function of $\delta = (n_2^2 - n_1^2)^{\frac{1}{2}}$ with $n_3 = 1.0$. The upper curves of τ/T indicate the percentage film-thickness change required to obtain a specific Δn. The approximation of Δn expressed in Eq. (16) is shown dashed for $n_1 = 1.45$.

Inspection of Eq. (15) indicates that the thickness change required to shift the mode index by Δn_{max} is made smaller by making either ξ large or δ small. However, Eq. (17) shows that δ small limits Δn_{max}, so that by choosing ξ large, we can reduce $\Delta\tau/\tau_{max}$ while increasing Δn_{max}. Because $\xi = (n_1^2 - n_3^2)^{\frac{1}{2}}$ and $\delta = (n_2^2 - n_1^2)^{\frac{1}{2}}$, large ξ is best obtained if n_3 is made small, indicating that the asymmetric thin-film guide for which $n_3 = 1.0$ (air) is most suitable for passive waveguide devices requiring high mode-index change with thickness.

Equations (15) and (17) are plotted in Fig. 5 as a function of δ and n_1 for the case $n_3 = 1.0$ along with the actual curves. Note that lower substrate indices favor larger values of Δn_{max} and that correspondingly greater thickness changes are required for a particular value of δ. The curves do not change rapidly with n_1, which is the reason for choosing δ as the independent variable instead of the film index n_2.

Mechanically robust thin-film glass lenses and prisms may be readily fabricated by sputtering through shaped masks or by sputter etching of predeposited films.[11] Vacuum evaporation of materials such as CeO_2, ZnS, and ZnO films is suitable for waveguides and passive components, except that optical losses in these materials restrict surface-wave propagation to centimeter distances. Thickness profiling of guiding films can produce practical lens components that have thin-film-equivalent f numbers as low as 8 for glass films and from 2 to 3 for high-index dielectrics.

FILM- AND SUBSTRATE-EMBEDDED LAYERS

Lens and prism devices can be fabricated by embedding appropriately shaped high-index layers into the guiding films or substrate regions, as shown in Fig. 6. This method has the advantage over profiled-thickness high-index films in that low-loss film materials can be used for the major part of an integrated system and lossy high-index materials can be confined to the regions of lenses. Addition of layers alters the modal field distributions as well as the propagation constants, so that field variations across an index gradient must be considered in order to determine scattered-power losses. Although the question of power loss is not treated here, it seems reasonable to expect that losses are minimized with the use of embedding techniques that provide for minimum distortion of the fields. We have computed transverse-field distributions for a variety of embedding configurations and show typical results for the central region of a lens in Fig. 6. The fields are evanescent within the waveguide, for these configurations. A sample lens that can be made by deposition techniques is also shown schematically in Fig. 6.

A numerical solution of Eq. (4) for the substrate-deposited film in a polyurethane guide 1.0λ thick is shown in Fig. 7 and illustrates how the mode index changes as a function of deposited-layer thickness τ, whereas the total thickness of the structure T remains constant. For example, with ZnS layers ($n_2=2.3$), mode-index changes of 0.5 are possible on a $1.0\lambda_0$ guide,

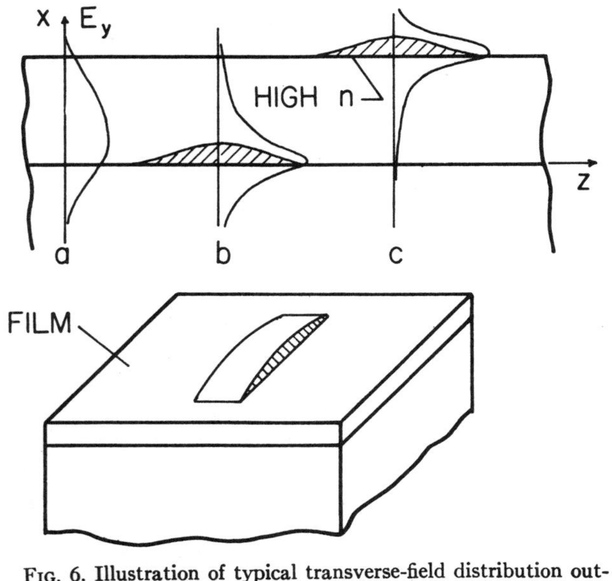

FIG. 6. Illustration of typical transverse-field distribution outside (a) and inside (b,c) the region of a film or substrate-embedded layer, and a schematic representation of thin-film lens made by deposition of high-index films.

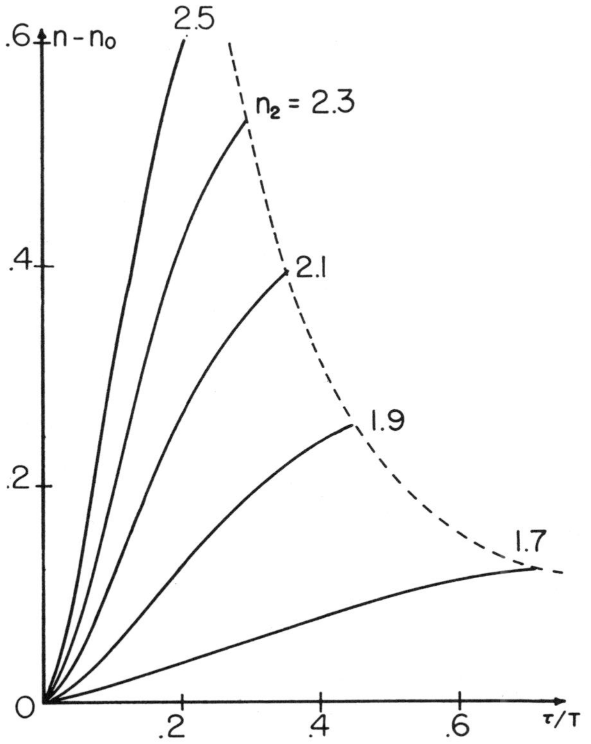

FIG. 7. The change of mode index $n-n_0$ for the two-film structure as a function of embedded layer thickness τ and refractive index n_2. The region to the left of dashed line is single mode, and $n_0=1.5249$ is the mode index in a guide of thickness $T=\lambda_0$ for $\tau=0$; $n_1=1.52$, $n_3=1.55$, and $n_4=1.0$.

but greater guide thickness T can produce even greater values of $(n-n_0)$ for a particular value of τ/T. By use of high-index dielectric materials, the embedding technique can yield lenses with f numbers near 2, depending on the materials used.

Although single-mode operation of the guide and device structure ensures that no higher-order propagating modes can be generated by slight irregularities, it may be possible to maintain essentially single-mode propagation on a multimode film by a judicious positioning of the high-index layer within the film. Clearly, a closer match between the transverse tangential fields at a and b in Fig. 6 results if the shaded layer is moved closer toward the center of the guiding film. Numerical results also show that a central location of the high-index film causes a greater change of the mode index than the same thickness of substrate or film-deposited layers. This is reasonable because the characteristics of a propagating mode are largely determined by the region to which most of the mode energy is confined.

THIN-FILM DIFFRACTION

By ruling finely spaced, shallow grooves in the surface of a film or by depositing strips of dielectric of lower refractive index on the film surface, as indicated in Fig. 8, thin-film diffraction gratings may be produced. The propagation constant of mode in such a structure undergoes periodic variations, which results in diffraction effects similar to the interaction of light with standing acoustic waves in bulk media. These phenomena have intriguing applications to a variety of thin-film spectral filter and mode-selection devices. Deposition of highly absorbing materials directly on the surface of a film quickly attenuates an optical surface wave, so that such layers may be used to form thin-film equivalents of amplitude masks and spatial filters, gratings, and linear Fresnel-zone lenses. An advantage of using deposited absorbing layers for spatial filters requiring continuously varying optical densities (as for optical-gradient or integral operators) is apparent because the absorbing layer may be easily shaped to produce any desired spatially varying absorption characteristic.

Propagation through a diffraction grating may be treated with a transfer function of the form of Eq. (9), if the length of the grating is not large compared with its period. In this case, the index $\hat{n}(y,z)$ is uniform in z and varies with y, in accordance with the mode propagation in the alternating regions. If alternate regions are absorptive, $\hat{n}(y,z)$ is complex. The diffracted field may be found in the Fraunhofer region from the one-dimensional Kirchhoff integral

$$U(\theta)=\int \exp\left\{ik_0\left[\int_{z_1}^{z_2} \hat{n}(y,z)dz+n_0 y \sin\theta\right]\right\} \times g_1(y,z_1)dy, \quad (18)$$

taken over the exit plane (y,z_2) of the grating. When the incident wave g_1 is plane, expansion of the first exponential term in a Fourier series with the periodicity of the grating results in a series of maxima of the integral in Eq. (18) at angles θ that correspond to the usual diffraction orders of a grating.

In more general and more readily fabricated gratings, the length of the grating is greater than the period, as shown in left-hand side of Fig. 8. Propagation along such a structure must then be treated in terms of modes of the Floquet form, $f_j(y)\,\exp i[\phi_{j,k}+k_0 n_j z]$, in which $f_j(y)$ is periodic in the period of the grating and $\phi_{j,k}$ is the phase from section to section. The amplitude of the modes is found from an integral over the product of $f_j(y)\,\exp i\phi_{j,k}$ and g_1 in the input-aperture plane. The phase $\phi_{j,k}$ is established by imposing periodic or other boundary conditions across the entire grating. The field produced by such a distributed grating generally results in many maxima in Eq. (18) as a result of the behavior of $\phi_{j,k}$ so that many diffracted beams are produced. When alternate regions are absorptive, the modes that are least attenuated are those that have most of their energy confined to the nonabsorptive regions, e.g., modes that are evanescent in the absorptive regions. Although these modes may not be strongly excited at the input to the grating when g_1 is a plane wave, poor grating-line resolution will result in their excitation.

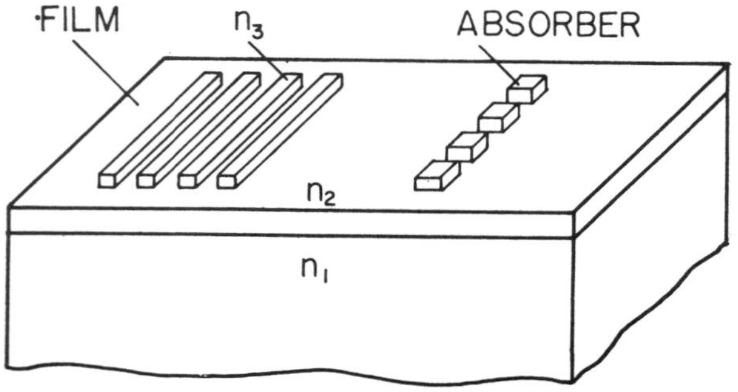

FIG. 8. Distributed thin-film diffraction gratings achieved by deposition of dielectric or absorbing materials.

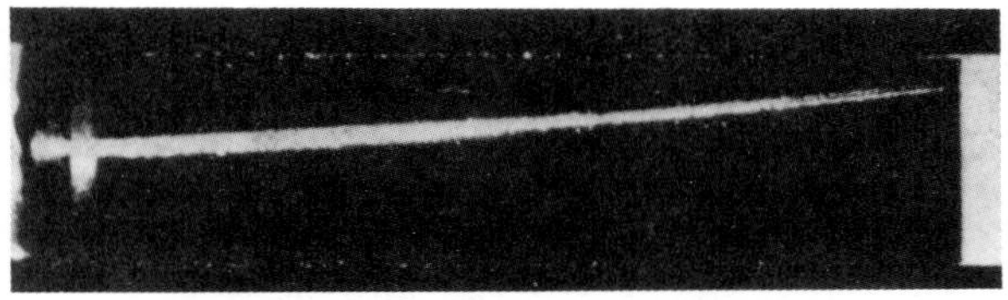

FIG. 9. Beam deflection on a glass film having transverse thickness gradient.

EXPERIMENTAL RESULTS

Films were fabricated by rf-sputtering high-refractive-index glass ($n=1.64$) in an argon atmosphere onto glass optical flats ($n=1.52$) to thicknesses ranging from 0.3 to 2 μm. High-index layers of ZnS and CeO_2 were thermally evaporated by methods described by Hass and Hall.[12,13] Coupling of a 6328-Å laser beam into the films with high-index glass prisms was achieved at over 50% efficiency; by coupling the energy out of the film with a second prism and observing the propagating mode spectrum, the approximate film thickness was determined.[4]

Figure 9 shows an example of surface-wave beam deflection on a film having a thickness gradient of about 0.4 μm/cm. The beam curves toward the thicker portion of the film, in which the mode index is increased and phase velocity decreased. A lens could be fabricated by profiling a portion of the film by either further deposition through shaped masks or by controlled etching techniques. Figure 10 indicates the basic operation of a thin-film lens fabricated by depositing a rectangular-profiled layer of CeO_2 about 0.08 μm thick at the center, on the surface of a glass film. The photograph was taken by superimposing a single beam that was moved between exposures; the beams intersect at a distance of 1.2 cm. Profiling of the layer is accomplished by evaporating the CeO_2 through a rectangular mask positioned at several distances about 1 cm below the

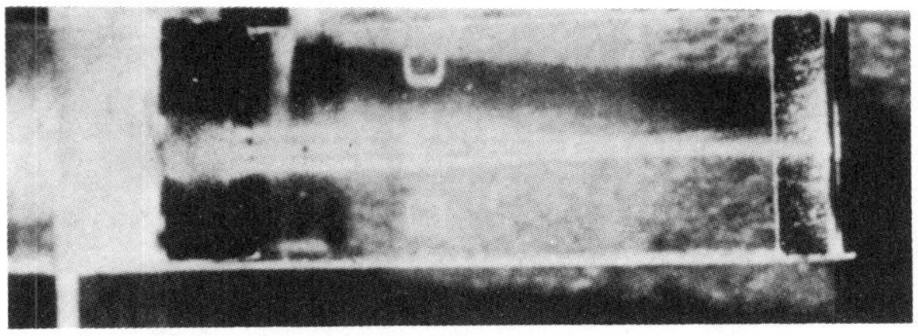

FIG. 10. Basic focusing of beam by film-surface-deposited CeO_2 lens.

substrate. More accurately controlled profiles have been obtained with a motor-driven mask assembly capable of producing film lens and prism layers of any desired thickness profile. Figure 11 is a photograph showing diffraction by a thin-film grating constructed by ruling 12-μm-wide lines separated by 12 μm in a 0.25-mm-wide Al strip deposited on the film surface. The diffraction pattern of the collimated input beam is a 5° diverging beam composed of several diffracted orders, which may

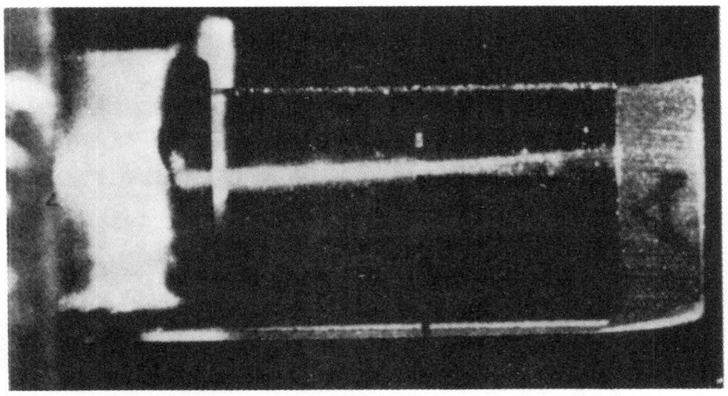

FIG. 11. Diffracted beam of a thin-film grating of 25-μ period ruled in an aluminum strip.

appear indistinct due to scattering. This pattern may be compared with a standard grating of similar slit width and periodicity, which produces a diffracted beam containing primarily zero, first, and second orders within a 4° cone.

A dielectric grating of the type shown in Fig. 8 was constructed on the surface of a glass film, about 2 μm thick, by developing the strip pattern in a thin layer of photoresist. The length of the grating along the strips

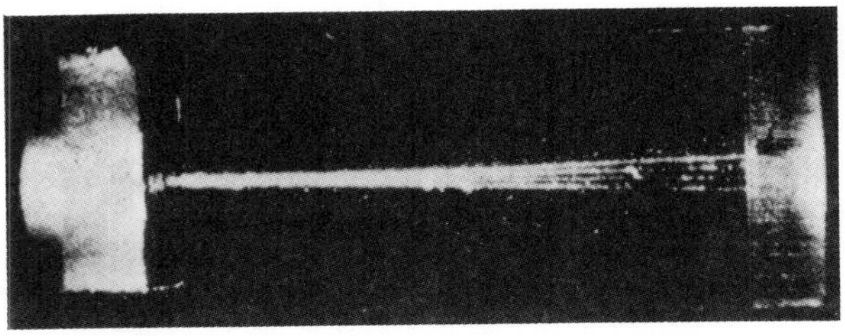

FIG. 12. Diffracted beams of a 1.55-mm-long dielectric grating of 19.2-μ period. Angular separation of the beams is 1.24°.

was 1.5 mm and consisted of 40 alternating lines and spaces each 9.6 μm wide. Propagation of the optical surface wave along such a grating, when the incident beam is nearly parallel to the grating lines, results in diffracted orders that are separated in angle by $\theta = \lambda_g/\Lambda$. The wavelength of the optical wave in the guide is λ_g, and Λ is the grating period. Figure 12 is a photograph showing the diffraction of a surface-wave beam by the long photoresist grating positioned just ahead of the coupling prism. A measurement of the angular separation of the diffracted beams in the film (1.24°) is in good agreement with the calculated value of θ.

CONCLUSION

The feasibility of passive integrated optical devices has been demonstrated; they are thin-film equivalents of conventional optical components and have potential application to integrated optical systems. Thin-film integrated optical devices can be produced in several different ways because of the greater degree of control over guided-wave propagation, in comparison to propagation in bulk media.

Methods of lens fabrication, including thickness profiling of guiding films and film- or substrate-embedding of high-index layers, have been discussed. The latter method has the advantage that low-loss guiding films may be used for most of the structure containing a lens, thus confining optical losses from slightly absorbing high-index layers to the lens region. Numerically computed solutions of the eigenvalue equations for typical one- and two-layer structures show that thin-film lenses with the low f numbers needed for integrated processors can be readily obtained. Experimental and theoretical aspects of distributed thin-film diffraction gratings have been discussed.

ACKNOWLEDGMENTS

The authors wish to thank W. R. Hill, R. Corwin, and P. Sulonen of Applied Physics Laboratory, University of Washington, for making available the vacuum-sputtering facilities used for fabricating the thin films.

APPENDIX

Assuming single-mode operation, we may derive Eq. (17) from Eq. (5) by determining n at $\tau_{co}^{(1)}$ because $\Delta n_{\max} = n(\tau_{co}^{(1)}) - n_1$. Substitution of $\tau_{co}^{(1)}$ of Eq. (5) and approximation of the first and second arctangents of Eq. (5) by $\pi/2$ and $\pi/2 - (n_2^2 - n^2/n^2 - n_1^2)^{\frac{1}{2}}$, respectively, give

$$\left(\frac{n_2^2 - n^2}{n^2 - n_1^2}\right)^{\frac{1}{2}} + \frac{1}{\delta}(n_2^2 - n^2)^{\frac{1}{2}} \tan^{-1}\frac{\xi}{\delta} + \frac{\pi}{\delta}(n_2^2 - n^2)^{\frac{1}{2}} = \pi. \quad \text{(A1)}$$

Multiplying through by $(n^2 - n_1^2)^{\frac{1}{2}} \sim (n_2^2 - n_1^2)^{\frac{1}{2}} = \delta$ and isolating the term containing n on one side of Eq. (A1) can be shown to yield

$$(n_2^2-n^2)^{\frac{1}{2}}=\pi\delta/[(1+\pi)+\tan^{-1}(\xi/\delta)]. \qquad \text{(A2)}$$

Expansion of $(n_2^2-n^2)^{\frac{1}{2}}$ in a Taylor series about n_1 and retention of two terms give

$$(n_2^2-n^2)^{\frac{1}{2}}\simeq\delta-(n_1/\delta)(n-n_1), \qquad \text{(A3)}$$

so that

$$\Delta n_{\max}=n-n_1\cong(\delta^2/n_1)[(1+\tan^{-1}\xi/\delta)/(1+\pi+\tan^{-1}\xi/\delta)]. \qquad \text{(A4)}$$

REFERENCES

* Presented at the Philadelphia Meeting of the Optical Society, April 1970. [J. Opt. Soc. Am. **60**, 725A (1970).] This work supported by NSF grant No. GK 10319.

[1] D. B. Anderson, J. J. Boyd, and J. D. McMullen, *Proceedings of the Symposium on Submillimeter Waves*, MRI Symposium Series, Vol. 20 (Polytechnic Institute of Brooklyn Press, New York, 1970).

[2] E. R. Schineller, Microwaves **7**, 77 (1968).

[3] S. E. Miller, Bell System Tech. J. **48**, 2059 (1969).

[4] J. H. Harris, R. Shubert, and J. N. Polky, J. Opt. Soc. Am. **60**, 1007 (1970).

[5] R. Shubert and J. H. Harris, IEEE **MTT16**, 1048 (1968).

[6] J. Kane and H. Osterberg, J. Opt. Soc. Am. **54**, 347 (1964).

[7] K. Weiser and Frank Stern, Appl. Phys. Letters **5**, 115 (1964).

[8] D. F. Nelson and J. McKenna, J. Appl. Phys. **38**, 4057 (1967).

[9] D. Marcuse, Bell System Tech. J. **49**, 273 (1970).

[10] P. K. Tien, R. Ulrich, and R. J. Martin, Appl. Phys. Letters **14**, 291 (1969).

[11] J. E. Goell and R. D. Standly, Bell System Tech. J. **48**, 3445 (1969).

[12] G. Hass, J. B. Ramsey, and R. Thun, J. Opt. Soc. Am. **48**, 324 (1958).

[13] J. F. Hall and W. F. Ferguson, J. Opt. Soc. Am. **45**, 714 (1955).

Geometrical Optics in Thin Film Light Guides

R. Ulrich and R. J. Martin

We consider thin film light guides consisting of a transparent film of high refractive index deposited on a substrate of lower index. The propagation of light in such a two-dimensional transmission medium can be described within the limits of geometrical optics by an effective index of refraction N. Its value depends on the film thickness. Therefore, a light beam in the thin film guide is refracted or totally reflected at a step of film thickness. We discuss these phenomena (Snell's law) and demonstrate them experimentally, using ZnS films on glass as guides. As applications, we show a thin film prism and thin film lenses for guided light beams. By properly choosing the film thicknesses at both sides of the step, one can obtain an unusually large positive or negative wavelength dispersion of the refraction or, if desired, achromatic refraction.

I. Introduction

The guidance of light in sheets or strips of a thin dielectric film has recently attracted increasing attention.[1–7] The interest in this optical technique is based on the expectation that many operations in future data processing and communication systems can be performed reliably and economically by integrated optical devices, employing surface-guided optical waves. The suggested applications may be divided into two categories. In one,[1] the light propagates along narrow strips of a high index film which are, in essence, optical fibers supported by a common substrate. In the other category[2–6] the second dimension of the film is large, too, and the light can propagate in two dimensions rather than only in one as on a strip or fiber.

In the present paper we report some experiments on this two-dimensional propagation in guides of the second category. The guides consist of a thin film of a transparent, high index optical material, deposited on a substrate of lower refractive index. We show that the light propagation on these surface waveguides can be described within the limits of geometrical optics by an effective index N of refraction of the guide. Its value depends, among other parameters, on the thickness W of the light-guiding film. In a film of nonuniform thickness $W(x,y)$ this index is a function $N(x,y)$ of position on the guide. A nonuniform guide, therefore, is an optically inhomogeneous propagation medium, and the light rays in it are curved (Fig. 1). We will see that by properly shaping the thickness profile $W(x,y)$ of the film one can build prisms, lenses, and similar optical elements for the guided optical waves. They are in many respects the equivalents of their three-dimensional counterparts in the usual *bulk* optics.

In Secs. II and III we report some experiments that illustrate the basic effects of refraction and of total reflection of a surface-guided optical wave. The application of these effects to the construction of thin film prisms and lenses is demonstrated in Sec. IV. The important question of the dispersion of the guides is discussed in Sec. V. Finally in an Appendix we derive the two-dimensional eikonal equation that governs the various geometrical optical phenomena discussed in other sections. In particular, we will show the validity of Snell's law for optical surface waves.

II. Refraction of a Guided Wave

A. Principle

We illustrate the basic effect of refraction of a guided wave in Figs. 1 and 2. Figure 1 shows a thin film light guide, consisting of a high index (n_1) dielectric film sandwiched between two other optical materials of lower refractive indices (n_0, n_2). The film has a uniform thickness W_I throughout the half-plane $y < 0$ (region I in Fig. 1), and a uniform, but larger, thickness W_{II} in the other half of the xy plane (region II). Near the common boundary $y = 0$ of the two regions the film thickness is tapered smoothly from W_I to W_{II} over a distance L_T that is much larger than the vacuum wavelength λ of the light to be guided. A well-collimated laser beam is coupled into the guide at some point in region I by a suitable coupler,[3,4,6] not shown in Fig. 1. The initial direction of propagation is chosen so that the guided beam is incident obliquely (angle of inci-

The authors are with Bell Telephone Laboratories, Inc., Holmdel, New Jersey 07733.

Received 21 April 1971.

Reprinted with permission from *Appl. Opt.*, vol. 10, pp. 2077–2085, Sept. 1971.

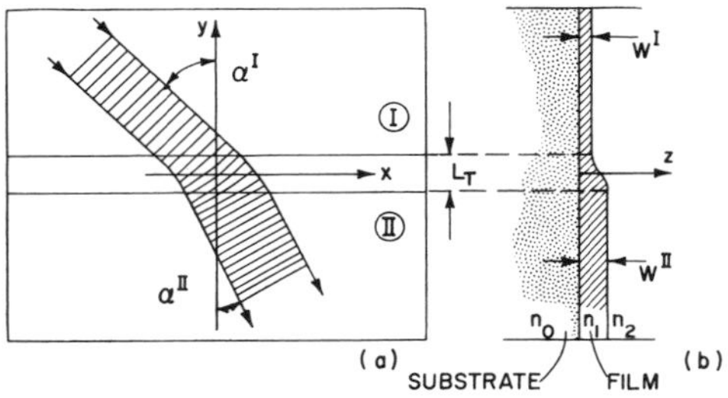

Fig. 1. A light beam, guided in a planar dielectric guide, is refracted at a step of the film thickness W. The difference in film thickness between the regions I and II causes their effective indices of refraction N^I, N^{II} to be different. (a) Top view; (b) side view.

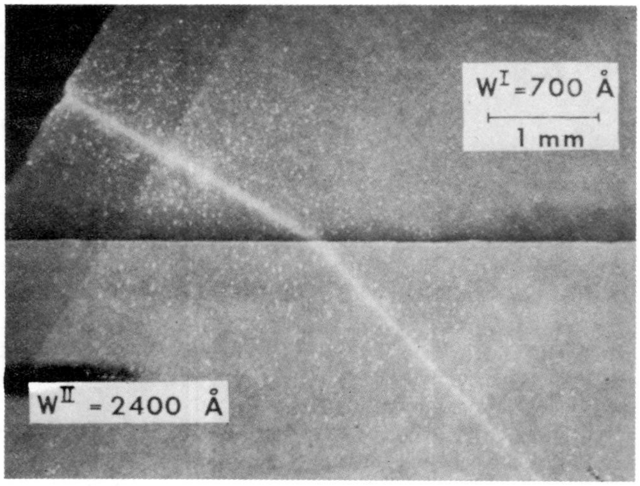

Fig. 2. Experimental demonstration of the refraction of a guided laser beam. The light guide is formed by a ZnS film vacuum-deposited on a glass substrate. The thicknesses W^I, W^{II} of the ZnS film are indicated. The He–Ne laser beam propagates as a TE(m = 0) mode.

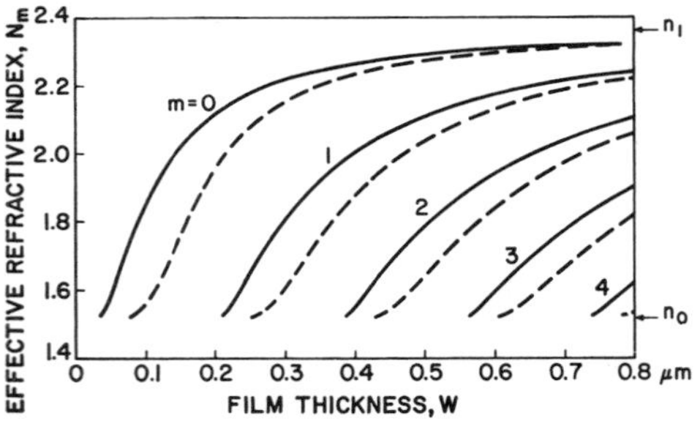

Fig. 3. The effective index of refraction N_m of a planar dielectric waveguide as a function of the film thickness W, given here for the example of a ZnS film on a glass substrate (n_0 = 1.51; n_1 = 2.35; λ = 6328 Å). The parameter at the curves is the mode number m. —— TE modes, ------ TM modes.

dence $\alpha^I \neq 0$) on the boundary. The beam will traverse the tapered region along a curved path and then proceed again as a straight, guided beam in the second region of the film. There, however, its direction of propagation is changed to a new angle $\alpha^{II} \neq \alpha^I$. This phenomenon must be interpreted as a refraction of the guided beam at the boundary of the two regions of different film thickness.

In the following we will discuss this refraction and some applications of it from a purely geometrical optical point of view. This means that we are interested mainly in the directions of propagation, i.e., in the optical rays, and not in intensity relations. The latter will be treated elsewhere. Therefore we are not concerned here with the radiation losses and the mode conversion that may occur at the step in film thickness.[7]

B. Effective Index of Refraction

As we will show theoretically in the Appendix, the refraction, not surprisingly, does obey Snell's law

$$N^I \sin\alpha^I = N^{II} \sin\alpha^{II}. \quad (1)$$

The parameter N appearing here has the meaning of an effective refractive index of the guide. Its value is related to the phase velocity v_{ph} of the guided wave by

$$N = c/v_{ph}, \quad (2)$$

where c is the velocity of light in vacuum. Equation (2) is exactly analogous to the relation between the phase velocity of a plane light wave in a bulk optical material and the bulk optical index.[8] In the thin film guide the situation is slightly more complicated by the fact that the guide can, in general, support several modes of propagation.[9] They can be distinguished[4] by a mode number m = 0,1,2,..., and by their polarization (TE or TM). Each mode has its own phase velocity, different from all other modes, and must therefore be characterized by its own effective index N_m.

The values of the indices N_m can be calculated from the bulk indices n_j, j = 0,1,2 of the materials forming the guide, from the wavelength and polarization of the light, and from the thickness W of the film.[2–4,7,9] As an illustration, we show here in Fig. 3 the behavior of the effective index N for the example of a light-guiding ZnS film on a glass substrate. We note from Fig. 3 that N is restricted for all modes to the range $n_0 < N < n_1$ between the bulk indices of substrate and film materials, assuming $n_2 \leq n_0$. A given mode m exists only if the film thickness W exceeds a certain *cutoff thickness* $W_{c,m}$. In the range of existence $W > W_{c,m}$, the effective index N_m of a mode is a monotonically increasing function of the film thickness W. This fact, which is basic to this paper, can also be derived analytically from the characteristic equation of the guide. We find from Eqs. (25)–(29) of Ref. 4 that

$$dN_m/dW = (N_m/W_{eq})[(n_1^2/N_m^2) - 1]. \quad (3)$$

Here, W_{eq} is a positive parameter which indicates the extent of the fields in z direction [$kW_{eq} = \mu$ with μ as defined in Eqs. (27) and (28) of Ref. 4]. Because of $N_m < n_1$ the slope Eq. (3) is positive definite. This means that the thicker region (II) in Fig. 1 has a higher index N than the thinner region (I). Consequently, according to Eq. (1), the beam in Fig. 1 should be refracted so that $\alpha^{II} < \alpha^{I}$. The experiments described below confirm this conclusion.

We note here that the angle α^{II} of refraction does not depend on the details of the thickness profile $W(y)$ of the film in the transition region in Fig. 1. However, this profile does affect the position of the refracted beam in the x direction, and also the losses caused by reflection, radiation, and mode conversion. A smoothly tapered transition will in general minimize these losses.

In the rest of this paper, we will suppress the mode number subscript m, assuming that initially only one mode was launched in the guide and ignoring conversion to other modes at the step in film thickness.

C. Experiment

We have observed the refraction of a guided wave experimentally (Fig. 2) in a guide that consists of a ZnS film vacuum-deposited on a glass substrate. The step in film thickness was produced by first coating the entire substrate with a film of uniform thickness $W_I \approx$ 700 Å. Region I was then masked off and a second layer of ZnS was deposited on top of the first one, increasing the film thickness in region II to $W_{II} \approx 2400$ Å. The mask was held at a distance of approximately 0.1 mm from the surface in this second evaporation. Surface diffusion and the natural angular spread of the incident vapor stream produced the smoothly tapered transition of width $L_T \approx 0.05$ mm between the regions I and II.

Figure 2 was photographed looking down normally at the plane of the guide. The two regions of different film thickness are discernible by their different reflectance. The beam of a He–Ne laser has been coupled into the TE ($m = 0$) mode of the guide at a point near the top left corner of Fig. 2 by means of a prism film coupler.[3,4] The path of the beam in the guide is clearly visible by the light that is scattered out of the guide at imperfections of the ZnS film and of the substrate. This scattering, in combination with true absorption, limited the total observable length of the beam in the guide to about 30 mm. From the point of view of application, a guide with such high losses would hardly be useful. In our experiments, however, whose purpose it was to demonstrate the basic effects, we found the strong scattering convenient for the observation of the guided beams. For practical applications much better guides are available,[10–12] whose attenuation is only a few percent per centimeter.

We have checked the validity of Snell's law [Eq. (1)] under various conditions. The values of N^I and N^{II} were measured with a prism film coupler,[3] and the angles α^I and α^{II} determined from photographs like Fig. 2. No deviation from Eq. (1) was found within the experimental accuracy, which was limited by the errors in determining the angles α. In the example of Fig. 2 we found $N^I = 1.678$ and $N^{II} = 2.165$ so that the relative index of region II with respect to region I is $N_R = N^{II}/N^I = 1.290$. The measured angles are $\alpha^I = 60.2°$ and $\alpha^{II} = 42.5°$. The resulting ratio $\sin\alpha^I/\sin\alpha^{II} = 1.286$ agrees satisfactorily with the expected value N_R given before.

In conclusion we can state that the phenomenon of refraction does exist for light waves guided in thin films. The refraction obeys Snell's law with an index N that is a function of the film thickness.

III. Reflection of a Guided Wave

When a guided wave is incident on the boundary between two regions of a guide having different indices $N^{II} \neq N^I$, some or all of the light can be expected to be reflected. We will show in this section that the reflection at a *tapered* step of film thickness (Fig. 1) either is extremely weak so that it can practically be neglected or is a total internal one, depending on the angle and side of incidence. In both situations the direction of the reflected beam is given by the reflection law of geometrical optics.

A. Influence of a Taper

We start by discussing the reflection of a thin film guided wave at an *abrupt* step of film thickness. Only the case of normal incidence ($\alpha = 0$) has been treated theoretically. Macruse[7] has shown that the power reflectance for a TE ($m = 0$) mode is given, under certain restrictions, by the familiar expression $R = (N^I - N^{II})^2/(N^I + N^{II})^2$, whereas for TM polarization a different, much more complicated expression holds. In either case, however, the reflectance of an abrupt step does not exceed a value of a few percent for typical index ratios N^{II}/N^I.

Maintaining normal incidence ($\alpha = 0$) of the beam, we can now consider a *tapered* step of the film thickness as the limiting case of a sequence of many small, abrupt steps. Because these are distributed more or less uniformly over a region L_T that is many wavelengths long, the waves reflected at the various steps all have different phases. Therefore they tend to cancel each other. This principle of impendance matching by smoothly tapering the transition between two different waveguides is well known in the microwave technique. A simple estimate shows that the power reflectance of a smoothly tapered step in film thickness, having a length $L_T \gg \lambda$, is reduced by a factor of the order $(kNL_T)^{-2}$ below the reflectance that the same step would have if it were abrupt. Here $k = 2\pi/\lambda$ is the vacuum propagation constant of the light. Even for the relatively short taper length of $L_T = 10\lambda$ this factor is as small as approximately 10^{-4}. Therefore, the normal reflectance of a tapered step may be considered negligible for most practical purposes.

For oblique incidence ($\alpha \neq 0$) we can only speculate that the reflectance of a thickness step behaves similarly to the reflectance for bulk optical waves at an index interface. For a TM mode this would mean that the reflectance rises monotonically when the angle of incidence is increased from $\alpha = 0$ to $\alpha \rightarrow \pi/2$. For TE polarization, on the other hand, we expect the reflectance to pass first through a Brewster-type minimum before it starts rising. Regardless of such details, how-

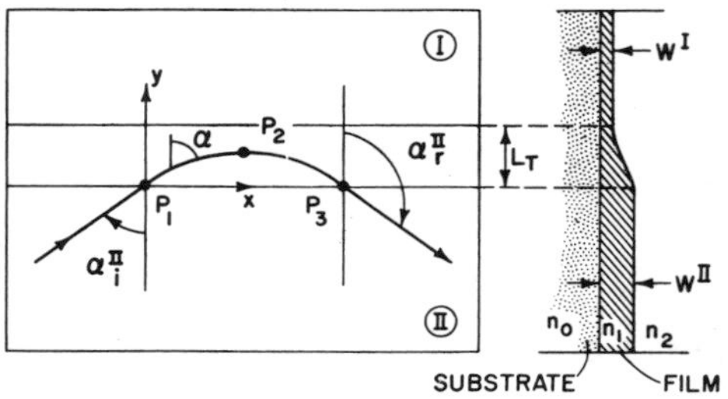

Fig. 4. The total reflection of a guided beam at a step of film thickness can be understood from the curvature of the rays inside the region L_T of nonuniform film thickness.

ever, our earlier considerations about the reflectance-lowering effect of a taper remain valid also for oblique incidence. This conclusion is supported experimentally by Fig. 2, where no indication of a reflected beam is observed. In summary, we can state that the tapered thickness step is equivalent to a very efficient antireflection coating of the graded index type.

B. Total Reflection

There is, of course, one very important exception to this rule. This is the case of total internal reflection. It exists if the guided beam is incident on the boundary line from the side of the higher effective index N^{II} (i.e., from the region of the thicker film W^{II} in Fig. 4) at a sufficiently large angle α_i^{II} of incidence,

$$\alpha_i^{II} > \alpha_c \text{ where } \alpha_c = \arcsin(N^I/N^{II}). \quad (4)$$

Thus if α_i^{II} exceeds the critical angle α_c, no refracted beam appears in the thin region W^I and the light beam is totally reflected back into region II.

The internal reflection of the guided beam at the region of tapered film thickness is a direct consequence of the law of refraction, Eq. (1). We recognize this by subdividing the taper into many small steps and applying Eq. (1) to each step individually (Fig. 4). The result is that the *numerical aperture* $\mathbf{NA} \equiv N \sin\alpha$ is constant along the beam, in exact analogy to the propagation of a bulk light beam through a stratified optical material. Beyond the point P_1 (Fig. 4), where N decreases, the beam azimuth α must therefore increase. Total reflection occurs if α reaches the value $\pi/2$ before the light has entered the thinner region I. In Fig. 4 we have $\alpha \rightarrow \pi/2$ at P_2. At that point the light propagates parallel to the lines of constant film thickness, and then it bends back toward the thicker region. At P_2 we have $N(P_2) = N^{II} \sin\alpha_i^{II} = \mathbf{NA}$, and the requirement that such point P_2 exist inside the tapered region leads directly to the condition (4) for total reflection.

From point P_3 on, the index is constant again, $N = N^{II}$, and the beam is straight. There we can identify α with the angle α_r^{II} of reflection, and from the constancy of $\mathbf{NA}$ we obtain the geometrical optical law of reflection for thin film guided beams,

$$\alpha_r^{II} = \pi - \alpha_i^{II}. \quad (5)$$

The total internal reflection is shown experimentally in Fig. 5. The guided light beam is incident from the upper left corner, propagating initially in the thin region. It enters the thicker region near the 90° corner, suffering a slight refraction. The beam is then incident on the horizontal boundary to the thinner region and is totally reflected. All experimental parameters (W^I, W^{II}, N^I, N^{II}, etc.) are the same here as in Fig. 2. Therefore, the critical angle is $\alpha_c^{II} = \arcsin(1/1.286) \approx 51°$. This angle is exceeded, as required, by the angle of incidence $\alpha_i^{II} \approx 63°$ in Fig. 5. The reflection is slightly diffuse because the edge of the thick region is not perfectly straight.

C. Total Reflection and Mode Cutoff

In the preceding discussion of total internal reflection we had tacitly assumed that the particular mode under consideration could exist both in regions I and II of the film and throughout the taper. This required that the film thickness be everywhere larger than the cutoff thickness $W_{C,m}$ of the particular mode. We will now discuss the situation where the film on one side of the boundary line is thinner than the cutoff thickness, so that the light cannot propagate there in the same mode in which it was incident on the boundary. A total reflection is still possible in this case, as we will see. The effective index N^I of the thinner region is no longer defined, however, and a new condition for total reflection must be found instead of Eq. (4).

A particular case of total reflection involving mode cutoff exists if $W^I = 0$, i.e., at the edge of region II. An example is given in Fig. 6. There the ZnS film covers only a part of the substrate. The rest of the surface had been masked during the vacuum deposition, producing a film with a smoothly tapered edge of width

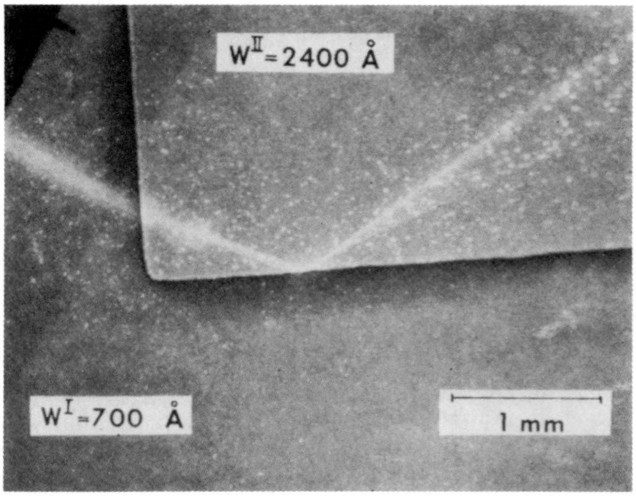

Fig. 5. Experimental demonstration of the total reflection of a guided laser beam at a step of film thickness. The beam is launched in the thin region, refracted when entering the thick region, and totally reflected at the boundary to the thin region. (ZnS on glass, TE_0 mode, $\lambda = 6328$ Å.)

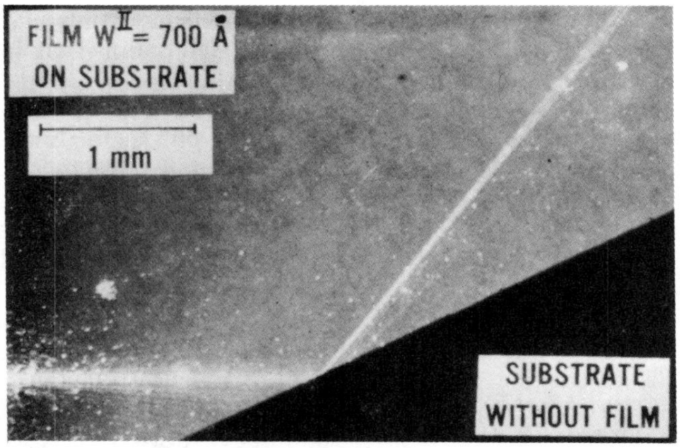

Fig. 6. Total reflection of a guided beam at the edge of the light-guiding film ($W^I = 0$).

$L_T \approx 0.05$ mm. As we see from Fig. 6, a total internal reflection does occur at this edge. It can be explained in exactly the same way as above, i.e., by a bending of the beam in the tapered region. The critical angle follows now from the condition that a point P_2 must exist at which $N(P_2) = \mathbf{NA} = N^{II} \sin\alpha_i^{II}$. The theory of the thin film guide[3,4,9] shows that perfect guidance is possible only when N exceeds the refractive index n_0 of the substrate, $N > n_0$. Applying this condition to the point P_2 leads to

$$\alpha_i^{II} > \arcsin(n_0/N^{II}) \tag{6}$$

as a necessary condition for any total internal reflection of a guided wave. Here N^{II} is the effective index of the thicker region and n_0 is the larger of the bulk indices of the two materials adjacent to the film (Fig. 1). According to its derivation, the necessary condition (6) is also a sufficient one if the film thickness on the thinner side is below the cutoff thickness $W^I < W_{C,m}$ of the incident mode. Thus $\alpha_c = \arcsin\,(n_0/N_m^{II})$ if the incident mode m is below cutoff in the thinner region I.

For smaller angles of incidence, when inequality (6) is violated, an interesting phenomenon occurs. The light, unable to propagate as a guided mode in region I, is radiated from the edge of the guide into the substrate. This effect has been studied recently by Tien and Martin.[13] They show that the effect can be employed for coupling into or out of the guide, provided a suitably large taper is used.

D. Totality of Reflection

Concluding this section on the reflection of a guided beam, we want to discuss under what conditions the internal reflection can truly be called a *total* one. This question arises because it is known that a guided light beam will lose some of its power by radiation[7] when it traverses a region of tapered film thickness, e.g., in the refraction experiment in Fig. 1. Do similar losses occur in the internal reflection ($\alpha_i^{II} > \alpha_c$) of a light beam guided in a thin film?

The answer is that no such losses occur if the step in film thickness is perfectly straight, i.e., if the contours of equal thickness (isohypses) are parallel, straight lines. This holds regardless of how steeply or smoothly the edge may be tapered.

We explain this with the help of Fig. 4. The straightness of the edge is characterized by the film thickness $W = W(y)$ being independent of x. When treating the reflection of a guided straight wave (the two-dimensional equivalent of the plane wave in bulk optics) at this step as an electromagnetic boundary value problem, it is recognized that all field components in the entire space have the common x dependence $\exp(ikxN^{II} \sin\alpha_i^{II})$ which is prescribed by the incident wave. Apart from this common factor, the problem is independent of x. Any radiation loss would have to appear in this treatment as an unguided continuum mode[14] propagating as $\exp[i(k_x x + k_y y + k_z z)]$ in the media n_0 or n_2 (see Fig. 1). However, a wave with the x dependence $\exp(ikxN^{II} \sin\alpha_i^{II})$ in these media is necessarily evanescent if the conditions (4) and (6) are satisfied, because $k_z = k[n_{0,2}^2 - (N^{II})^2]^{\frac{1}{2}}$ becomes imaginary. The evanescent wave cannot carry power away from the guide, and therefore the internal reflection is truly a total one. Losses of the mentioned type must be expected, however, if the edge is not straight but curved[15] or irregular.

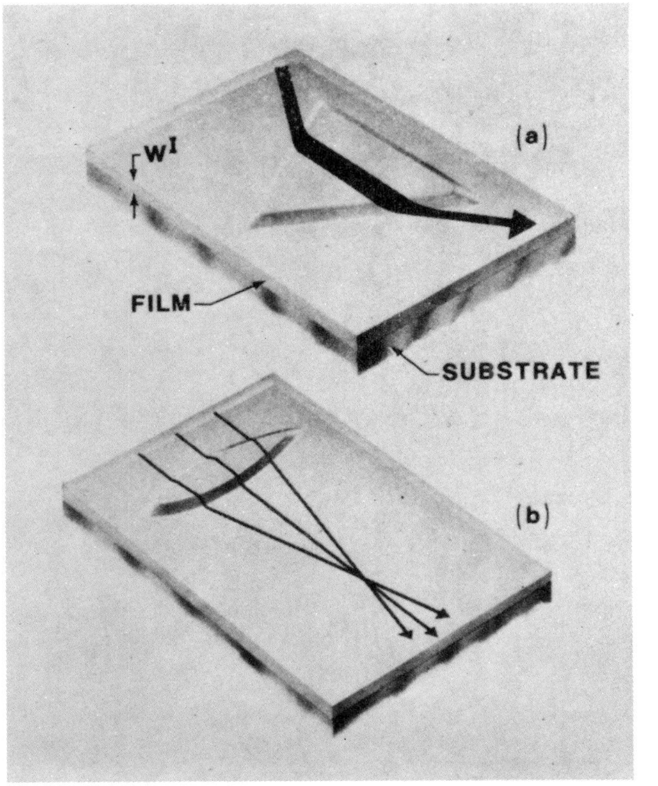

Fig. 7. Thin film optical elements like a prism (a) or lens (b) are formed by suitably shaping the boundary lines between the regions of thin and thick film thickness.

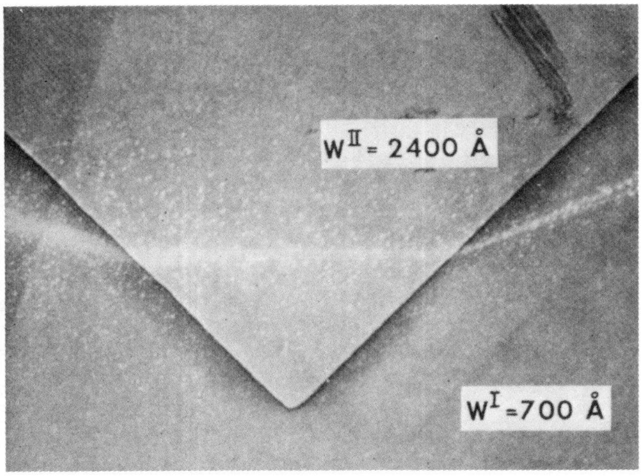

Fig. 8. A thin film prism deflecting a guided laser beam. This is the same prism as in Fig. 5, but the direction of the incident beam has been changed.

IV. Application

Prisms and lenses are the basic components of bulk optics. There they are formed by properly shaping the surfaces of dielectric bodies, e.g., of pieces of glass. Similarly we can form prisms and lenses for surface-guided light by properly shaping in a thin film guide the boundaries of a region of modified refractive index N, which means here of modified thickness W. We will discuss these possibilities in the following.

A. Thin Film Prism

The thin film prism is shown schematically in Fig. 7(a), and Fig. 8 is a photograph of a light beam in an experimental prism. The prism consists of a triangular region of increased film thickness. The transition to the surrounding area of thinner film thickness is tapered smoothly to reduce the radiation loss at these boundaries. In Fig. 8 the region inside the 90° sector forms the prism. We see the light beam being refracted, in the usual way, both when it is entering and leaving this sector. The deflection angle δ of the beam and its dispersion $d\delta/d\lambda$ can be calculated from the standard formalism for bulk prisms if the relative index $N_R = N^{II}/N^I$ is used.

Such a thin film prism may find application as the dispersive element in a thin film prism spectrometer to analyze the frequency spectrum of a light beam guided in any selected mode m of the film. Alternatively, we may use the prism to spatially separate the light of a single frequency and single initial direction that is propagating in the various modes of a film. This is possible because the effective index N is a function not only of frequency but also of the mode number m and of the polarization. Finally, the prism may serve simply to change, without dispersion, the direction of a guided beam. For this application it is important that the deflection angle δ can be made achromatic, as will be shown later on.

B. Thin Film Lenses

A thin film lens for guided waves is obtained when the region of modified thickness has curved boundaries, e.g., as indicated in Fig. 7(b). The photograph in Fig. 9 shows such a lens focusing an initially parallel laser beam into a spot. The guided beam appears broken off into many pencil beams because each one of the numerous imperfections in the guide casts a long shadow down the beam. The focal length of this lens is approximately 2 mm.

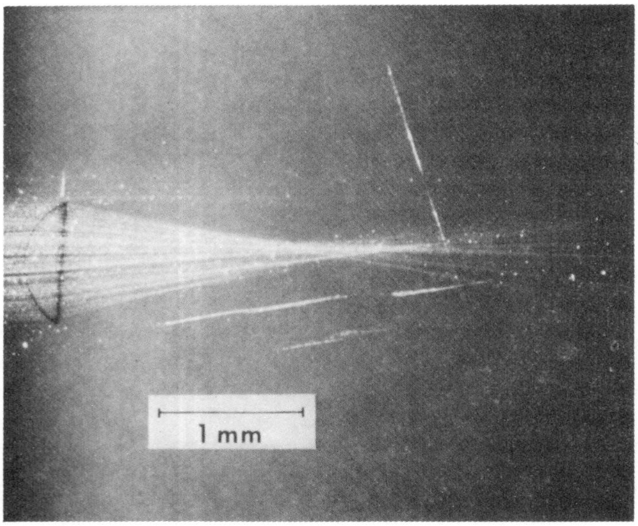

Fig. 9. A thin film lens collimating a guided laser beam. The film thickness is increased in the lens-shaped area. The lens is slightly tilted with respect to the incident beam; the resulting aberrations are apparent.

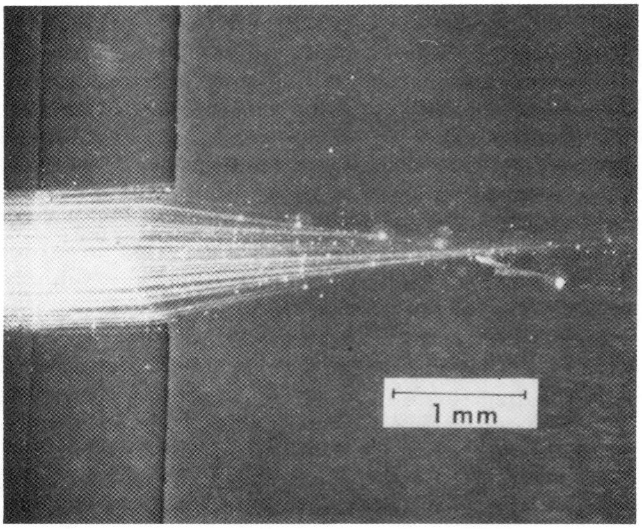

Fig. 10. Another collimating thin film lens. The lens has a planoconcave shape; its effective index is *lower* than that of the surrounding region, because the film is thinner in the lens region than in its surrounding.

The lens shown in Fig. 10 is also a focusing lens, although it is planoconcave instead of planoconvex as in Fig. 9. The lens is focusing because its refractive index N^{II} has been reduced relative to the index N^{I} of the surrounding area by making the film thinner in the region of the lens than on the rest of the substrate, so that $N_R < 1$. Here again, the standard formulas hold for the calculation of the geometrical-optical characteristics of the lens (focal points, cardinal points, aberrations), provided the relative index N_R of the lens is used.

V. Dispersion of Thin Film Guides

The wavelength dispersion of the refractive index N plays an important role for the applications mentioned above. We will show here that this dispersion can be controlled within wide limits by a choice of the parameters of the guide, particularly of the film thickness W.

The dispersion of the relative index N_R can be expressed as

$$d\ln N_R/d\lambda = (d\ln N^{II}/d\lambda) - (d\ln N^{I}/d\lambda). \quad (7)$$

It is equal to the difference of the logarithmic dispersions of the effective index $N(W,\lambda)$ of the guide, taken at the two thicknesses W^{II} and W^{I}. We illustrate this relation by Fig. 11, which is a modified version of the well-known dispersion vs refractive index representation of optical materials. Each bulk material can be characterized by a point in this figure. Because $\ln N$ is chosen as abscissa, the horizontal separation of two materials in this figure gives the logarithm of their relative refractive index, $\ln N_R = \ln N^{II} - \ln N^{I}$. Similarly, the vertical separation of the two materials gives the logarithmic dispersion [Eq. (7)] of their relative index N_R. A thin film light guide of given thickness is likewise represented by a point in Fig. 11. When the film thickness W is varied, the point moves along a line that connects the points of the bulk substrate and film materials. The thickness has been indicated in Fig. 11 as the parameter along the curve.

We notice from Fig. 11 that the dispersion of the guide can be considerably higher than the dispersion of the materials of which the guide has been made. The extra dispersion is intrinsic to the mechanism of guiding.

From Fig. 11 we can now read off directly the relative index N_R and the dispersion [Eq. (7)] of a step in film thickness in a thin film guide. The two film thicknesses are represented, e.g., by the points I and II. We recognize from Fig. 11 that by a proper choice of these points (i.e., of the film thicknesses W^{I} and W^{II}) we can simultaneously obtain specified values of the relative index N_R and of the dispersion of N_R. In particular it is possible to obtain refraction without dispersion (e.g., points I and III, representing W^{I} = 440 Å, W^{III} = 3000 Å). This is important for achromatizing prisms and lenses. On the other hand, large positive and negative dispersions are possible by choosing one of the film thicknesses very small or very large and the other one near the maximum of the dispersion curve in Fig. 11.

VI. Conclusions

We have shown that the basic laws of geometrical optics, those of refraction and reflection, are valid also in thin film optical waveguides. In principle, therefore, it is possible to translate many of the bulk optical elements and iustruments into thin film form. In this process one loses, of course, one dimension of their performance. A thin film microscope, for example, would have only a one-dimensional line of view. Yet various useful applications are conceivable for such devices in the field of analog optical data processing, as was pointed out by Shubert and Harris.[2]

Despite the many properties which thin film optical elements and their bulk equivalents have in common, they differ in a number of important points:

(1) In bulk optics, all elements (prisms, lenses) are ordinarily immersed in a common medium of propagation (vacuum, air, even water) whose index is *lower* than that of the elements (glass). In thin film optical elements, the effective index N_{med} of the common immersion medium can be made larger or smaller than the index N_{el} of the element with about equal ease, simply by choosing the proper thickness distribution $W(x,y)$ of the film. This was illustrated above with the planoconvex and planoconcave lenses of Figs. 9 and 10, both being positive. The freedom of choosing $N_{el} \gtrless N_{med}$ may simplify the design of some thin film optical systems.

(2) In bulk optics for the visible, the refractive indices used in practice are limited to the range n = 1.45–2.00. In addition, of course, n = 1 is used for the immersion. In thin film optics of the type discussed above, the need for a supporting substrate limits the range of available indices to $N \gtrsim 1.45$ at its lower end, excluding the case $N = 1$. At the upper end it might be expected that the range of possible effective indices N will extend out higher than in bulk optics, because there are high index materials, e.g., Ta_2O_5 ($n \approx 2.3$), that can be readily deposited as thin film guides,[11] but could not be used

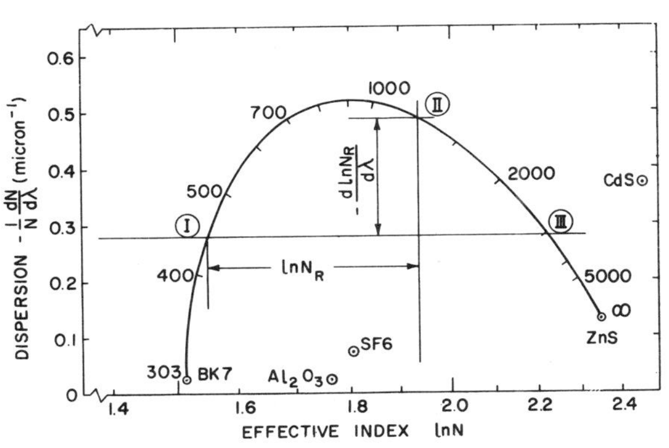

Fig. 11. This diagram shows the refractive index and dispersion at λ = 6328 Å of selected optical materials and of a thin film guide (ZnS film on BK7 glass, TE_0 mode). The parameter along the curve is the film thickness in angstroms. Abscissa in logarithmic scale, ordinate linear scale. BK7 and SF6 are designations of some Schott optical glasses.

economically in bulk form. At present, however, the technical problems of fabricating low-loss films, particularly of high index, are still considerable.

(3) The refractive index n and its dispersion $dn/d\lambda$ are material constants in bulk optics. Therefore, hundreds of optical materials (glasses) have been developed to satisfy the requirements of optical engineers in these respects. In thin film optics, as we have seen, both the relative index N_R and its dispersion are parameters that can be designed to match given specifications within wide limits, simply by choosing the thicknesses W^I and W^{II}.

(4) Bulk optical elements are preferably antireflection-coated, especially if they have a high refractive index. Thin film optical elements can be made reflectionless simply by smoothly tapering their thickness profile at their boundaries. Instead of reflection losses, however, the thin film optical elements show some losses by radiation and mode conversions at their boundaries. Therefore, a taper is desirable anyway to reduce these losses, although in order to be noticeably effective, it it must be long.[7]

(5) If, on the other hand, a partial reflection at the boundary of an optical element is a desirable feature, the step in index must be very abrupt, within a small fraction of a wavelength. In bulk optics such a step is obtained simply by polishing. In thin film light guides, sufficiently sharp steps of film thickness may be produced by photolithographic techniques if pressed to their ultimate limit of resolution. More promising appears to us the use of electron-beam lithographic techniques for this purpose.

We have discussed here only variations of the film thickness W as a means of locally modifying the effective index N of the guide, because this method appears to be technically the simplest. Another possibility for modifying N is by varying the composition of the guide, e.g., by changing locally[2,8] one or more of the indices n_0, n_1, n_2. Such embedding of a lens into the film or cladding without thickness changes appears technically difficult to us, however, because it might introduce irregularities along the boundary lines. An interesting alternative[16] has come to our attention since completion of this manuscript. This is the deposition of a second film of refractive index n_3 on top of the main film (index n_1) in suitably shaped areas. For $n_3 = n_1$, the situation is exactly the one discussed here, and for $n_3 < n_1$ it is still very similar. However, if the second film has a higher index than the main film, $n_3 > n_1$, and is thick enough, it may take over as the guiding film and leave to the main film only the role of a cladding. This method of varying N has the advantage that the higher losses, which are usually associated with high index films, are present only along the short sections of the rays inside the lenses or prisms.

Appendix

Two-Dimensional Wave Propagation

When light is guided by a thin dielectric film (Fig. 1), the electromagnetic field energy is in z direction sharply confined to the film and its immediate vicinity. In the other two dimensions (x,y) of the film, the fields can propagate, but not entirely unrestrictedly. Rather, the thickness profile $W(x,y)$ of the film determines which paths of propagation in the xy plane are possible and which are not. The allowed ones we call two-dimensional optical rays, in exact analogy to three-dimensional rays in bulk optics. Below we will show that these rays are governed by equations which are formally identical to those of bulk optics. Therefore, all conclusions from these equations, i.e., the laws of geometrical optics, must hold also for the thin film guides.

Propagation in a Uniform Guide

The modes of propagation in a guide of uniform thickness have been described in the literature.[2-4,9,14] Because the modes are orthogonal, it is sufficient here to consider propagation of monochromatic light in only one of them, representative for all others. The general, three-dimensional wave equation can be separated into a two-dimensional wave equation, describing the propagation in the xy plane

$$\bar{\nabla} V(x,y) + k^2N^2V(x,y) = 0, \tag{8}$$

and into an ordinary differential equation which determines the z dependence of the particular mode. The two equations are linked by the separation constant N^2. Its value must be found as the eigenvalue of the ordinary differential equation under the proper boundary conditions at the surfaces of the film. Therefore, N depends on the polarization and on the thickness W, as was discussed above in connection with Fig. 3.

With N known, the wave equation (8) can be solved. In this equation $V(x,y)$ is the local peak (in z direction) amplitude[4] of the transverse field component of the mode under consideration, and $\bar{\nabla} = (\partial/\partial x, \partial/\partial y)$ is the two-dimensional *nabla* operator. The general solutions of Eq. (8) are plane or cylindrical waves of phase velocity c/N, propagating along straight rays.

Propagation in a Nonuniform Guide

In a film with nonuniform thickness the propagation of light may locally be described as a superposition of modes of a uniform guide, having the film thickness $W(x,y)$. Through its dependence on that thickness $W(x,y)$ the effective index N has become a function $N(x,y)$ of the position in the xy plane. The propagation of the field $V(x,y)$ is then also governed by a two-dimensional wave equation. Compared to Eq. (8), however, this wave equation contains additional terms proportional to $\bar{\nabla} W$ and to all higher derivatives of $W(x,y)$. The effect of these terms is to couple all modes of the guide to each other, thus describing the mode conversion and radiation losses caused by the nonuniformity of thickness. We postulate now that the thickness variations are so smooth that the variation of the effective index $N(x,y)$ over a distance of one wavelength is small,

$$|k\bar{\nabla} N| = |(\partial N/\partial W)k\bar{\nabla} W| \ll 1. \tag{9}$$

Under this condition the complicated wave equation reduces to Eq. (8), and the dependence $N = N_m(W)$ of the index on film thickness is the same as for a uniform guide.

Eikonal and Ray Equations

We are interested now in the geometrical optical properties of the field $V(x,y)$. We therefore separate any amplitude variations by introducing (following, e.g., Born and Wolf[17]) the eikonal $\mathbf{S}(x,y)$ as real, scalar function of x,y. Its implicit definition is

$$V(x,y) = A(x,y) \exp[ik\mathbf{S}(x,y)], \quad (10)$$

with a real amplitude function $A(x,y)$. On inserting Eq. (10) into the wave Eq. (8), and retaining only the leading terms in the geometrical optical limit $k \to \infty$, we obtain the eikonal equation

$$[\nabla\mathbf{S}(x,y)]^2 = N^2(x,y). \quad (11)$$

This equation is independent of the amplitude $A(x,y)$. For a guide of given thickness profile $W(x,y)$ and a given mode the index profile $N(x,y)$ can be calculated for all points $W > W_C$, so that Eq. (11) can be solved, in principle. From the definition Eq. (10) it follows that the curves $\mathbf{S}(x,y) =$ constant define all possible sets of wavefronts. Their orthogonal trajectories are the light rays in this two-dimensional transmission medium. We denote by $\mathbf{r} = \{x,y\}$ the position vector of a general point on the ray and by s the distance measured along the ray. Then $\mathbf{s} = d\mathbf{r}/ds$ is the unit vector in the direction of the ray, and the differential equation of the rays can be written as

$$(d/ds)[N(\mathbf{r})\mathbf{s}(\mathbf{r})] = \nabla N(\mathbf{r}). \quad (12)$$

This differential equation describes in full generality the optical rays in an inhomogeneous, two-dimensional propagation medium.

The eikonal Eq. (11) and the ray Eq. (12) are formally identical with the corresponding equations in bulk optics,[17] except that here the effective index N appears instead of the bulk index. Therefore, the propagation of a guided mode along a nonuniform film obeys the same laws of geometrical optics as ordinary light propagation in a bulk optical medium whose index is given by $N(x,y)$. In particular, Fermat's principal and Snell's law must hold.

Examples

It is illustrative to discuss Eq. (12) for some simple cases. We note first that in a uniform guide, $N =$ constant, Eq. (12) yields the result that all rays are straight lines. Next we calculate the curvature κ of a general ray. It can be written as[17]

$$\kappa \equiv \rho^{-1} = (1/N)(\partial N/\partial W)|\mathbf{s} \times \nabla W|, \quad (13)$$

where ρ is the radius of curvature. We see from this equation that a ray in the guide does not bend in a tapered region if its direction coincides with the direction of the steepest slope ∇W of the film profile. A ray that propagates perpendicular to that direction, i.e., along a direction of constant thickness, is curved strongest. A simple consideration, using the positive definiteness [Eq. (3)] of $\partial N/\partial W$, shows that the ray is bent always toward the thicker region (higher N) of the guide.

We now obtain the law of refraction simply by applying the ray Eq. (12) to the situation in Fig. 1. The film thickness W and the index N are independent of x. Therefore $N \cdot s_x = N \sin\alpha$ must be constant along the ray, which implies the validity of Snell's law.

This paper is based on material presented at the Devices Research Conference, Rochester, New York, June 1969.

References

1. E. S. Miller, Bell Syst. Tech. J. **48,** 2059 (1969).
2. R. Shubert and J. H. Harris, IEEE Trans. Microwave Theory Techn. **MTT-16,** 1048(1968).
3. P. K. Tien, R. Ulrich, and R. J. Martin, Appl. Phys. Lett. **14,** 291 (1969); P. K. Tien and R. Ulrich, J. Opt. Soc. Am. **60,** 1325 (1970).
4. R. Ulrich, J. Opt. Soc. Am. **60,** 1337 (1970).
5. L. Kuhn, M. L. Dakks, P. F. Heidrich, and B. A. Scott, Appl. Phys. Lett. **17,** 265 (1970).
6. M. L. Dakks, L. Kuhn, B. A. Scott, and P. F. Heidrich, Appl. Phys. Lett. **16,** 523 (1970); H. Kogelnik and T. Sosnowski, Bell Syst. Tech. J. **49,** 1602 (1970); J. H. Harris, R. Shubert, and N. Polky, J. Opt. Soc. Am. **60,** 1007 (1970).
7. D. Marcuse, Bell Syst. Tech. J. **49,** 273, 919 (1970).
8. In Ref. 2, Eqs. (20)–(22), the effective index is defined erroneously as the reciprocal of Eq. (2) here.
9. J. Kane and H. Osterberg, J, Opt. Soc. Am. **54,** 347 (1964); E. A. J. Marcatili, Bell Syst. Tech. J. **48,** 2071 (1969).
10. J. E. Goell and R. D. Standley, Bell Syst. Tech. J. **48,** 3445 (1969).
11. D. H. Hensler, J. D. Cuthbert, P. K. Tien, and R. J. Martin, Appl. Opt. **10,** 1037 (1971).
12. R. Ulrich and H. P. Weber, BTL; to be published.
13. P. K. Tien and R. J. Martin, Appl. Phys. Lett. **18,** 398 (1971).
14. D. Marcuse, Bell Syst. Tech. J. **48,** 3187 (1969).
15. E. A. J. Marcatili, Bell Syst. Tech. J. **48,** 2103 (1969).
16. R. Shubert and J. H. Harris, J. Opt. Soc. Am. **61,** 154 (1971).
17. M. Born and E. Wolf, *Principles of Optics* (New York Pergamon, 1965).

Part 3
Waveguide Papers

A Circular-Harmonic Computer Analysis of Rectangular Dielectric Waveguides

By J. E. GOELL

(Manuscript received April 8, 1969)

This paper describes a computer analysis of the propagating modes of a rectangular dielectric waveguide. The analysis is based on an expansion of the electromagnetic field in terms of a series of circular harmonics, that is, Bessel and modified Bessel functions multiplied by trigonometric functions. The electric and magnetic fields inside the waveguide core are matched to those outside the core at appropriate points on the boundary to yield equations which are then solved on a computer for the propagation constants and field configurations of the various modes.

The paper presents the results of the computations in the form of curves of the propagation constants and as computer generated mode patterns. The propagation curves are presented in a form which makes them refractive-index independent as long as the difference of the index of the core and the surrounding medium is small, the case which applies to integrated optics. In addition to those for small index difference, it also gives results for larger index differences such as might be encountered for microwave applications.

I. INTRODUCTION

It is anticipated that dielectric waveguides will be used as the fundamental building blocks of integrated optical circuits. These waveguides can serve not only as a transmission medium to confine and direct optical signals, but also as the basis for circuits such as filters and directional couplers.[1] Thus, it is important to have a thorough knowledge of the properties of their modes.

Circular dielectric waveguides have received considerable attention because circular geometry is commonly used in fiber optics.[2–5] In many integrated optics applications it is expected that waveguides will consist of a rectangular, or near rectangular, dielectric core embedded in a dielectric medium of slightly lower refractive index. The modes

Reprinted with permission from *Bell Syst. Tech. J.*, vol. 48, pp. 2133–2160, Sept. 1969.

for this geometry are more difficult to analyze than those of the metallic rectangular waveguide because of the nature of the boundary.

Marcatili, using approximations based on the assumption that most of the power flow is confined to the waveguide core, has derived in closed form the properties of a rectangular dielectric waveguide.[6] In his solution, fields with sinusoidal variation in the core are matched to exponentially decaying fields in the external medium. In each region only a single mode is used. The results of this method are obtained in a relatively simple form for numerical evaluation.

The properties of the principal mode of the rectangular dielectric waveguide have been studied by Schlosser and Unger using a high-speed digital computer.[7] In their approach the transverse plane was divided into regions, as shown in Fig. 1, and rectangular coordinate solutions assumed in each of the regions. The longitudinal propagation constant was then adjusted so that a field match could be achieved at discrete points along the boundary. This method gives results which, theoretically, are valid over a wider range than Marcatili's, but with a significant increase in computational difficulty. One shortcoming of the method is that for a given mode, as the wavelength increases the field extent increases, so, in the limit it becomes increasingly difficult to match the fields along the boundaries between regions [1] and [2] and between regions [2] and [3].

A variational approach has been undertaken by Shaw and others.[8] They assume a test solution with two or three variable parameters in the core. From this test solution, the fields outside the core are then derived and the parameters are varied to achieve a consistent

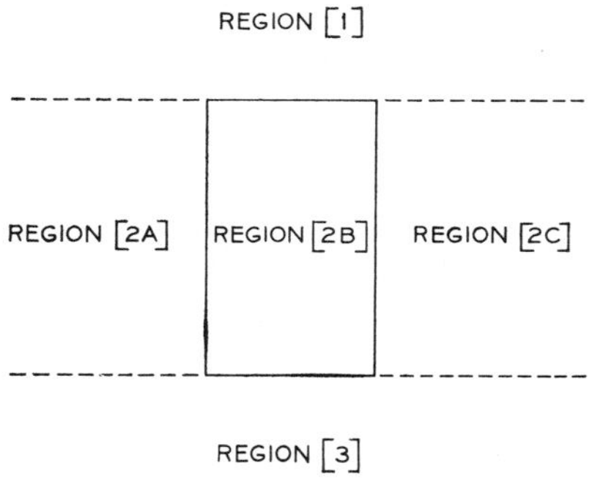

Fig. 1 — Matching boundaries for rectangular mode analysis.

solution. This approach, like that of Schlosser, requires involved computations. Also, it has the disadvantage that the test function must be assumed in advance. In addition, some of his preliminary results do not show the proper behavior for the limiting cases (waveguide dimensions which are very large or very small compared with the wavelength).

In the present analysis the radial variation of the longitudinal electric and magnetic fields of the modes are represented by a sum of Bessel functions inside the waveguide core and by a sum of modified Bessel functions outside the waveguide core. Solutions are found by matching the fields along the perimeter of the core. Thus, the matching boundary is not a function of the waveguide parameters, so the computational complexity does not increase with wavelength.

Section II discusses the underlying theory of the circular-harmonic analysis of rectangular dielectric waveguides. This is followed by a description of computational techniques and special graphical methods of presentation used. Section III is divided into three parts, the first describing the accuracy of the computations, the second describing field patterns, and the third presenting propagation curves.

II. DERIVATION OF EQUATIONS

The waveguide considered here consists of a rectangular core of dielectric constant, ϵ_1, surrounded by an infinite medium of dielectric constant, ϵ_0. Both media are assumed to be isotropic, and have the permeability of free space, μ_0. Figure 2 shows the coordinate systems (rectangular and cylindrical) and rod dimension used in this paper. The direction of propagation is in the $+z$ direction (towards the observer).

In cylindrical coordinates the field solutions of Maxwell's equations take the form of Bessel functions and modified Bessel functions multiplied by trigometric functions, and their derivatives. In order for propagation to take place in the z direction, the field solutions must be Bessel functions in the core and modified Bessel functions outside. Since Bessel functions of the second kind have a pole at the origin and modified Bessel functions of the first kind a pole at infinity, the radial variation of the fields is assumed to be a sum of Bessel functions of the first kind and their derivatives inside the core and a sum of modified Bessel functions and their derivatives outside the core.

In cylindrical coordinates, the z components of the electric and magnetic fields are given by

$$E_{z1} = \sum_{n=0}^{\infty} a_n J_n(hr) \sin (n\theta + \varphi_n) \exp [i(k_z z - \omega t)] \tag{1a}$$

and

$$H_{z1} = \sum_{n=0}^{\infty} b_n J_n(hr) \sin (n\theta + \psi_n) \exp [i(k_z z - \omega t)] \tag{1b}$$

inside the core, and by

$$E_{z0} = \sum_{n=0}^{\infty} c_n K_n(pr) \sin (n\theta + \varphi_n) \exp [i(k_z z - \omega t)] \tag{1c}$$

and

$$H_{z0} = \sum_{n=0}^{\infty} d_n K_n(pr) \sin (n\theta + \psi_n) \exp [i(k_z z - \omega t)] \tag{1d}$$

outside the core, where ω is the radian frequency and k_z the longitudinal propagation constant. The transverse propagation constants are given by

$$h = (k_1^2 - k_z^2)^{\frac{1}{2}} \tag{2a}$$

and

$$p = (k_z^2 - k_0^2)^{\frac{1}{2}} \tag{2b}$$

where $k_1 = \omega(\mu_0 \epsilon_1)^{\frac{1}{2}}$ and $k_0 = \omega(\mu_0 \epsilon_0)^{\frac{1}{2}}$. The terms J_n and K_n are the nth order Bessel functions and modified Bessel functions, respectively, and ψ_n and φ_n are arbitrary phase angles.

The transverse components of the fields are given by[9]

$$E_r = \frac{ik_z}{k^2 - k_z^2} \left[\frac{\partial E_z}{\partial r} + \left(\frac{\mu_0 \omega}{k_z r} \right) \frac{\partial H_z}{\partial \theta} \right] \tag{3a}$$

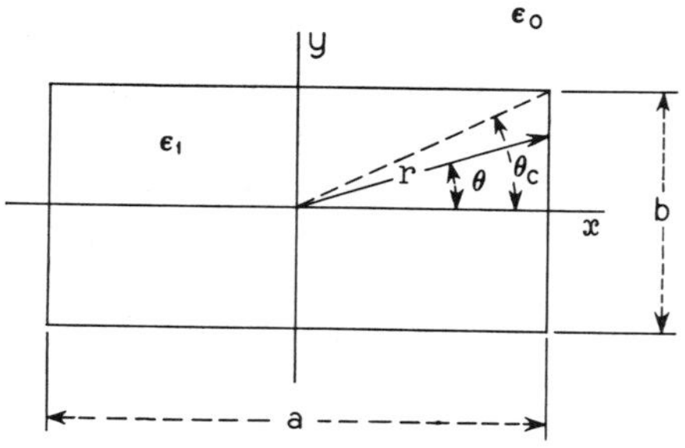

Fig. 2 — Dimensions and coordinate system.

$$E_\theta = \frac{ik_z}{k^2 - k_z^2}\left[\frac{1}{r}\frac{\partial E_z}{\partial \theta} - \left(\frac{\mu_0\omega}{k_z}\right)\frac{\partial H_z}{\partial r}\right] \tag{3b}$$

$$H_r = \frac{ik_z}{k^2 - k_z^2}\left[-\left(\frac{k^2}{\mu_0\omega k_z r}\right)\frac{\partial E_z}{\partial \theta} + \frac{\partial H_z}{\partial r}\right] \tag{3c}$$

$$H_\theta = \frac{ik_z}{k^2 - k_z^2}\left[\left(\frac{k^2}{\mu_0\omega k_z}\right)\frac{\partial E_z}{\partial r} + \frac{1}{r}\frac{\partial H_z}{\partial \theta}\right], \tag{3d}$$

where k can be either k_1 or k_0.

Finally, the component of the electric field tangent to the rectangular core is given by

$$E_t = \pm(E_r \sin\theta + E_\theta \cos\theta) \quad \begin{array}{c} -\theta_c < \theta < \theta_c \\ \pi - \theta_c < \theta < \pi + \theta_c \end{array} \tag{4a}$$

or

$$E_t = \pm(-E_r \cos\theta + E_\theta \sin\theta), \quad \begin{array}{c} \theta_c < \theta < \pi - \theta_c \\ \pi + \theta_c < \theta < -\theta_c \end{array}, \tag{4b}$$

where θ_c is the angle which a radial line to the corner in the first quadrant makes with the x axis. Similar expressions exist for the tangential magnetic field.

2.1 *Effects of Symmetry*

Since the waveguide is symmetrical about the x axis the fields must be either symmetric or antisymmetric about this axis. This is true because the structure is invarient under 180° rotations and therefore the field patterns must be invarient under a 180° rotation, except for sign. From this and the fact that $\partial/\partial\theta$ appears in each of equations (3), it is evident that two types of modes must exist, the first type with $\varphi_n = 0$ and $\psi_n = \pi/2$ and the second type with $\varphi_n = \pi/2$ and $\psi_n = \pi$.

Similarly, the field functions must also be symmetric or antisymmetric about the y axis. Suppose, for example, E_{z0} exhibits a sinusoidal angular dependence about $\theta =$ (E_{z0} is odd about the x axis). Then, letting $\alpha = \theta - \pi/2$, equation (1c) can be put in the form

$$E_{z0} = \sum_{n=0}^{\infty} c_n K_n(pr)(\sin n\alpha \cos n\pi/2 + \cos n\alpha \sin n\pi/2). \tag{5}$$

For E_{z0} to be purely symmetric about $\alpha = 0$ (the y axis), all n must be odd; for E_{z0} to be antisymmetric about $\alpha = 0$ all n must be even.

Since similar results apply for cosinusoidal variation of E_{z0} about $\theta = 0$, and all other field functions as well, any given mode must consist of either even harmonics or odd harmonics.

From the preceeding analysis it is evident that if the matching points are selected symmetrically about both the x and y axes, then, except possibly for sign, every point will have an equivalent point in each quadrant. Therefore, the field matching need only be performed in one quadrant. Thus, the use of the symmetry of the structure not only reduces the number of constants required to calculate the properties of a given mode by a factor of four, it also decreases the number of points to achieve a given degree of accuracy by the same factor.

2.2 *Selection of Matching Points*

As mentioned in Section 2.1, the matching point locations should be symmetrical about the x and y axes. For the odd harmonic cases, the points used to compute the results to be presented in Section III were $\theta_m = (m - 1/2)\pi/2N;\ m = 1, \cdots, N$, where N was the number of space harmonics.

The choice of points for the even harmonic cases was more complicated since simultaneous existence of an $n = 0$ harmonic for both the TE and TM circular modes is inconsistent with the waveguide symmetries. Thus, if the maximum n for both the TE and TM solutions are equal, the total number of coefficients to be found will be $4N - 2$ rather than $4N$ as in the previous case.

The method of choosing points for the even harmonic modes used for the computation of the results of Section III was to pick the points for the field components with even symmetry about $\theta = 0$ to be $\theta_m = (m - 1/2)\pi/2N;\ m = 1, 2, \cdots, N$, and for the field components with odd symmetry about $\theta = 0$ to be $\theta_m = (m - N - 1/2)\pi/2(N - 1)$; $m = N + 1, N + 2, \cdots, (2N - 1)$ for cases with unity aspect ratio, $(a/b = 1)$. For aspect ratios other than unity, all points were chosen according to the first formula, except that the first and last points for the odd z component were omitted.

2.3 *Formulation of Matrix Elements*

The coefficients of equation (1) were found by matching the tangential electric and magnetic fields along the boundary of the waveguide core. Since each type of field consists of both longitudinal and transverse components, four types of matching equations exist.

To facilitate computer analysis the matching equations were put in matrix form. The matching equations in matrix form for the longitudinal field components are

$$E^{LA}A = E^{LC}C \tag{6a}$$

for the electric field and

$$H^{LB}B = H^{LD}D \tag{6b}$$

for the magnetic field. For the transverse fields the matrix matching equations are given by

$$E^{TA}A + E^{TB}B = E^{TC}C + E^{TD}D \tag{6c}$$

for the electric field and

$$H^{TA}A + H^{TB}B = H^{TC}C + H^{TD}D \tag{6d}$$

for the magnetic field. The A, B, C, and D matrices are N element column matrices of the a_n, b_n, c_n, and d_n mode coefficients, respectively. The elements of the $m \times n$ matrices E^{LA}, E^{LC}, H^{LB}, H^{LD}, E^{TA}, E^{TB}, E^{TC}, E^{TD}, H^{TA}, H^{TB}, H^{TC}, and H^{TD} are given by

$$e_{mn}^{LA} = JS, \tag{7a}$$

$$e_{mn}^{LC} = KS, \tag{7b}$$

$$h_{mn}^{LB} = JC, \tag{7c}$$

$$h_{mn}^{LD} = KC, \tag{7d}$$

$$e_{mn}^{TA} = -k_z(\mathbf{J}'SR + \mathbf{J}CT), \tag{7e}$$

$$e_{mn}^{TB} = k_0 Z_0(\mathbf{J}SR + \mathbf{J}'CT), \tag{7f}$$

$$e_{mn}^{TC} = k_z(\mathbf{K}'SR + \mathbf{K}CT), \tag{7g}$$

$$e_{mn}^{TD} = -k_0 Z_0(\mathbf{K}SR + \mathbf{K}'CT), \tag{7h}$$

$$h_{mn}^{TA} = \epsilon_r k_0(\mathbf{J}CR - J'ST)/Z_0\ , \tag{7i}$$

$$h_{mn}^{TB} = -k_z(\mathbf{J}'CR - \mathbf{J}ST), \tag{7j}$$

$$h_{mn}^{TC} = -k_0(\mathbf{K}CR - \mathbf{K}'ST)/Z_0\ , \tag{7k}$$

$$h_{mn}^{TD} = k_z(\mathbf{K}'CR - \mathbf{K}ST), \tag{7l}$$

where

$$Z_0 = (\mu_0/\epsilon_0)^{\frac{1}{2}},$$

$$\epsilon_r = \epsilon_1/\epsilon_0\ ,$$

$$\left.\begin{aligned} S &= \sin (n\theta_m + \varphi) \\ C &= \cos (n\theta_m + \varphi) \end{aligned}\right\} \quad \text{or} \quad \begin{aligned} \varphi &= 0 \\ \varphi &= \pi/2 \end{aligned} ,$$

$$J = J_n(hr_m), \qquad K = K_n(pr_m),$$

$$J' = J_n'(hr_m), \qquad K' = K_n'(pr_m),$$

$$\mathbf{J} = \frac{nJ_n(hr_m)}{h^2 r_m}, \qquad \mathbf{K} = \frac{nK_n(pr_m)}{p^2 r_m},$$

$$\mathbf{J}' = \frac{J_n'(hr_m)}{h}, \qquad \mathbf{K}' = \frac{K_n'(pr_m)}{p},$$

and

$$\left.\begin{aligned} R &= \sin \theta_m \\ T &= \cos \theta_m \\ r_m &= (a/2) \cos \theta_m \end{aligned}\right\} \theta < \theta_c , \qquad \left.\begin{aligned} R &= -\cos \theta_m \\ T &= \sin \theta_m \\ r_m &= (b/2) \sin \theta_m \end{aligned}\right\} \theta > \theta_c .$$

For $\theta = \theta_c$, the boundary at the corner was assumed to be perpendicular to the radial line connecting it to the origin, so for this case $R = \cos (\theta_m + \pi/4)$, $T = \cos (\theta_m - \pi/4)$, and $r_m = (a^2 + b^2)^{\frac{1}{2}}/4$.

2.4 *Mode Designation*

Unlike metallic waveguides, the field patterns of dielectric waveguides are sensitive to refractive index difference, wavelength, and aspect ratio. This complicates the problem of finding a reasonably descriptive mode designation scheme.

For rectangular metallic waveguides, the accepted approach is to designate the modes as TE (or H) and TM (or E), and to specify the number of field maxima in the x and y directions with a double subscript. When there is no variation the subscript 0 is used.

Since the rectangular dielectric waveguide modes are neither pure TE nor pure TM, a different scheme must be used. The scheme adopted is based on the fact that in the limit, for large aspect ratio, short wavelength, and small refractive index difference, the transverse electric field is primarily parallel to one of the transverse axes. Modes are designated as E^y_{mn} if in the limit their electric field is parallel to the y axis and as E^x_{mn} if in the limit their electric field is parallel to the x axis. The m and n subscript are used to designate the number of maxima in the x and y directions, respectively.†

† This scheme agrees with that used by Marcatili in Ref. 6.

2.5 *Electric and Magnetic Field Function Differences*

For a hollow metallic waveguide where pure TE and TM modes can exist, it is evident from equation (3) that E_r and H_θ have similar transverse variations as do E_θ and H_r, so that the impedance is independent of position. Furthermore, the transverse electric and magnetic fields are perpendicular and the power flow, Re $\{E \times H^*\}$, does not change sign anywhere across the waveguide.

By examination of equation (3), it is clear that for the mixed modes of the dielectric waveguide, the field functions are not similar and the impedance is a function of position. In order for the transverse fields E_t and H_t to be perpendicular,

$$E_t \cdot H_t = E_r H_r + E_\theta H_\theta = 0. \tag{8}$$

Now, from equation (3)

$$E_t \cdot H_t = \frac{k_z^2 - k^2}{k_z^2} \left(\frac{\partial H_z}{\partial r} \frac{\partial E_z}{\partial r} + \frac{1}{r^2} \frac{\partial H_z}{\partial \theta} \frac{\partial E_z}{\partial \theta} \right). \tag{9}$$

Thus, E_t and H_t are not necessarily perpendicular. Finally, since the transverse variations of E_t and H_t are not the same, the electric field and magnetic field can change sign at different points, which results in negative power flow.†

Three special cases exist where the electric and magnetic fields, and the impedance, have the same positional dependence, and where the power flow does not change sign across the waveguide:

(*i*) in one of the regions if the propagation constant is approximately equal to the bulk propagation constant of that region, that is, if $k \approx k_1$ or $k \approx k_0$,

(*ii*) everywhere in the limit for small refractive index difference, since case *i* will then hold in both regions, and

(*iii*) everywhere for circular symmetry of both the structure and the modes.

2.6 *Normalization*

The arguments of the Bessel and modified Bessel functions are given by $hr = (k_1^2 - k_z^2)^{\frac{1}{2}} r$ and $pr = (k_z^2 - k_0^2)^{\frac{1}{2}} r$, respectively. The first argument can be put in the form

$$hr = [k_1^2 - k_0^2 - p^2]^{\frac{1}{2}} r. \tag{10}$$

† This unusual property has also been observed for helices.[10] Presumably, if loss were included there would be a radial component of power to feed the reverse flow, and the lossless case can be thought of as the limit of the lossy case.

Letting

$$\mathcal{P}^2 = \frac{(k_z/k_0)^2 - 1}{n_r^2 - 1}, \tag{11}$$

and

$$\mathcal{R} = rk_0(n_r^2 - 1)^{\frac{1}{2}}, \tag{12}$$

where

$$n_r = (k_1/k_0)^{\frac{1}{2}} \tag{13}$$

is the index of refraction of the core relative to the outer medium, gives

$$pr = \mathcal{P}\mathcal{R} \tag{14}$$

and

$$hr = \mathcal{R}(1 - \mathcal{P}^2)^{\frac{1}{2}}. \tag{15}$$

The curves of the propagation constant given in Section III are drawn in terms of $\mathcal{P}^2$ and $\mathcal{B}$, where

$$\mathcal{B} = \frac{2b}{\lambda_0}(n_r^2 - 1)^{\frac{1}{2}} \tag{16}$$

and $\lambda_0 = 2\pi/k_0$. Since $\mathcal{R}$ is proportional to $1/(n_r^2 - 1)^{\frac{1}{2}}$ and $\mathcal{P}$ and $\mathcal{B}$ are proportional to $(n_r^2 - 1)^{\frac{1}{2}}$, the use of $\mathcal{P}^2$ and $\mathcal{B}$ as plotting variables eliminates the explicit dependence of the Bessel and modified Bessel function arguments on the refractive indices of the media.

Examination of the matching equations, equations (6), reveals that ϵ_r appears in the H^{TA} term. However, since ϵ_r appears as a multiplicative factor in H^{TA}, for sufficiently small values the normalized propagation constant, $\mathcal{P}^2$, is independent of ϵ_r.

The normalized propagation constant, $\mathcal{P}^2$, has two additional properties which make its use convenient. First, its range of variation is on the interval (0, 1). Second, for $n_r \approx 1$,

$$\mathcal{P}^2 \approx \frac{k_z/k_0 - 1}{\Delta n_r}, \tag{17}$$

where $\Delta n_r = n_r - 1$; so for small n_r, $\mathcal{P}^2$ is proportional to $k_z - k_0$. The latter property is the reason that $\mathcal{P}^2$ rather than $\mathcal{P}$ was used as a plotting variable.

2.7 *Method of Computation*

2.7.1 *Propagation Constant*

Equation (6) yields $4N$ simultaneous homogeneous linear equations for the a_n, b_n, c_n, and d_n for the odd modes and $4N$-2 equations for the even modes, using the matching points previously described. The equations can be combined to form a single matrix equation

$$[Q][T] = 0, \tag{18}$$

where

$$Q = \begin{bmatrix} E^{LA} & 0 & -E^{LC} & 0 \\ 0 & H^{LB} & 0 & -H^{LD} \\ E^{TA} & E^{TB} & -E^{TC} & -E^{TD} \\ H^{TA} & H^{TB} & -H^{TC} & -H^{TD} \end{bmatrix}$$

and the column matrix

$$[T] = \begin{bmatrix} A \\ B \\ C \\ D \end{bmatrix}.$$

All of the quantities in the matrices $[Q]$ and $[T]$ are themselves matrices as defined by equations (1), (6), and (7).

In order for a nontrivial solution to equation (18) to exist

$$\text{Det } [Q] = 0. \tag{19}$$

The normalized propagation constant, $\mathcal{P}^2$, was found by substituting test values into equation (19). First, values of $\mathcal{P}^2$ evenly distributed in the interval (0, 1) were substituted to crudely locate the roots. Then, Newton's method was used to find the roots to the desired accuracy.[11] Generally, one Newton approximation was used to find $\mathcal{P}^2$ for the propagation curves and about ten Newton's approximations when $\mathcal{P}^2$ was to be used to calculate field plots.

Both the simple method of triangulation[12] and the more complicated Gauss pivotal condensation method[13] were used to evaluate the determinant, the former for almost all cases and the latter for a few cases when roundoff error was apparent because the value of the determinant was not a smooth function of $\mathcal{P}^2$. In all cases double precision arithmetic was used. For five space harmonics, about 0.1 second of IBM 360/65 computing time was required for each value of $\mathcal{P}^2$ to evaluate the determinant using the triangulation method.

Due to the wide dynamic range of the coefficients, steps had to be taken to prevent underflow and overflow of the computer and to reduce the effects of roundoff. Multiplying a row or column of the matrix by a finite constant is equivalent to multiplying the determinant by that constant. Thus, any row or column of the determinant can be multiplied by a positive function without shifting its zeroes.

A detailed theory giving the "best functions" can be derived. However, since a "brute force" method was used, the more sophisticated method, which was not used because it would have required a substantial increase in the complexity of the program logic, is not discussed. It was found that multiplying the Bessel function terms by $h^2d/|J_n(hb)|$ and the modified Bessel function terms by $p^2d/k_n(pb)$, where d is the average of the waveguide dimensions, kept the variation of the terms "under control." A further simplification was made by setting Z_0 to unity, which does not shift the zeroes of the determinant because if the H_t rows are multiplied by Z_0, then if Z_0 appears in a column, it will appear in a similar manner in every element of the column.

2.7.2 *Mode Configurations*

The electric and magnetic fields were calculated for representative cases from equation (3). To find the a_n, b_n, c_n, and d_n coefficients, k_z was first found from equation (19). Its value was then substituted into equation (18). By setting one of the elements of the T column matrix to unity, all of the other elements were then found by standard matrix techniques.[13]

Several approaches were used to obtain information that could be used to derive the field patterns. These included computation of the field components along radial cuts of the waveguide cross section, computer generated isoclines giving the direction of the electric field, and computer generated mode pictures.

The isoclines and pictures were drawn using a simulated Stromberg Carlson SC-4020 cathode ray tube plotter, which is capable of generating points and lines on a 1024 × 1024 grid.† A single quadrant was used for the isoclines and intensity picture since the results for all quadrants are identical except for orientation. In general, the dimensions were scaled so that the long dimension of the rectangular waveguide core extended over 80 percent of the displayed width. All figures were plotted at the points $(20m, 20n)$, where m and n take on all integer values from 0 to 49.

Isocline drawings were made by drawing a line at each of the coordinate points parallel to the electric field at that point (all lines

† An SC-4060 plotter was used to simulate the SC-4020 plotter to take advantage of previously existing programs.

had the same length). The isocline drawings were used as working tools to derive the field line drawings in Section III.

In order to draw pictures of mode patterns, the power density was calculated at each of the points to be plotted. The square root of the power density was then normalized to the square root of the peak power density and quantized into 21 levels. About each point in the picture, a portion of the figure shown in Fig. 3 was then plotted, starting at 1 and going to the point corresponding to the appropriate quantized level (except at the points where the quantized power was zero where no plotting was done). Since the size of the cathode ray tube spot is approximately equal to the line spacing in the figure, the plotted figures are filled in. Therefore, the light passed by these figures is approximately equal to the power density to be represented. For small index difference, since the power density is proportional to the square of the transverse electric field, the dynamic range of the pictures (in terms of the electric field) is 400.

Starting with the single quadrant pictures, complete pictures were generated by making quadruple exposures of the microfilm. In general, about 30 to 60 seconds of IBM 360/65 computing time were required for each picture.

III. RESULTS OF COMPUTATION

This section gives the computed results. Section 3.1 discusses accuracy. This is followed by a discussion of field plots and mode

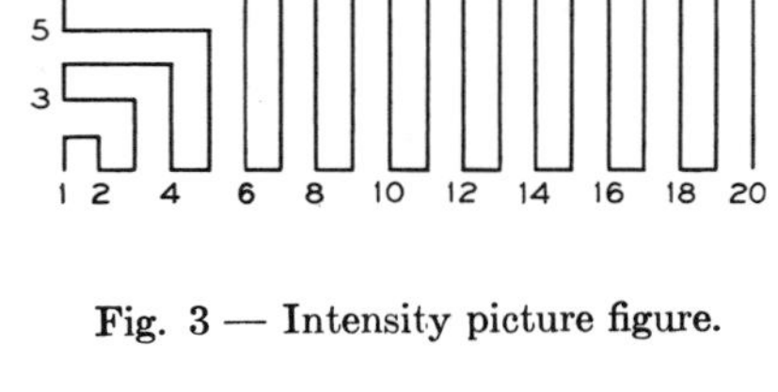

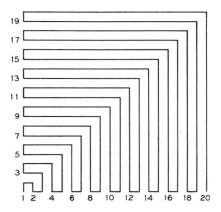

Fig. 3 — Intensity picture figure.

Table I—Sample Accuracy Results

Number of Harmonics Used	$\mathcal{P}^2$			
	$a/b = 1$	$a/b = 2$	$a/b = 3$	$a/b = 4$
3	0.714	0.811	0.820	0.828
4	0.713	0.811	0.820	0.819
5	0.715	0.808	0.819	0.813
6	0.714	0.808	0.822	0.820
7	0.715	0.808	0.820	0.813
8	0.715	0.807	0.820	0.814
9	0.715	0.807	0.823	0.815
Variation	0.2%	0.4%	0.4%	1.5%

pictures in Section 3.2. Finally, curves of the propagation constant for a variety of conditions are presented in Section 3.3.

3.1 *Accuracy*

Numerous test runs were made in order to obtain an estimate of the accuracy of the computed results. The results of several of these runs are given in Table I for the first mode with $\mathcal{B} = 2$. The numbers at the bottom of the table represent the total variation for a given aspect ratio taken as a percentage of the full range possible (one).

For small aspect ratios, it is clear that the convergence is very rapid. However, for larger aspect ratios the convergence is not as good. For example, the variation for an aspect ratio of four is 1.5 percent (taken as a percentage of the full range of variation). For this case, from the table and from the limit for infinite aspect ratio[14] which is an upper bound for $\mathcal{P}^2$, it appears the error is about 3 percent. This error is achieved with a relatively small number of harmonics and can only be improved by using a prohibitively large number of harmonics on a computer which carries more significant digits than the one which was available for this study. However, since solutions exist for an infinite aspect ratio, the decrease in accuracy for the large aspect ratio of the circular-harmonic method is not a serious problem.

Computations similar to those for Table I were performed to obtain an estimate of the upper bound of the accuracy of the cases presented in Section 3.3. From these calculations, it is believed that all of the data to be presented in the following sections is accurate to 1 percent, except for the results of calculations using even harmonics for aspect ratios other than unity which are believed to be accurate to better than 2 percent. In general, accuracy decreases as the mode order increases, although not monotonically.

The results of the circular-harmonic analysis and of Marcatili's analysis agree.[6] In the regions where his method and the circular-harmonic method are both theoretically valid, the agreement is well within the tolerances given above. To avoid duplication, the reader is directed to his curves for a comparison.

The effect of the number of harmonics used in the field patterns is of some interest. This question has not been explored in great detail; however, a few comparisons of intensity pictures for different numbers of circular harmonics were made. In general, it was found that five harmonics were sufficient to give a good representation of the modes that this paper presents. An example of this is given in Fig. 4, comparing the E^y_{11} mode intensity patterns for five and nine harmonics. For the results which follow, five circular harmonics were used.

3.2 *Mode Configurations*

Figure 5 shows intensity pictures for the first six modes for unity aspect ratio, $\mathcal{B} = 3$, and an index difference of 0.01. Figure 6 gives similar data for an aspect ratio of two and $\mathcal{B} = 2$. For both, the plots are arranged in ascending order of cutoff frequency. All of the pictures are for E^y_{mn} modes. These pictures are virtually indistinguishable from the corresponding E^x_{mn} modes so both sets are not presented. In general, for small index differences the E^y_{mn} and E^x_{mn} can be considered to be near duals, that is, to have identical field patterns except that the electric and magnetic fields are interchanged.

The field distribution patterns for the modes of Figs. 5 and 6 are more complicated than those for the rectangular metallic waveguide

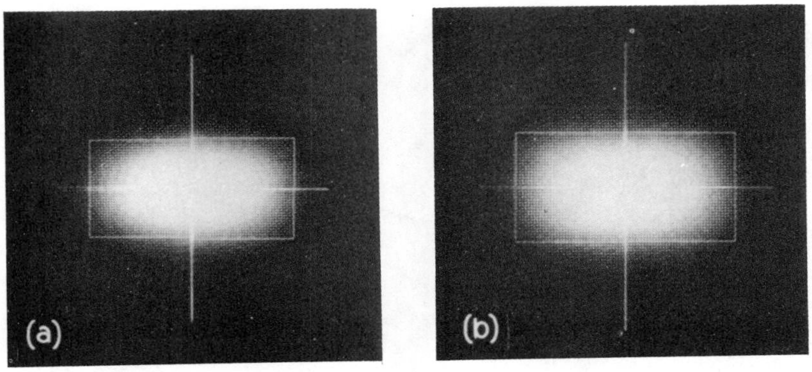

Fig. 4 — Intensity for the E^y_{11} mode for $a/b = 2$, $\mathcal{B} = 2$, and $\Delta n_r = .01$: (a) for five harmonics and (b) for nine harmonics.

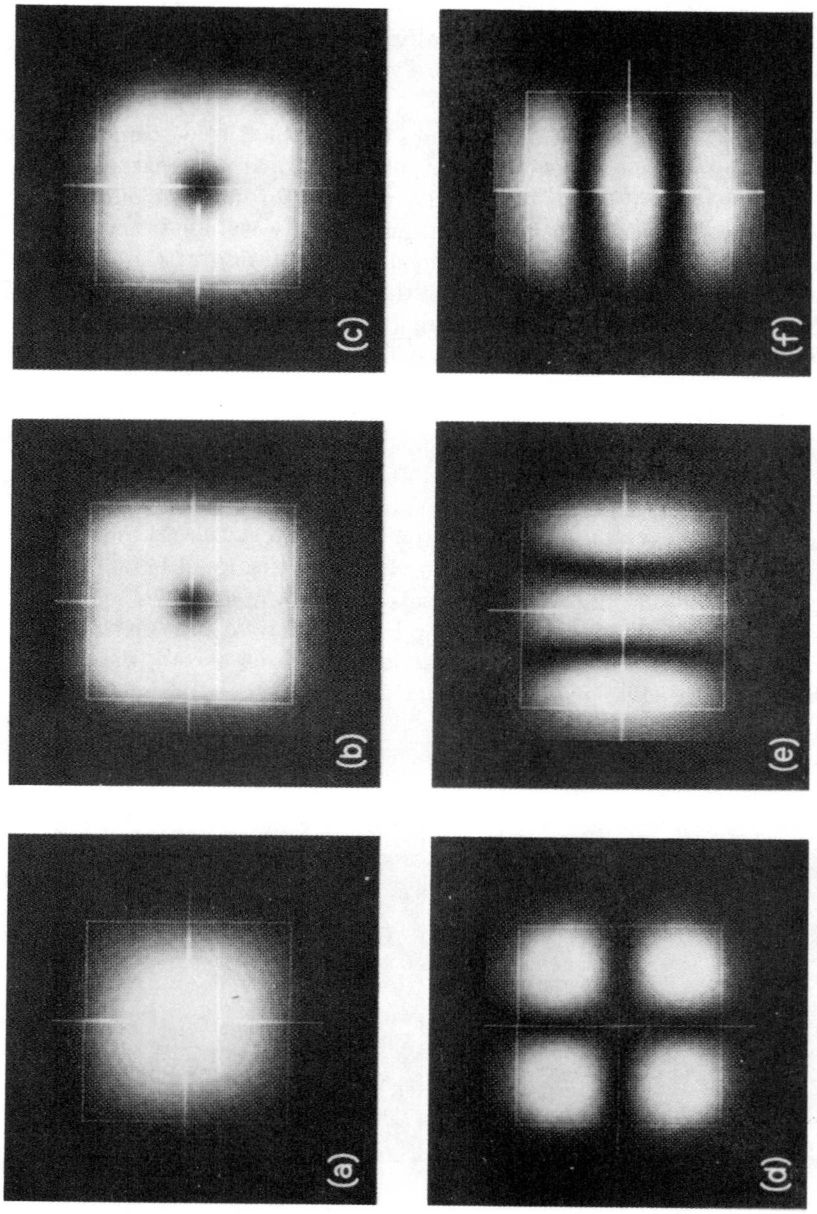

Fig. 5 — Intensity for some E^y_{mn} modes with unity aspect ratio, $\mathcal{B} = 3$, and $\Delta n_r = 0.01$: (a) E^y_{11}, (b) E^y_{21}, (c) E^y_{12}, (d) E^y_{22}, (e) E^y_{31}, and (f) E^y_{13}.

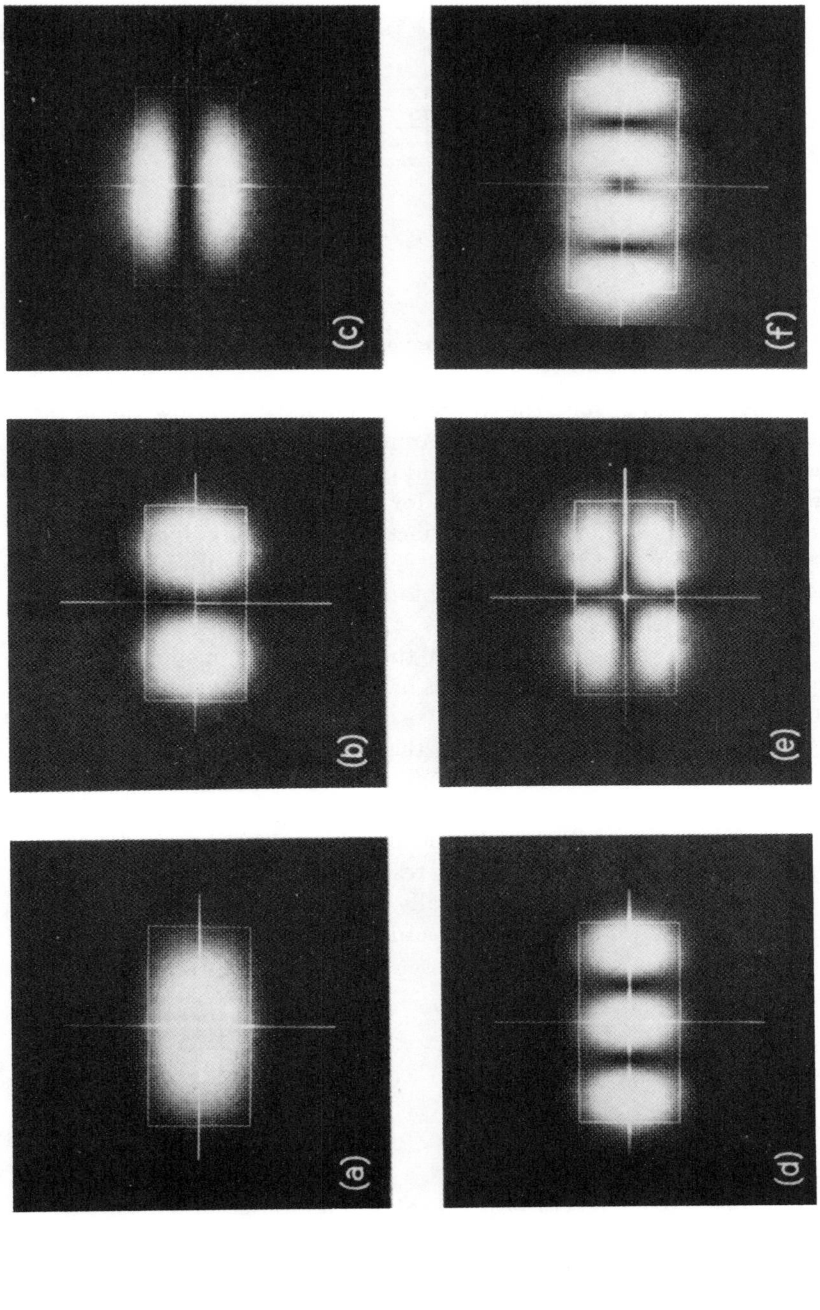

Fig. 6 — Intensity picture for some E^y_{mn} modes with $a/b = 2$, $\mathcal{B} = 2$, and $\Delta n_r = 0.01$: (a) E^y_{11}, (b) E^y_{21}, (c) E^y_{12}, (d) E^y_{31}, (e) E^y_{22}, and (f) E^y_{41}.

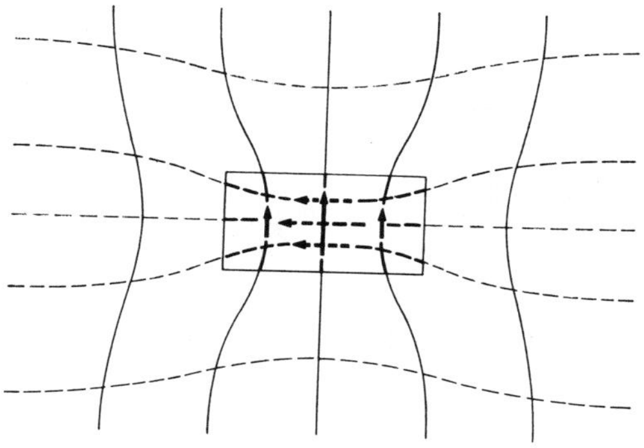

Fig. 7— Field configuration of the E^y_{11} mode.

since they extend beyond the waveguide boundary and, in general, their shape is dependent on waveguide parameters other than shape. The E^x_{11} and E^y_{11} modes have the simplest field patterns. Figure 7 shows the electric and magnetic field orientations for the E^y_{11} mode. In this figure and the following ones, there are heavy lines in the regions of high field intensity and light lines in regions of low field intensity. Only E^y_{mn} modes are shown since the E^x_{mn} modes can be obtained by interchanging the electric and magnetic field vectors.

Figure 8 shows the field lines for the E^y_{21} and E^y_{12} modes for a large aspect ratio. (For $a/b \rightarrow \infty$ the fields have the appearance of rectangular metallic waveguide modes.) However, as the aspect ratio approaches unity, the E^y_{12} and E^x_{21} modes and the E^y_{21} and E^x_{12} modes couple and shift to the patterns shown in Fig. 9. Most of the change takes place with the aspect ratio close to unity.

Figures 10, 11, and 12 show the field configurations for the E^y_{22} mode, the E^y_{31} mode, and the E^y_{13} mode, respectively. The field patterns of these modes do not change drastically with the aspect ratios.

Figure 13a shows an intensity picture of the E^y_{32} mode and Figure

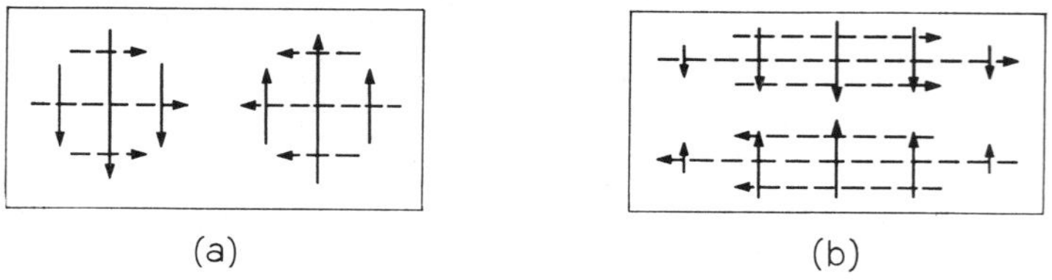

Fig. 8—Field configurations for the (a) E^y_{21} and (b) E^y_{12} modes far from cutoff.

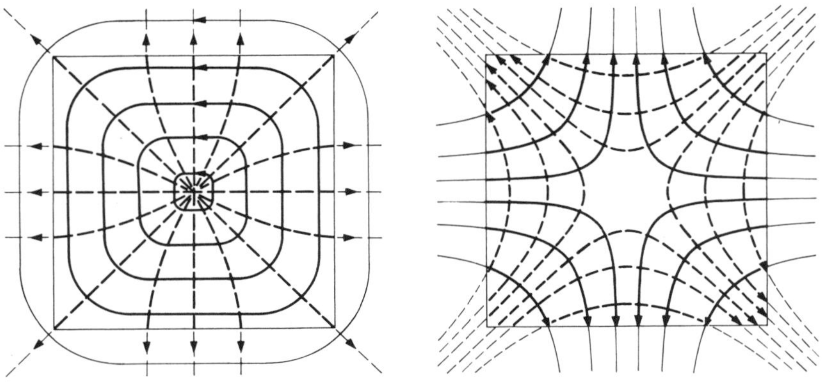

Fig. 9 — Field configurations for the square (a) E_{21}^{y} and (b) E_{12}^{y} modes.

13b its field pattern for unity aspect ratio. The field pattern inside the core is similar to a sum of the TE_{23} and TE_{32} of metallic waveguide, shown in Fig. 13c and d, respectively. Figure 13a demonstrates that the circular-harmonic analysis can generate complex field patterns with a relatively small number of harmonics.

Figures 14 and 15 show the variation of the intensity distribution with $\mathcal{P}^2$ for the E_{11}^{y} and E_{21}^{y} modes, respectively. As one would expect, for small values of $\mathcal{P}^2$ the radial extent of both modes increases very rapidly as $\mathcal{P}^2$ decreases. It is of significance, however, that most of the energy is contained within the waveguide core, even for relatively small values of $\mathcal{P}^2$ and Δn. Thus, Marcatili's assumption that very little energy propagates in the region of the corners is valid over a wide range.

3.3 *Propagation Curves*

In all cases of computed propagation curves, the normalized waveguide height $\mathcal{B}$, as given in equation (11), is plotted on the horizontal

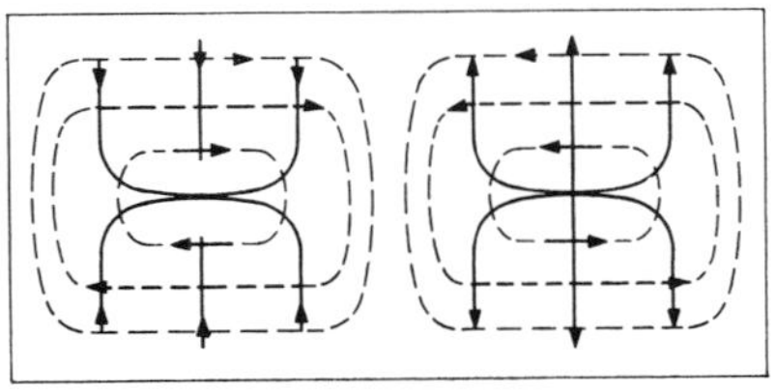

Fig. 10 — Field configuration of the E_{22}^{y} mode.

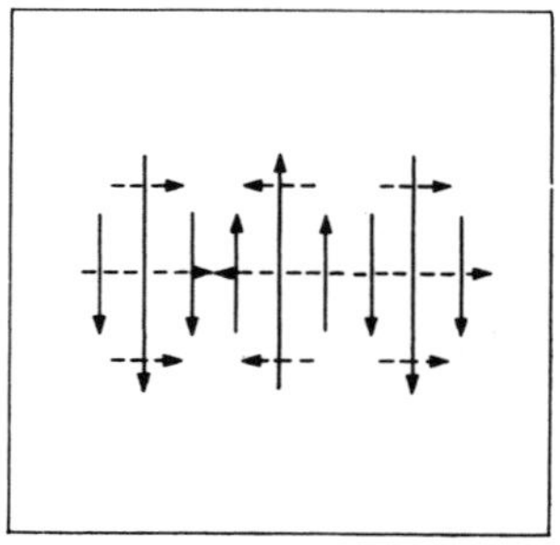

Fig. 11 — Field configuration of the E^y_{31} mode.

axis and the normalized propagation constant, $\mathcal{P}^2$, given in equation (16), along the vertical axis.

Figure 16 shows the case of vanishing index difference for an aspect ratio of one. The first 16 modes are shown. For this case the following six degenerate groups exist

$$E^y_{11}\ ,\ E^x_{11}$$
$$E^y_{12}\ ,\ E^x_{12}\ ,\ E^y_{21}\ ,\ E^x_{21}$$
$$E^y_{31}\ ,\ E^x_{13}$$
$$E^x_{31}\ ,\ E^y_{13}$$
$$E^x_{22}\ ,\ E^y_{22}$$
$$E^y_{32}\ ,\ E^x_{23}\ ,\ E^y_{23}\ ,\ E^x_{23}\ .$$

In addition, the E^y_{31} and the E^x_{31} modes are almost degenerate except

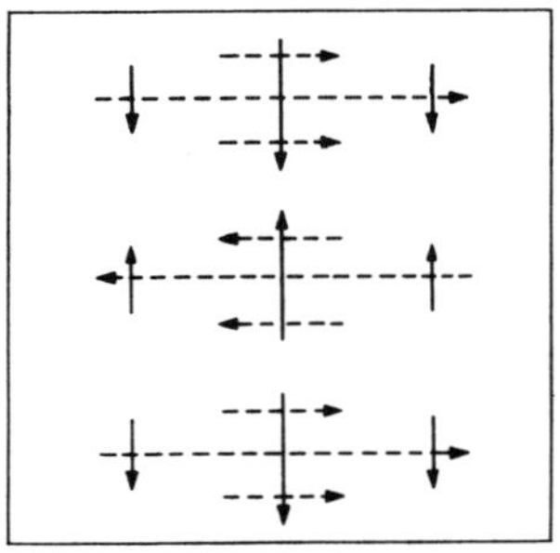

Fig. 12 — Field configuration of the E^y_{13} mode.

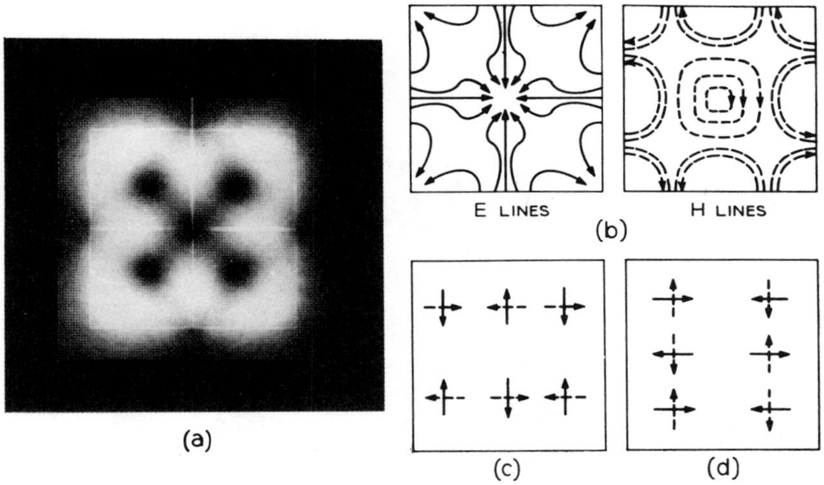

Fig. 13 — The E^y_{32} mode for unity aspect ratio: (a) intensity, (b) field configuration, (c) TE_{32}, and (d) TE_{23}.

near cutoff. The splitting of these modes can be accounted for by the differences of the field patterns shown in Fig. 11 and 12. Since the E^x_{31} mode reversals occur along the direction of the electric field lines, the electric field for this mode must have a larger longitudinal field component than for the E^y_{31} mode.

All degeneracies, except the $E^y_{mn} - E^x_{mn}$, are broken by a change in the aspect ratio as demonstrated in Fig. 17, which is drawn for the first 12 modes of a waveguide of aspect ratio 2. One interesting feature of this curve is the mode crossing of the E^y_{31} and E^y_{12} modes. Crossings of this type, which cannot occur in metallic waveguides, are possible because the field functions are frequency dependent. Qualitatively, it can be explained by noting that field reversals must take place in the core, therefore constraining the central lobe of the E^y_{31} more than any of the E^y_{12} mode lobes as cutoff is approached. Far from cutoff, however, all fields are well constrained and the E^y_{31} mode has a larger propagation constant than the E^y_{12} mode, as it does for the similar metallic waveguide mode with an aspect ratio of 2.

The effect of finite index difference on the modes can be observed by comparing Fig. 16, which is computed for unity aspect ratio and a vanishing index difference, with Fig. 18, which is computed for unity aspect ratio and a 0.5 index difference. The curves for modes whose

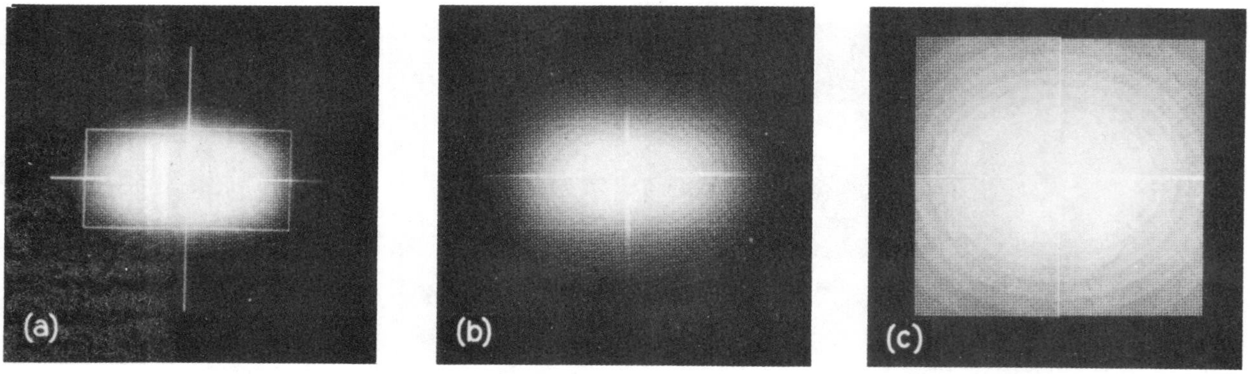

Fig. 14 — Intensity pictures of the E_{11}^{y} mode for (a) $\mathcal{P}^2 = 0.81$, (b) $\mathcal{P}^2 = 0.50$, and (c) $\mathcal{P}^2 = 0.02$.

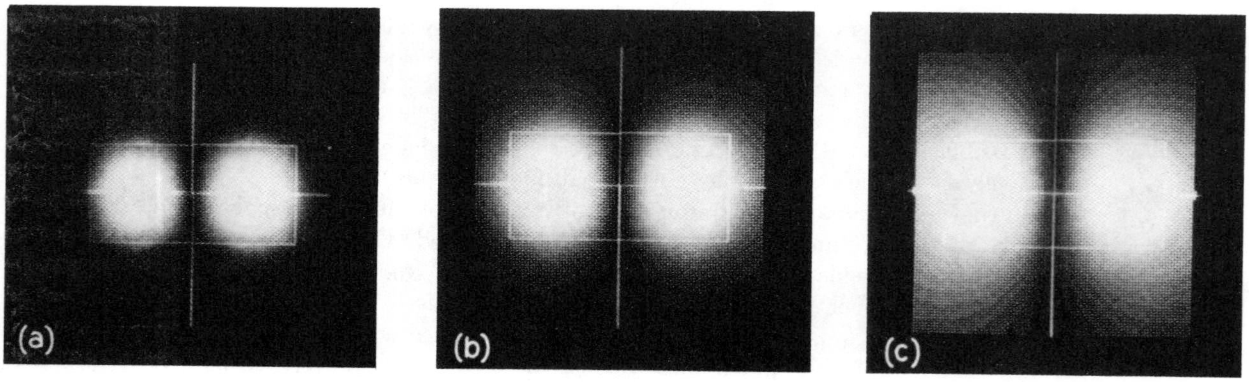

Fig. 15 — Intensity pictures of the E_{21}^{y} mode for (a) $\mathcal{P}^2 = 0.76$, (b) $\mathcal{P}^2 = 0.31$, and $\mathcal{P}^2 = 0.04$.

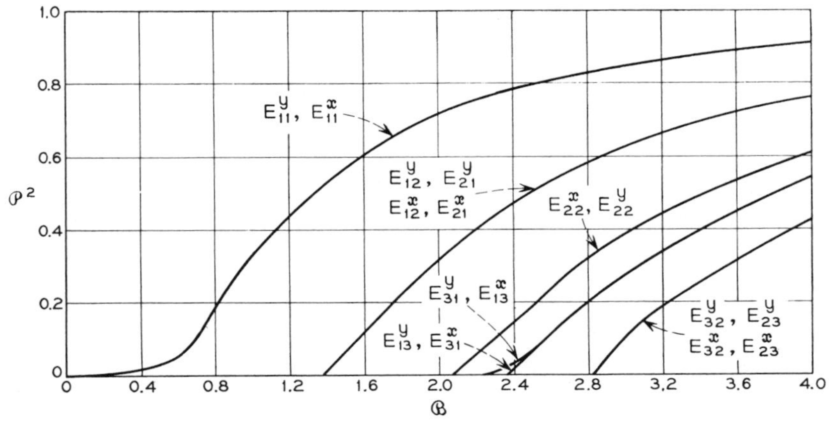

Fig. 16 — Propagation curves for the first 16 modes for unity aspect ratio and $\Delta n_r \rightarrow 0$.

field lines reverse direction across the origin are no longer degenerate, but those whose field lines do not reverse still are degenerate. For all degeneracies to be split, there must exist a finite index difference as well as an aspect ratio other than unity. Figure 19 illustrates one such case.

The effect of index difference on the degenerate principal modes for unity aspect ratio is examined in Fig. 20. The curve shows both a low and high index difference limit. In the range of interest for optical

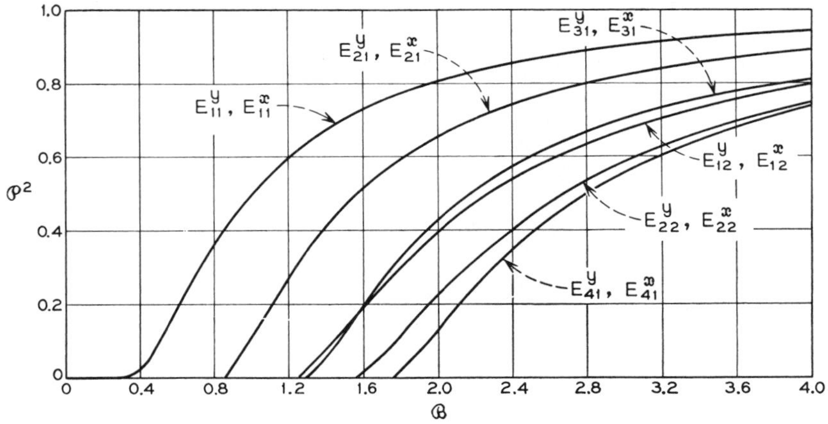

Fig. 17 — Propagation curves for the first 12 modes for $a/b = 2$ and $\Delta n_r \rightarrow 0$.

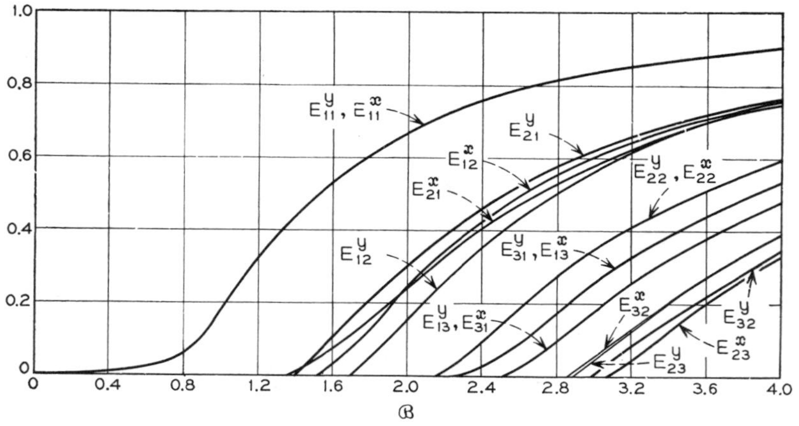

Fig. 18 — Propagation curves for the first 16 modes for unity aspect ratio and $\Delta n_r = 0.5$.

circuits (0 − 0.1) the vanishing difference curve is an excellent approximation. The greatest changes occur in the 0.1 − 10 range, which is the range of interest for some microwave problems.

Figure 21 presents the computed results for the effect of index changes on the principal modes for an aspect ratio of 2. The effect is much stronger on the E^y_{11} mode than the E^x_{11} mode. In fact, the effect on the E^x_{11} mode is comparatively small, except near cutoff.

The effect of aspect ratio on the principal modes is demonstrated for

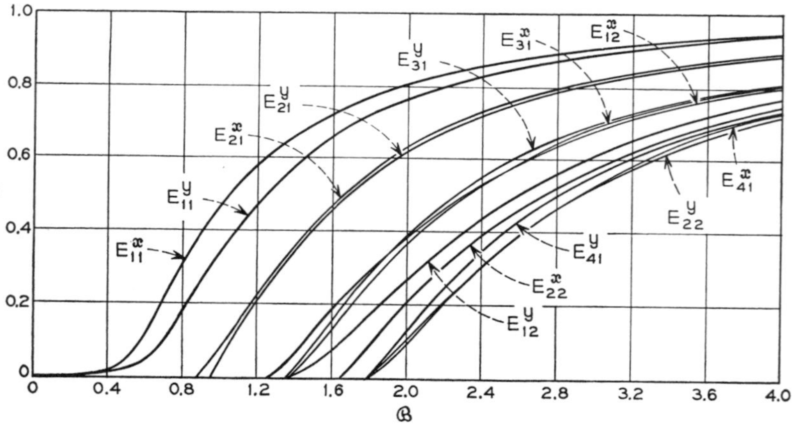

Fig. 19 — Propagation curves for the first 12 modes for $a/b = 2$ and $\Delta n_r = 0.5$.

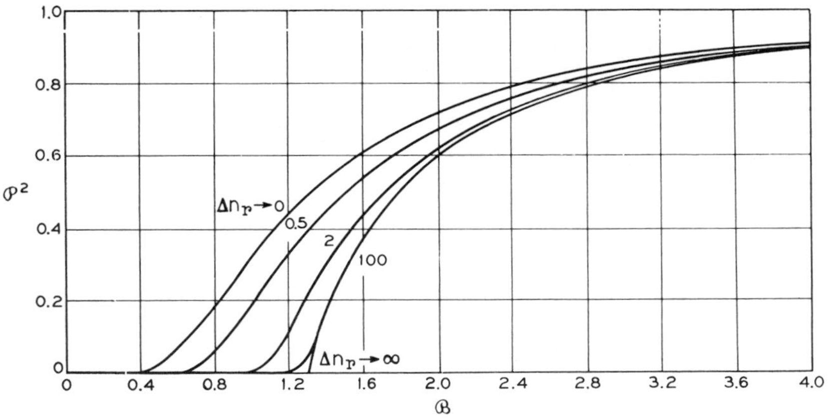

Fig. 20 — E^y_{11} and E^x_{11} mode propagation curves for several values of Δn_r with unity aspect ratio.

vanishing index difference in Fig. 22. The curve for infinite aspect ratio was obtained from the exact analysis of the slab case.[14]

IV. CONCLUSIONS

The results of the computations show that the circular harmonic method for analyzing rectangular dielectric waveguides gives excel-

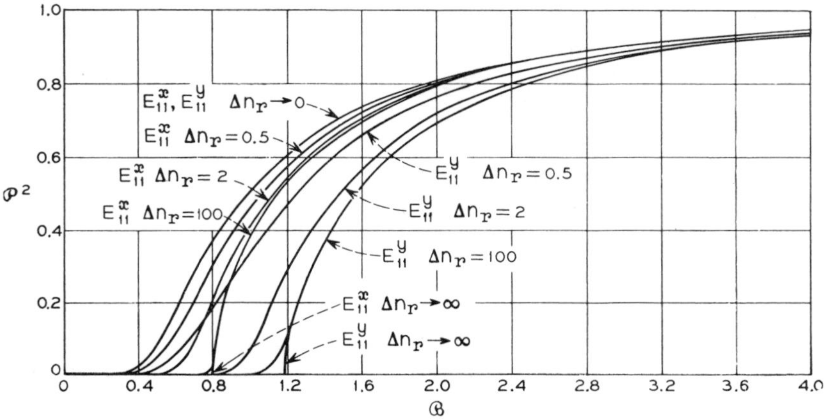

Fig. 21 — E^y_{11} and E^x_{11} mode propagation curves for several values of Δn_r with $a/b = 2$.

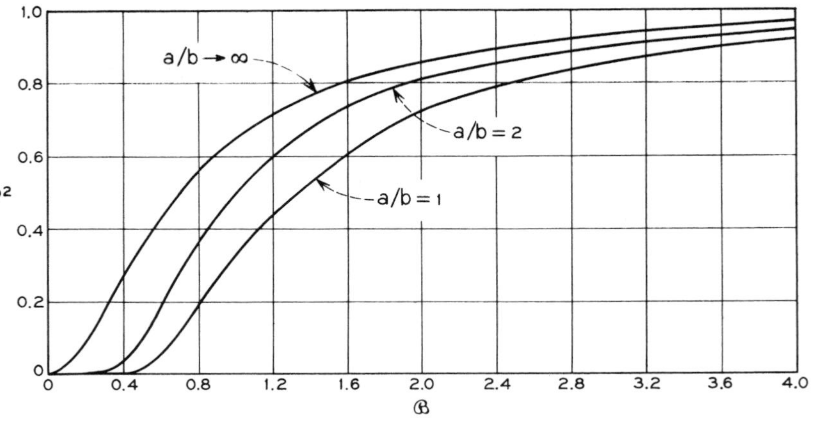

Fig. 22 — E^y_{11} and E^x_{11} mode propagation curves for several values of a/b with $\Delta n_r \rightarrow 0$.

lent results for waveguides of moderate aspect ratio. The convergence of the computed results was rapid and the results are in agreement with those of Marcatili's in the regions where his approximations apply. Furthermore, the results compare very well with Schlosser's curves for the principal mode.

Comparison of the results presented here with Marcatili's show that the two methods give values of the normalized propagation constant, $\mathcal{P}^2$, which are within a few percent for $\mathcal{P}^2 > 0.5$. Thus for $\mathcal{P}^2$ in this range his method is to be preferred since the calculations required are much simpler. However, for $\mathcal{P}^2 < 0.5$, and when it is desired to differentiate between modes for some of the near degenerate cases, another method must be used.

The circular harmonic analysis is attractive for small $\mathcal{P}^2$ because of the nature of the matching boundary. For large refractive index difference and moderate $\mathcal{P}^2$ both the method presented here and the one presented by Scholosser can be used.

V. ACKNOWLEDGMENTS

The author wishes to express his appreciation to T. Li and E. A. J. Marcatili for their valuable suggestions, to Mrs. C. L. Beattie for her aid in writing the plotting program, and to Mrs. E. Kershbaumer for her aid in writing the program for computing the propagation constants.

REFERENCES

1. Miller, S. E., "Integrated Optics: An introduction," this issue, pp. 2059–2069.
2. Kapany, N. S., *Fiber Optics,* New York: Academic Press, 1967, pp. 36–80.
3. Bracey, M. F., Cullen, A. L., Gillespie, E. F. F., and Staniforth, J. A., "Surface Wave Research in Sheffield," I.R.E. Trans. Antennas and Propagation, *AP-7,* No. 10 (December 1959), pp. 219–225.
4. Snitzer, E., "Cylindrical Dielectric Waveguide Modes," J. Opt. Soc. of Amer., *51,* No. 5 (May 1961), pp. 491–498.
5. Snitzer, E., and Osterberg, H., "Observed Dielectric Waveguide Modes in the Visible Spectrum," J. Opt. Soc. of Amer., *51,* No. 5 (May 1961), pp. 491–505.
6. Marcatili, E. A. J., "Dielectric Rectangular Waveguide and Directional Coupler for Integrated Optics," this issue, pp. 2071–2102.
7. Schlosser, W., and Unger, H. G., "Partially Filled Waveguides and Surface Waveguides of Rectangular Cross-Section," *Advances in Microwaves,* New York: Academic Press, 1966, pp. 319–387.
8. Shaw, C. B., French, B. T., and Warner, C. III, "Further Research on Optical Transmission Lines," Sci. Rep. No. 2, Contract AF449 (638)-1504 AD 625 501, Autonetics Report No. C7-929/501, pp. 13–44.
9. Stratton, J. A., "Electromagnetic Theory," New York: McGraw-Hill, 1941, p. 361.
10. Laxpati, S. R., and Mittra, R., "Energy Considerations in Open and Closed Waveguides," IEEE Trans. Antennas and Propagation, *AP-13,* No. 6 (November 1965), pp. 883–890.
11. Hamming, R. W., *Numerical Analysis for Scientists and Engineers,* New York: McGraw-Hill, 1962, pp. 81–82.
12. Freed, B. H., "Algorithm 41," Revision Evaluation of Determinant Comm. ACM, *6,* No. 9 (September 1963), p. 520.
13. "System/360 Scientific Subroutine Package," IBM, White Plains, N. Y., H20-0205-2, pp. 179–182.
14. Collin, R. E., "Field Theory of Guided Waves," New York: McGraw-Hill, 1960, pp. 480–495.

Dielectric Rectangular Waveguide and Directional Coupler for Integrated Optics

By E. A. J. MARCATILI

(Manuscript received March 3, 1969)

We study the transmission properties of a guide consisting of a dielectric rod with rectangular cross section, surrounded by several dielectrics of smaller refractive indices. This guide is suitable for integrated optical circuitry because of its size, single-mode operation, mechanical stability, simplicity, and precise construction.

After making some simplifying assumptions, we solve Maxwell's equations in closed form and find, that, because of total internal reflection, the guide supports two types of hybrid modes which are essentially of the TEM *kind polarized at right angles. Their attenuations are comparable to that of a plane wave traveling in the material of which the rod is made.*

If the refractive indexes are chosen properly, the guide can support only the fundamental modes of each family with any aspect ratio of the guide cross section. By adding thin lossy layers, the guide presents higher loss to one of those modes. As an alternative, the guide can be made to support only one of the modes if part of the surrounding dielectrics is made a low impedance medium.

Finally, we determine the coupling between parallel guiding rods of slightly different sizes and dielectrics; at wavelengths around one micron, 3-dB directional couplers, a few hundred microns long, can be achieved with separations of the guides about the same as their widths (a few microns).

I. INTRODUCTION

Proposals have been made for dielectric waveguides capable of guiding beams in integrated optical circuits very much as waveguides and coaxials are used for microwave circuitry.[1–3] Figure 1 shows the basic geometries for these waveguides. The guide is a dielectric rod of refractive index n immersed in another dielectric of slightly smaller refractive index $n(1 - \triangle)$; both are in contact with a third dielectric which may be air (Fig. 1a) or a dielectric of refractive index $n(1 - \triangle)$,

Reprinted with permission from *Bell Syst. Tech. J.*, vol. 48, pp. 2071–2102, Sept. 1969.

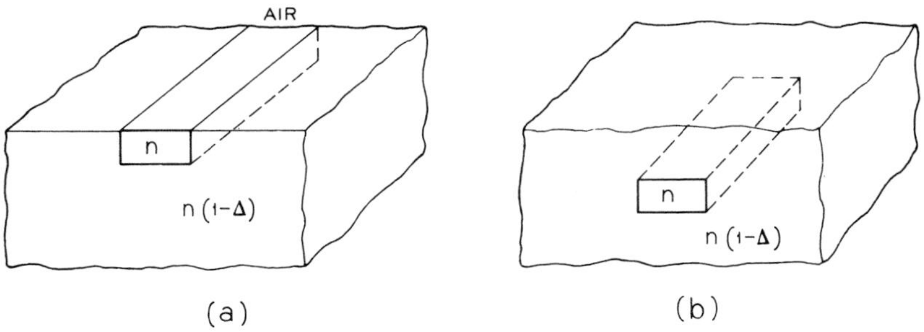

Fig. 1 — Dielectric waveguides for integrated optical circuitry.

(Fig. 1b). These geometries are attractive not only because of simplicity, precision of construction, and mechanical stability, but also because by choosing Δ small enough, single-mode operation can be achieved with transverse dimensions of the guide large compared with the free space wavelengths, thus relaxing the tolerance requirements.

Even though in a real guide the cross section of the guiding rod is not exactly rectangular and the boundaries between dielectrics are not sharply defined, as in Fig. 1, it is worth finding the characteristics of the modes in the idealized structure and the requirements to make it a single-mode waveguide.

Furthermore, directional couplers made by bringing two of those guides close together, Fig. 2, may become important circuit components.[1,2] In this paper we study the transmission through such a coupler; the modes in a single guide result as a particular case, when the separation between the two guides is so large that the coupling is negligible. Through use of a perturbation technique, we also find the coupler properties when the two guides are slightly different.

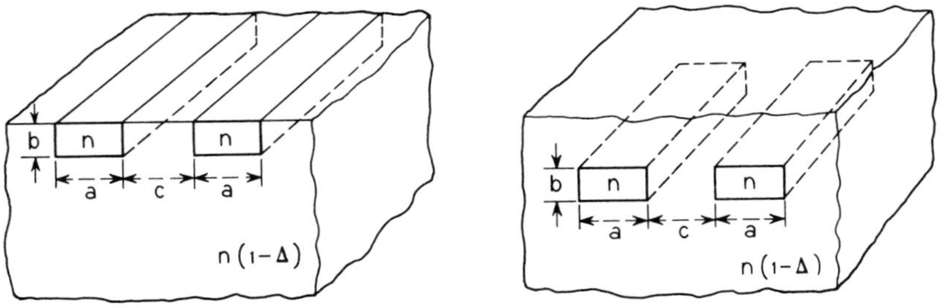

Fig. 2 — Directional couplers.

The guiding properties of the rectangular cross section guide immersed in a single dielectric are compared with those derived through computer calculations by Goell.[4] Similarly, the coupling properties of two guides of square cross section immersed in a single dielectric are compared with those of two guides of circular cross section derived by Jones and by Bracey and others.[5,6] In both comparisons agreement is quite good.

II. FORMULATION OF THE BOUNDARY VALUE PROBLEM

For analysis, we redraw in Fig. 3 the cross section of the coupler subdivided in many areas. Nine of the areas have refractive indexes n_1 to n_5; we do not specify the refractive indexes in the six shaded areas. The reasons for these choices will become obvious.

A rigorous solution to this boundary value problem requires a computer;[4,7] nevertheless, it is possible to introduce a drastic simplification which enables one to get a closed form solution. This simplification arises from observing that, for well-guided modes, the field decays exponentially in regions 2, 3, 4, and 5; therefore, most of the power travels in regions 1, a small part travels in regions 2, 3, 4, and 5, and even less travels in the six shaded areas. Consequently, only a small error should be introduced into the calculation of fields in regions 1 if one does not properly match the fields along the edges of the shaded areas.

The matching made only along the four sides of regions 1 can be achieved assuming simple field distribution. Thus the field components in regions 1 vary sinusoidally in the x and y direction; those in 2 and 4 vary sinusoidally along x and exponentially along y; and those in regions 3 and 5 vary sinusoidally along y and exponentially along x. The propagation constants k_{x1}, k_{x2}, and k_{x4} along x in media 1, 2, and

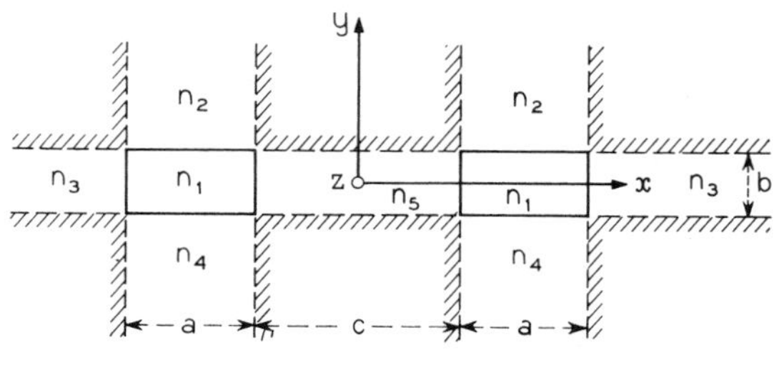

Fig. 3 — Coupler cross section subdivided for analysis.

4 are identical and independent of y. Similarly, the propagation constants k_{y1}, k_{y3}, and k_{y5} along y in the regions 1, 3, and 5 are also identical and independent of x.

In the appendix we calculate these propagation constants and find, as expected, that all the modes are hybrid and that guidance occurs because of total internal reflection. Nevertheless, because of another approximation which consists of choosing the refractive indexes n_2, n_3, n_4, and n_5 slightly smaller than n_1, total internal reflection occurs only when the plane wavelets that make a mode impinge on the interfaces at grazing angles.* Consequently, the largest field components are perpendicular to the axis of propagation; the modes are essentially of the TEM kind and can be grouped in two families, E^x_{pq} and E^y_{pq}. The main field components of the members of the first family are E_x and H_y, while those of the second are E_y and H_x. The subindex p and q indicate the number of extrema of the electric or magnetic field in the x and y directions, respectively. Naturally, E^x_{11} and E^y_{11} are the fundamental modes; we concentrate on them as we discuss the transmission properties of different structures.

III. GUIDE IMMERSED IN SEVERAL DIELECTRICS

The guide immersed in several dielectrics (Fig. 4a) is derived from Fig. 3 by choosing

$$c = \infty. \tag{1}$$

It supports a discrete number of guided modes which we group in two families E^x_{pq} and E^y_{pq} plus a continuum of unguided modes.[8,9]

3.1 *The E^y_{pq} Modes*

The main transverse field components of the E^y_{pq} modes are E_y and H_x. They are depicted in solid and broken lines, respectively, in Fig. 4a for the fundamental mode E^y_{11}. Within the guiding rod each component varies sinusoidally both along x and along y. Outside the guide each component decays exponentially. Such functional dependence is given in equation (38) and depicted in Fig. 4b. We assume $n_2 \neq n_3 \neq n_4 \neq n_5$; consequently the field distributions are not symmetric with respect to the planes $x = 0$ and $y = 0$. In Fig. 5a we assume $n_2 = n_4$ and $n_3 = n_5$; the E^y_{pq} modes depicted are either symmetric or antisymmetric with respect to the same planes. These modes look similar to those in laser

* This approximation is not very demanding. Even when n_1 is 50 percent larger than n_2, n_3, n_4, and n_5, the results are valid.

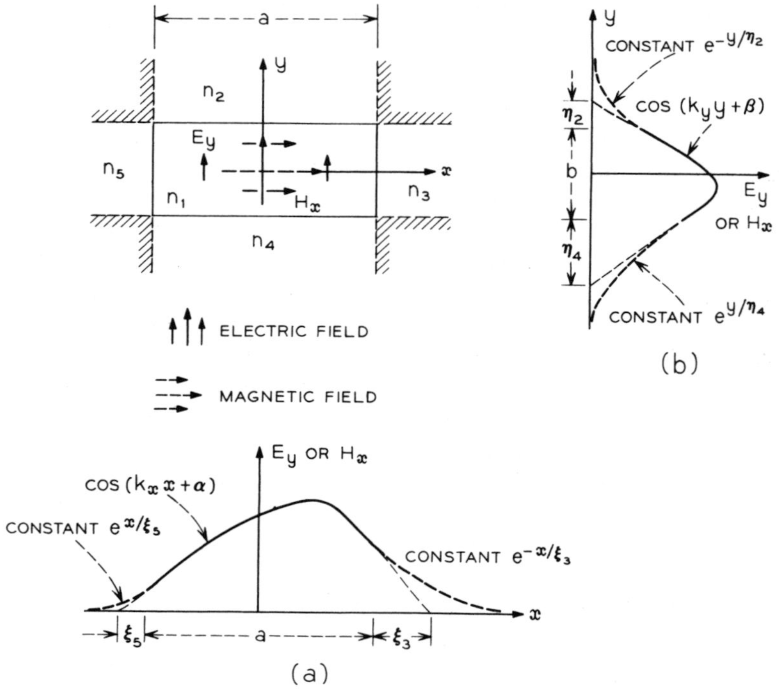

Fig. 4 — Guide immersed in different dielectrics: (a) cross section and (b) field distribution of the fundamental mode E_{11}^y.

cavities with rectangular flat mirrors, but our nomenclature is different.[10] The subindexes p and q indicate the number of extrema each component has within the guide.

Now we describe these modes quantitatively by reproducing the propagation constants found for each medium in Section A.1 of the appendix. Let us call k_z the axial propagation constant and $k_{x\nu}$ and $k_{y\nu}$ the transverse propagation constants along the x and the y directions, respectively, in the νth medium ($\nu = 1, 2, \cdots 5$). Furthermore, let us call

$$k_\nu = kn_\nu = \frac{2\pi}{\lambda} n_\nu \tag{2}$$

the propagation constant of a plane wave in a medium of refractive index n_ν and free-space wavelength λ.

According to equations (39) through (52)

$$k_z = (k_1^2 - k_x^2 - k_y^2)^{\frac{1}{2}} \tag{3}$$

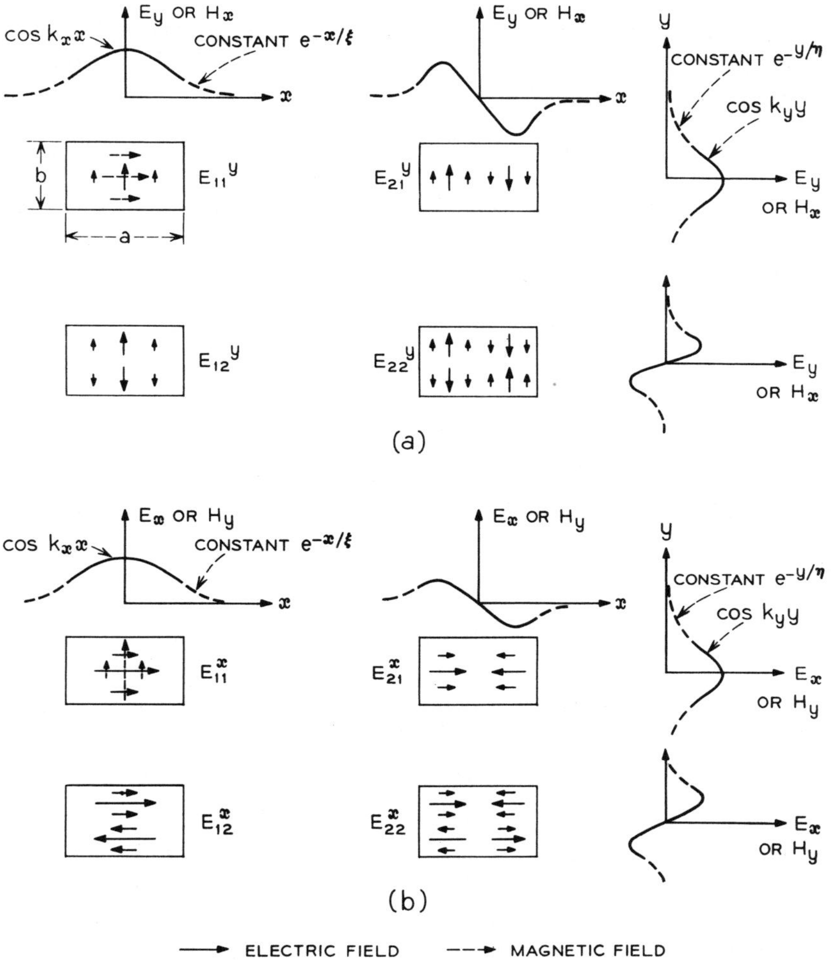

Fig. 5 — (a) Field configuration of $E_{pq}{}^y$ modes. (b) Field configuration of $E_{pq}{}^x$ modes.

in which

$$k_x = k_{x1} = k_{x2} = k_{x4} \tag{4}$$

and

$$k_y = k_{y1} = k_{y3} = k_{y5}\,. \tag{5}$$

This means that the fields in media 1, 2, and 4 have the same x

dependence and similarly those in media 1, 3, and 5 have identical y dependence. These transverse propagation constants are solutions of the transcendental equations:

$$k_x a = p\pi - \tan^{-1} k_x \xi_3 - \tan^{-1} k_x \xi_5 \tag{6}$$

$$k_y b = q\pi - \tan^{-1} \frac{n_2^2}{n_1^2} k_y \eta_2 - \tan^{-1} \frac{n_4^2}{n_1^2} k_y \eta_4 \tag{7}$$

in which

$$\xi_{\substack{3\\5}} = \frac{1}{\left| k_{x\substack{3\\5}} \right|} = \frac{1}{\left[\left(\dfrac{\pi}{A_{\substack{3\\5}}} \right)^2 - k_x^2 \right]^{\frac{1}{2}}} \tag{8}$$

$$\eta_{\substack{2\\4}} = \frac{1}{\left| k_{y\substack{2\\4}} \right|} \quad \frac{1}{\left[\left(\dfrac{\pi}{A_{\substack{2\\4}}} \right)^2 - k_y^2 \right]^{\frac{1}{2}}} \tag{9}$$

and

$$A_{2,3,4,5} = \frac{\pi}{(k_1^2 - k_{2,3,4,5}^2)^{\frac{1}{2}}} = \frac{\lambda}{2(n_1^2 - n_{2,3,4,5}^2)^{\frac{1}{2}}}. \tag{10}$$

In the transcendental equations (6) and (7), a and b are the transverse dimensions of the guiding rod, and the $\tan^{-1}$ functions are to be taken in the first quadrant.

What are the physical meanings of $\xi_{\substack{3\\5}}$, $\eta_{\substack{2\\4}}$, and $A_{2,3,4,5}$? The amplitude of each field component in medium 3 (Fig. 4) decreases exponentially along x. It decays by $1/e$ in a distance $\xi_3 = 1/\left| k_{x3} \right|$. Similarly ξ_5, η_2, and η_4 measure the "penetration depths" of the field components in media 5, 2, and 4, respectively.

The meaning of A_2 is the following. Consider a symmetric slab derived from Fig. 4 by choosing $a = \infty$ and $n_2 = n_4$. The maximum thickness for which the slab supports only the fundamental mode is A_2.

Expressions (3), (8), and (9) contain k_x and k_y, which are solutions of the transcendental equations (6) and (7). These cannot be solved exactly in closed form. Nevertheless, for well-guided modes, most of the power travels within medium 1, implying

$$\left(\frac{k_x A_{\substack{3\\5}}}{\pi} \right)^2 \ll 1 \quad \text{and} \quad \left(\frac{k_y A_{\substack{2\\4}}}{\pi} \right)^2 \ll 1. \tag{11}$$

It is possible then to solve those transcendental equations in closed, though approximate, form. Their solutions are

$$k_x = \frac{p\pi}{a}\left(1 + \frac{A_3 + A_5}{\pi a}\right)^{-1} \tag{12}$$

$$k_y = \frac{q\pi}{b}\left(1 + \frac{n_2^2 A_2 + n_4^2 A_4}{\pi n_1^2 b}\right)^{-1}. \tag{13}$$

For large a and b, the electrical width, $k_x a$, and the electrical height, $k_y b$, of the guide are close to $p\pi$ and $q\pi$, respectively.

Substituting equations (12) and (13) in equations (3), (8), and (9), we obtain explicit expressions for k_z, ξ_3, ξ_5, η_2, and η_4:

$$k_z = \left[k_1^2 - \left(\frac{\pi p}{a}\right)^2\left(1 + \frac{A_3 + A_5}{\pi a}\right)^{-2} - \left(\frac{\pi q}{b}\right)^2\left(1 + \frac{n_2^2 A_2 + n_4^2 A_4}{\pi n_1^2 b}\right)^{-2}\right]^{\frac{1}{2}} \tag{14}$$

$$\xi_{\substack{3\\5}} = \frac{A_{\substack{3\\5}}}{\pi}\left[1 - \left\{\frac{pA_{\substack{3\\5}}}{a}\frac{1}{1 + \frac{A_3 + A_5}{\pi a}}\right\}^2\right]^{-\frac{1}{2}} \tag{15}$$

$$\eta_{\substack{2\\4}} = \frac{A_{\substack{2\\4}}}{\pi}\left[1 - \left\{\frac{qA_{\substack{2\\4}}}{b}\frac{1}{1 + \frac{n_2^2 A_2 + n_4^2 A_4}{\pi n_1^2 b}}\right\}^2\right]^{-\frac{1}{2}}. \tag{16}$$

3.2 *The E_{pq}^x Modes*

Except for the fact that the main transverse components are E_x and H_y, the E_{pq}^x modes are qualitatively similar to the E_{pq}^y modes (Fig. 5b); they differ quantitatively. Distinguishing with bold-face type the symbols corresponding to E_{pq}^x modes, the axial propagation constant and the "penetration depth" in media 2, 3, 4, and 5 are, according to equations (60), (63), and (64),

$$\mathbf{k}_z = (k_1^2 - \mathbf{k}_x^2 - \mathbf{k}_y^2)^{\frac{1}{2}} \tag{17}$$

$$\boldsymbol{\xi}_{\substack{3\\5}} = \frac{1}{\left|\mathbf{k}_{x\substack{3\\5}}\right|} = \frac{1}{\left[\left(\frac{\pi}{A_{\substack{3\\5}}}\right)^2 - \mathbf{k}_x^2\right]^{\frac{1}{2}}} \tag{18}$$

$$\boldsymbol{\eta}_{\substack{2\\4}} = \frac{1}{\left|\mathbf{k}_{y\substack{2\\4}}\right|} = \frac{1}{\left[\left(\frac{\pi}{A_{\substack{2\\4}}}\right)^2 - \mathbf{k}_y^2\right]^{\frac{1}{2}}} \tag{19}$$

in which $\mathbf{k}_x$ and $\mathbf{k}_y$ are solutions of the transcendental equations

$$\mathbf{k}_x a = p\pi - \tan^{-1} \frac{n_3^2}{n_1^2} \mathbf{k}_x \xi_3 - \tan^{-1} \frac{n_5^2}{n_1^2} \mathbf{k}_x \xi_5 \tag{20}$$

$$\mathbf{k}_y b = q\pi - \tan^{-1} \mathbf{k}_y \mathbf{n}_2 - \tan^{-1} \mathbf{k}_y \mathbf{n}_4 \,. \tag{21}$$

The approximate closed form solutions of these equations are

$$\mathbf{k}_x = \frac{p\pi}{a} \left(1 + \frac{n_3^2 A_3 + n_5^2 A_5}{\pi n_1^2 a} \right)^{-1} \tag{22}$$

and

$$\mathbf{k}_y = \frac{q\pi}{b} \left(1 + \frac{A_2 + A_4}{\pi b} \right)^{-1} . \tag{23}$$

Substituting these expressions in equations (17), (18), and (19), we derive the explicit results:

$$\mathbf{k}_z = \left[k_1^2 - \left(\frac{\pi p}{a} \right)^2 \left(1 + \frac{n_3^2 A_3 + n_5^2 A_5}{\pi n_1^2 a} \right)^{-2} - \left(\frac{\pi q}{b} \right)^2 \left(1 + \frac{A_2 + A_4}{\pi b} \right)^{-2} \right]^{\frac{1}{2}} \tag{24}$$

$$\xi_{\substack{3\\5}} = \frac{A_{\substack{3\\5}}}{\pi} \left[1 - \left\{ \frac{p A_{\substack{3\\5}}}{a} \frac{1}{1 + \dfrac{n_3^2 A_3 + n_5^2 A_5}{\pi n_1^2 a}} \right\}^2 \right]^{-\frac{1}{2}} \tag{25}$$

$$\mathbf{n}_{\substack{2\\4}} = \frac{A_{\substack{2\\4}}}{\pi} \left[1 - \left\{ \frac{q A_{\substack{2\\4}}}{b} \frac{1}{1 + \dfrac{A_2 + A_4}{\pi b}} \right\}^2 \right]^{-\frac{1}{2}} . \tag{26}$$

If

$$\frac{1}{n_1} \left\{ n_1 - n_{\substack{2\\3\\4\\5}} \right\} \ll 1,$$

these results coincide with those in equations (14), (15), and (16), indicating that the E^x_{pq} and E^y_{pq} modes become degenerate.

3.3 *Examples*

The axial propagation constants k_z and $\mathbf{k}_z$, given in equations (3) and (17) and properly normalized, have been plotted in Figs. 6a through k as a function of the normalized height of the guide

$$\frac{b}{A_4} = \frac{2b}{\lambda} (n_1^2 - n_4^2)^{\frac{1}{2}}$$

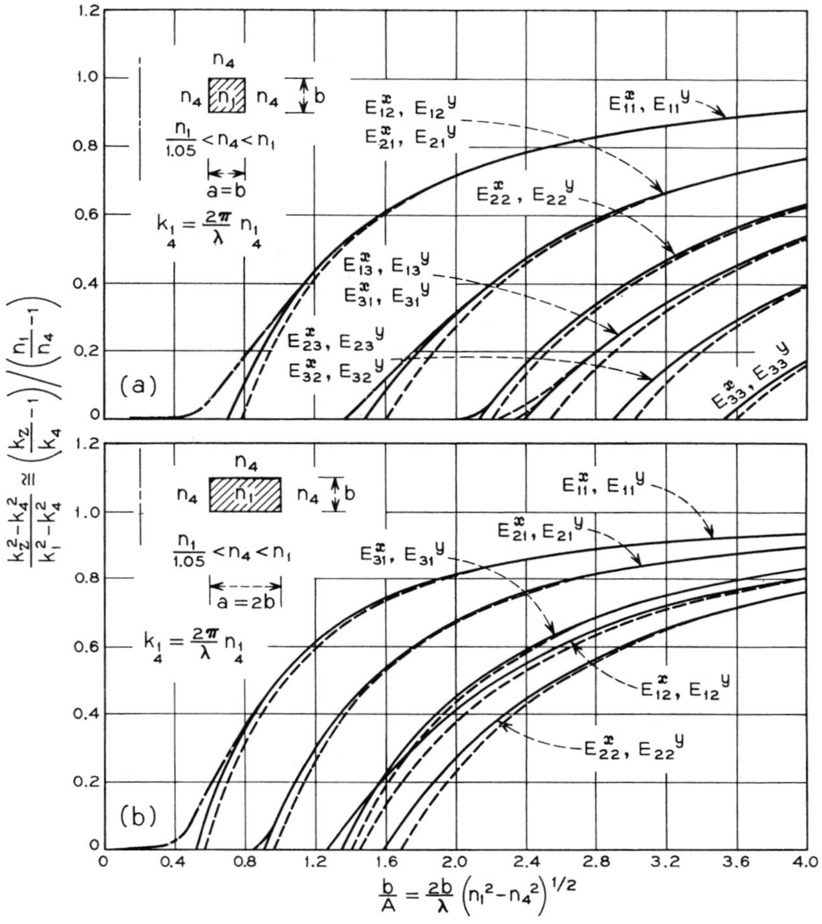

Fig. 6 — Propagation constant for different modes and guides. ———— transcendental equation solutions; — — — — closed form solutions; —·—·— Goell's computer solutions of the boundary value problem.

for several geometries and surrounding media.* The ordinate in each of these figures is

$$\frac{k_z^2 - k_4^2}{k_1^2 - k_4^2};$$

it varies between 0 and 1. It is 0 when $k_z = k_4$, that is, when the guide

* In these figures we use the same symbol k_z for both the $E_{pq}{}^y$ and the $E_{pq}{}^x$ modes.

is so small that the mode under consideration becomes unguided or, in other words, the "penetration depth" in medium 4 is ∞. It is 1 when the guide is so large that $k_z = k_1$, which means that all the field travels within the guiding rod and the "penetration depths" in media 2, 3, 4, and 5 are zero.

The solid curves have been obtained using the exact numerical solutions of the transcendental equations (6), (7), (20), and (21); for the transverse propagation constants k_x and k_y; the dashed lines have been derived using the closed form approximations (12), (13), (22), and (23). In Figs. 6a, 6b, 6e, and 6f, for comparison, we have also included the dotted-dashed lines which are the results obtained by Goell as computer solutions of the boundary value problem.[4]

The three solutions coincide even for moderately large values of b.

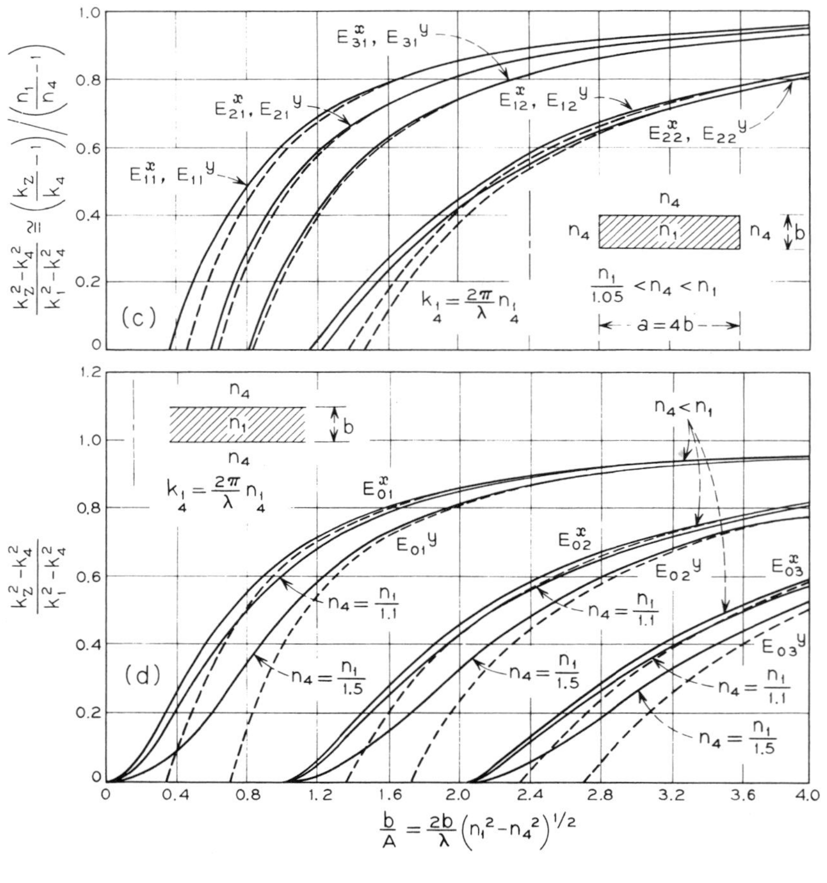

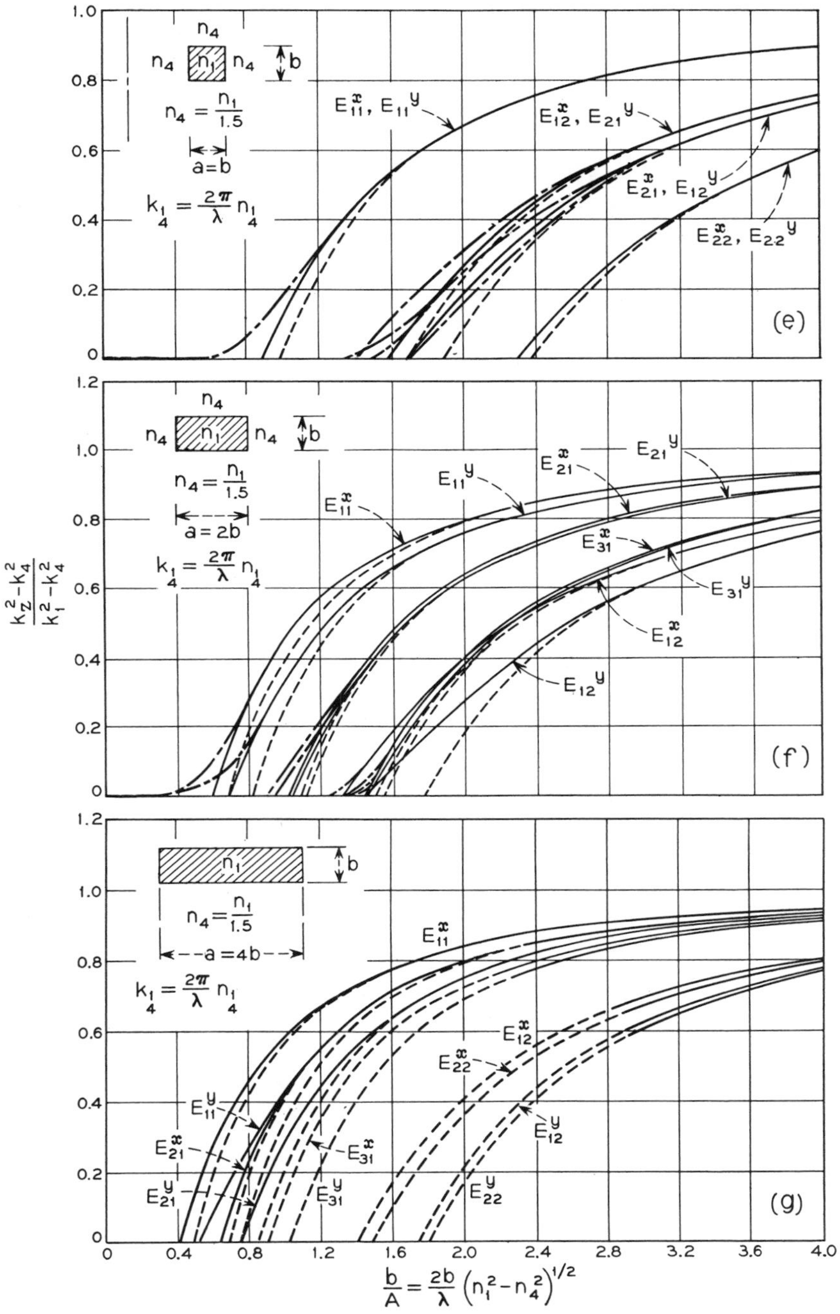
(e)
n_4
n_1
b
$n_4 = \frac{n_1}{1.5}$
a=b
$k_{1,4} = \frac{2\pi}{\lambda} n_{1,4}$
E_{11}^x, E_{11}^y
E_{12}^x, E_{21}^y
E_{21}^x, E_{12}^y
E_{22}^x, E_{22}^y
(f)
a=2b
E_{11}^x
E_{11}^y
E_{21}^x
E_{21}^y
E_{31}^x
E_{31}^y
E_{12}^x
E_{12}^y
(g)
a=4b
E_{22}^x
E_{22}^y
E_{31}^y
$\frac{k_z^2 - k_4^2}{k_1^2 - k_4^2}$
$\frac{b}{A} = \frac{2b}{\lambda}\left(n_1^2 - n_4^2\right)^{1/2}$

Thus, for a guide and mode for which

$$\frac{k_z^2 - k_4^2}{k_1^2 - k_4^2} \geqq 0.5,$$

the closed form approximation is within a few percent of the exact value. This gives us confidence to use our results in guides with an aspect ratio $a/b > 2$, in guides surrounded by several dielectrics and in directional couplers for which there are no computer calculations available.

The largest discrepancy between our results and Goell's occurs for

$$\frac{k_z^2 - k_4^2}{k_1^2 - k_4^2} \cong 0$$

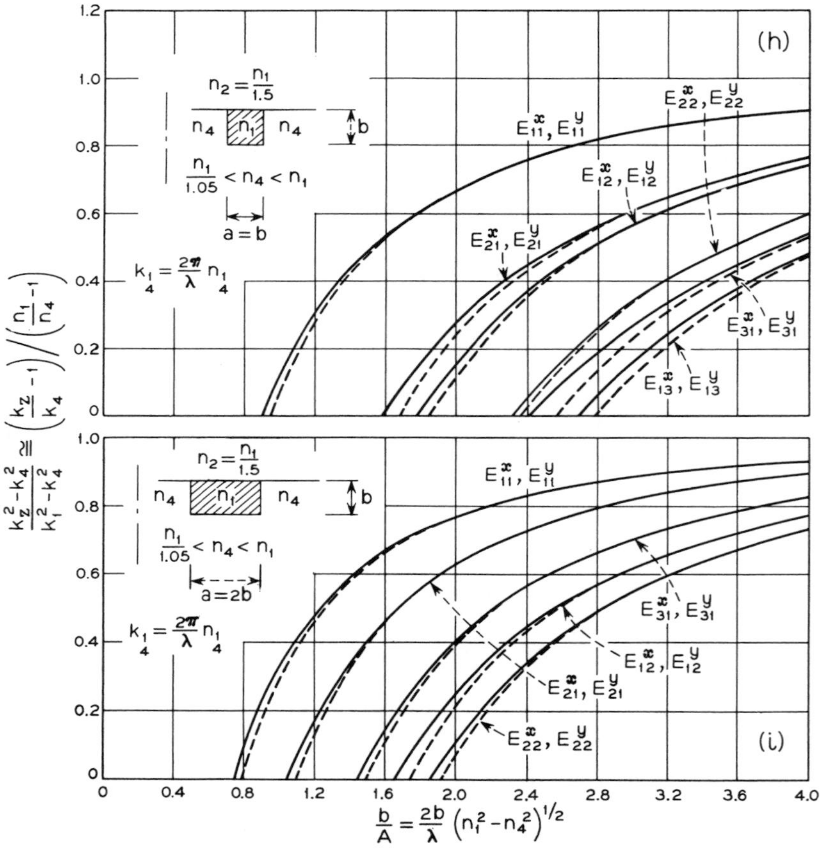

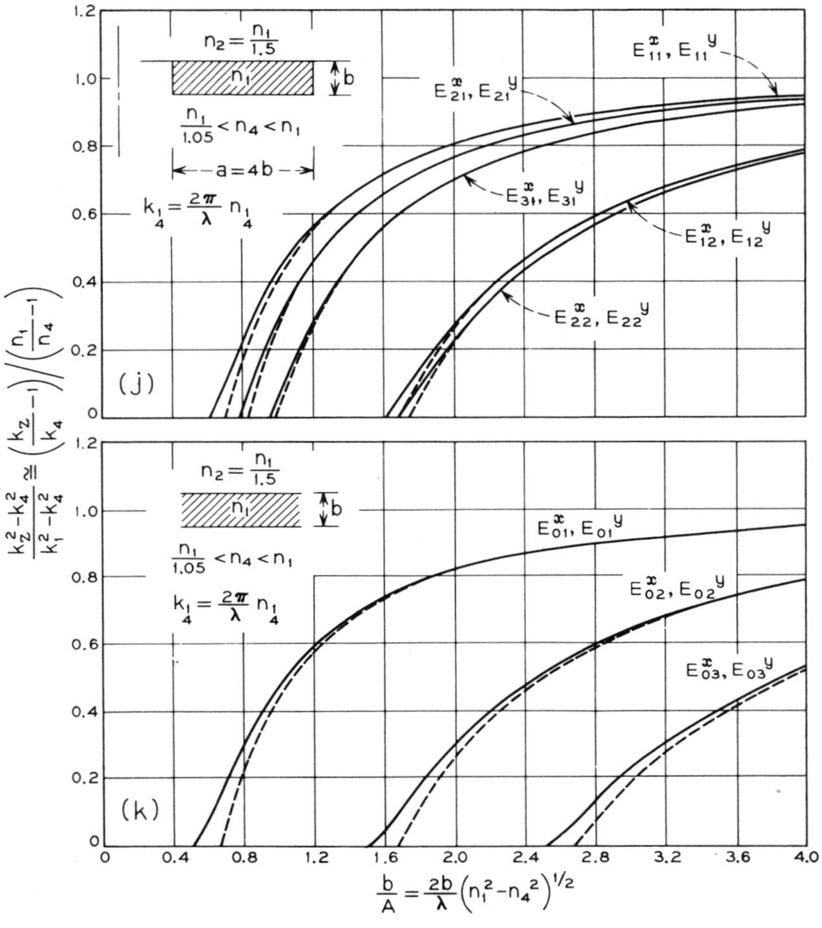

and especially for the fundamental modes E^x_{11} and E^y_{11}. Our approximate theory is incapable of predicting the fact that these modes remain guided no matter how small the guide's cross section.

Figures 6a through d cover the cases of rectangular guides totally embedded in a single dielectric of slightly lower refractive index. For all practical purposes, given p and q, the E^x_{pq} and E^y_{pq} modes are degenerate, and the square cross section provides the widest separation between modes.

Figures 6e through g also consider rectangular guides embedded in a single dielectric, but the external refractive index is 1.5 times smaller than the internal one. A glass rod immersed in air is an example. The substantial difference of refractive indexes breaks the degeneracy for any rectangular cross section. Rectangular waveguides as in Fig. 1a, with three sides in contact with slightly lower refractive indexes and the fourth side in contact with air, are covered in Fig. 6h through k.

The approximate dispersion relation (14) for E^y_{pq} modes, in a rectangular guide surrounded by four different dielectrics, has been put in graphical form in Fig. 7 by plotting the equivalent equation

$$p^2X + q^2Y = 1 \tag{27}$$

in which

$$X = \left(\frac{\pi}{a}\right)^2\left(1 + \frac{A_3 + A_5}{\pi a}\right)^{-2}(k_1^2 - k_z^2)^{-1} \tag{28}$$

and

$$Y = \left(\frac{\pi}{b}\right)^2\left(1 + \frac{n_2^2A_2 + n_4^2A_4}{\pi n_1^2 b}\right)^{-2}(k_1^2 - k_z^2)^{-1}. \tag{29}$$

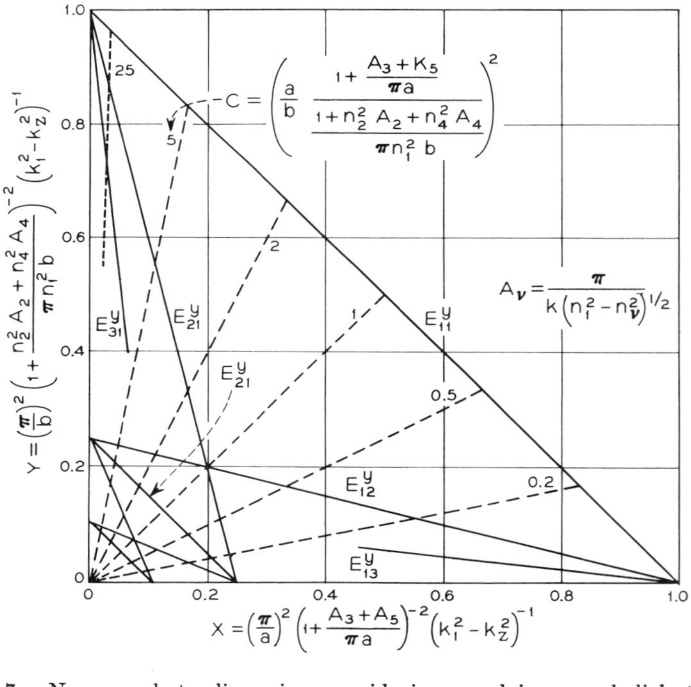

Fig. 7 — Nomograph to dimension a guide immersed in several dielectrics in such a way that it supports any prescribed number of modes.

The curves plotted for different values of p and q are straight lines (solid lines); since the values of X and Y are physically meaningful when they are positive, the plots are kept within the first quadrant.

In Fig. 7 the dotted lines depict the equation

$$\frac{Y}{X} = \left[\frac{a}{b} \frac{1 + \dfrac{A_3 + A_5}{\pi a}}{1 + \dfrac{n_2^2 A_2 + n_4^2 A_4}{\pi n_1^2 b}} \right]^2 = C. \tag{30}$$

Given any guide, we can calculate C which is a function of the dimensions, refractive indexes, and wavelength. The corresponding dotted line intersects all the solid lines representing the different modes. The abscissa or ordinate of each intersection yields, after some algebra, the propagation constant k_z of each particular mode. If the resulting k_z is smaller than the smallest k_ν, that mode is not guided.

Another way of using the graph is this: Suppose one wants a guide with such dimensions that at a given wavelength only the E_{11}^y mode is supported. Picking $k_z = k_{\nu\min}$, any combination of n_1, n_2, n_3, n_4, n_5, a, and b represented by a point within the triangle limited by the solid lines E_{11}^y, E_{12}^y, and E_{21}^y will satisfy the proposed single-mode requirement.

In the graph it is enough to substitute a by b and everything we said about E_{pq}^y modes is applicable to E_{pq}^x modes.

Figures 6a through k have been used to determine dimensions for several guides. All of them have the maximum dimensions compatible with exclusive guidance of the E_{11}^x and E_{11}^y modes. The results are collected in Table I.

In general, the geometry with $n_2 < n_4$ requires a larger waveguide cross section than with $n_2 = n_4$. This means reducing the refractive index on one side of the guide reduces its ability to guide. The explanation of this paradox is found in the known fact that a symmetric slab indeed guides "better" than an asymmetric one. Comparing, for example, Figs. 6d and 6k, in which the solid curves have been drawn solving Maxwell's equations exactly, the E_{p1}^x and E_{p1}^y modes can be guided by the symmetric slab (Fig. 6d) no matter how small the thickness b; there is a minimum thickness required for the asymmetric slab (Fig. 6k) to guide the same modes.[9]

Consider the guide immersed in a single dielectric. In general, the guide's height b is inversely proportional to

$$\frac{1}{(n_1^2 - n_4^2)^{\frac{1}{2}}}.$$

TABLE I—TYPICAL DIMENSIONS FOR SEVERAL GUIDES*

	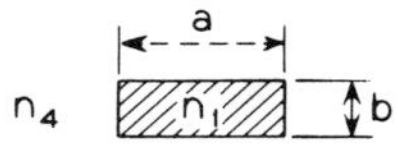				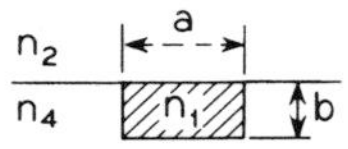		
	$\frac{n_1}{n_4} = 1.001$	$\frac{n_1}{n_4} = 1.01$	$\frac{n_1}{n_4} = 1.05$	$\frac{n_1}{n_4} = 1.5$	$\frac{n_1}{n_2} = 1.5$; $\frac{n_1}{n_4} = 1.001$	$\frac{n_1}{n_2} = 1.5$; $\frac{n_1}{n_4} = 1.01$	$\frac{n_1}{n_2} = 1.5$; $\frac{n_1}{n_4} = 1.05$
a = b	15.3†	4.9	2.25	0.92	17.7	5.6	2.6
a = 2b	19	6.1	2.8	1.21	23.2	7.4	3.4
a = 4b	26.8	8.5	3.8	1.37	34.9	11	4.9

* Dimensions are for guides capable of supporting only the fundamental modes E_{11}^x and E_{11}^y.
† All numbers in the table must be multiplied by λ/n_1.

For $n_1 = 1.5$, $n_4 = 1$, and $\lambda = 1\mu$, the largest guide height corresponds to the square cross section, and $b = a = 0.61\mu$. This dimension may be too small and difficult to control. The tolerance requirements may be relaxed by choosing $n_1 - n_4 \ll 1$. Nevertheless, this difference cannot be made arbitrarily small because the guide loses its ability to negotiate sharp bends.[11]

In all these examples the fundamental modes E_{11}^x and E_{11}^y are almost degenerate, so symmetry imperfections of the guide tend to couple these modes. A lossy layer, added to one of the interfaces between guiding rod and surrounding dielectrics, should attenuate the mode with polarization parallel to that interface. As an alternative, the guide can be made to support only the fundamental mode E_{11}^y by substituting medium 2 with a low impedance medium such as a dielectric with large refractive index or a metal.

An example of such a guide and the propagation constant of its modes are shown in Fig. 8. By choosing

$$a < \frac{0.7\lambda}{(n_1^2 - n_4^2)^{\frac{1}{2}}}$$

only the E_{11}^y mode is guided. If the metal is not perfect, there is power leakage into the low impedance medium. The smaller that impedance, the smaller the leakage.

Guides for integrated optics may be easier to build with $a/b \gg 1$. We can use Fig. 7 to design a guide of arbitrary dimensions a and b which is

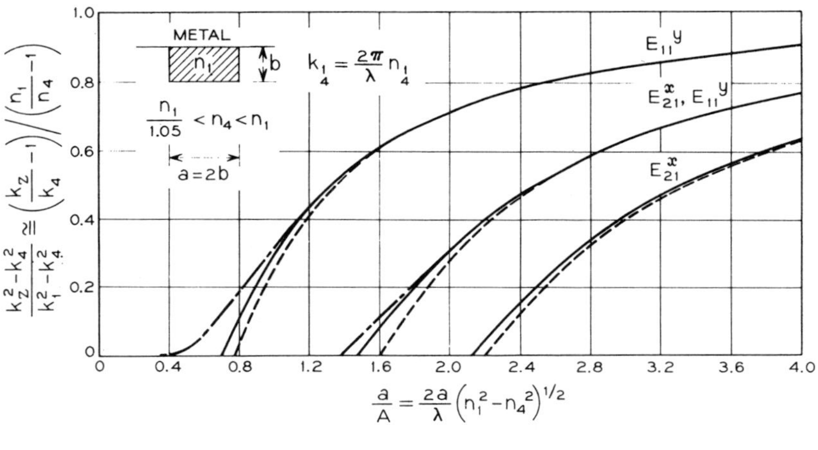

Fig. 8 — Propagation constant for modes in a guide surrounded by metal and dielectrics. ———— transcendental equation solutions; — — — — closed form solutions; —·—·— Goell's computer solutions of the boundary value problem.

still capable of supporting only the E_{11}^x and E_{11}^y modes. An as example, let us calculate what the values

$$n_3 = n_5 = n_1(1 + \Delta) \quad \text{and} \quad n_2 = n_4 = n_1(1 + \Delta')$$

should be, assuming

$$\Delta,\ \Delta' \ll 1, \quad \text{and} \quad \frac{a}{b} = 5.$$

Choosing

$$\left(\frac{\Delta'}{\Delta}\right)^{\frac{1}{2}} = \frac{a}{b} = 5, \tag{31}$$

one derives from Fig. 7

$$C = \left(\frac{a}{b}\right)^2 = 25.$$

The curve corresponding to $C = 25$ has been plotted as a dotted line in Fig. 7. It intercepts the E_{21}^y line at

$$Y = \left[\frac{b}{\pi} + \frac{1}{\pi k n_1}\left(\frac{2}{\Delta'}\right)^{\frac{1}{2}}\right]^{-2} (k_1^2 - k_z^2)^{-1} = 0.88.$$

In this expression, by making

$$k_z = k n_1(1 - \Delta),$$

the guide supports only the E_{11}^y and E_{11}^x modes; its height is then

$$b = 1.66 \frac{\lambda}{n_1(\Delta')^{\frac{1}{2}}}. \tag{32}$$

We can choose b arbitrarily by the proper selection of $\triangle'$.

For

$$\lambda = 1\mu \; n_1 = 1.5, \text{ and } b = 5\mu,$$

from equations (31) and (32) we obtain

$$a = 25\mu, \; \Delta = 0.002, \text{ and } \Delta' = 0.05.$$

IV. DIRECTIONAL COUPLER

In general, the directional coupler can transmit E^x_{pq} and E^y_{pq} modes; but if the sides a and b of the guides are selected small enough, only the fundamental modes E^x_{11} and E^y_{11} are guided. Let us concentrate on the E^y_{11} mode. The coupler guides two kinds of E^y_{11} modes: one is symmetric (Fig. 9c) while the other is antisymmetric (Fig. 9d). Both are essentially TEM modes with main field components E_y and H_x. The electric and magnetic field intensity profiles for both modes are depicted qualitatively in Figs. 9b, c, and d.

Ignoring the small effects introduced by the loose coupling, the electrical width $k_x a$ and height $k_y b$ of each guide, as well as the field penetrations $\underset{5}{\xi_3}$ and $\underset{4}{\eta_2}$, coincide with those of the guide described in Section III. Similar reasoning applies to the E^x_{11} mode.

The coupling coefficient K between the two guides and the length L necessary for complete transfer of power from one to the other are, according to equations (56) and (59),[12]

$$-iK = \frac{\pi}{2L} = 2 \frac{k_x^2 \xi_5}{k_z a} \frac{\exp(-c/\xi_5)}{1 + k_x^2 \xi_5^2}. \tag{33}$$

For E^y_{pq} modes, k_z and ξ_5 are given in equations (3) and (8), and k_x is the solution of equation (6). Similarly, for E^x_{pq} modes, k_z, ξ_5, and k_x are obtained from equations (17), (18), and (20). As expected, the coupling decreases exponentially with the ratio c/ξ_5 between the guide's separation and the field penetration in medium 5.

The normalized coupling coefficient

$$\frac{|K|\, a}{\left[1 - \left(\frac{n_5}{n_1}\right)^2\right]^{\frac{1}{2}}} \frac{k_z}{k_1} = \frac{\pi}{2} \frac{a}{L} \frac{1}{\left[1 - \left(\frac{n_5}{n_1}\right)^2\right]^{\frac{1}{2}}} \frac{k_z}{k_1}$$

$$= 2\left(\frac{k_x A_5}{\pi}\right)^2 \left[1 - \left(\frac{k_x A_5}{\pi}\right)^2\right]^{\frac{1}{2}} \exp\left\{-\pi \frac{c}{A_5}\left[1 - \left(\frac{k_x A_5}{\pi}\right)^2\right]^{\frac{1}{2}}\right\} \tag{34}$$

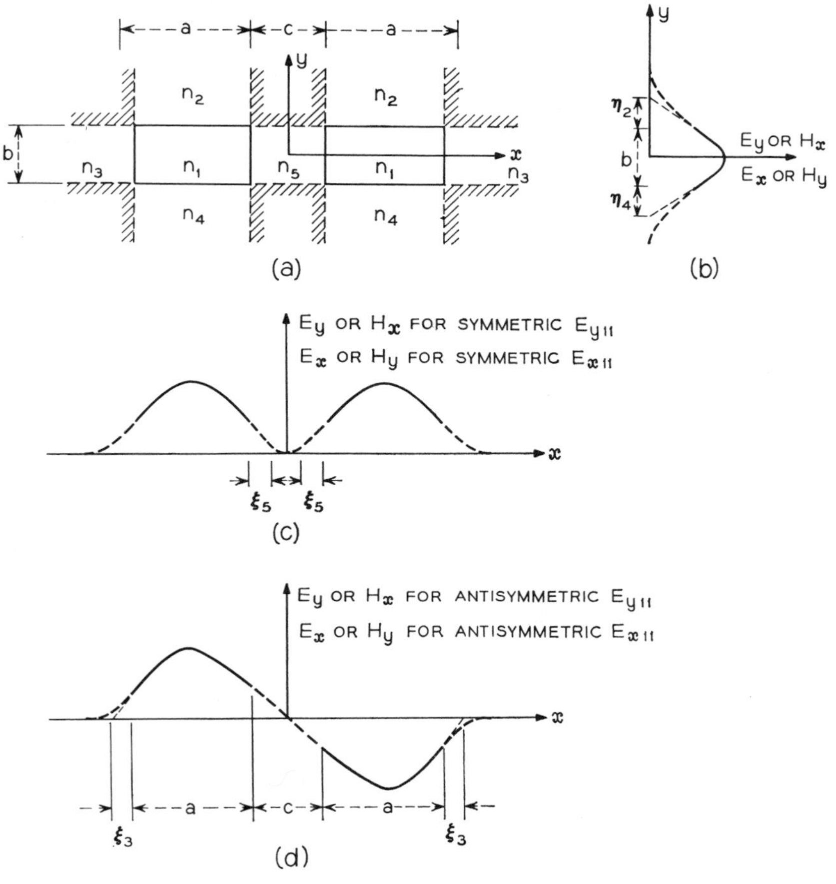

Fig. 9 — Directional coupler immersed in several dielectrics: (a) cross section, (b), (c), and (d) field distributions.

derived from equation (33) by substituting ξ_5 for its value given in equation (8) has been plotted in Fig. 10 for the E^y_{1q} mode, assuming $n_3 = n_5$ and n_1/n_5 is arbitrary. The solid and dotted lines were obtained using the exact solution of (6) and the approximate expression (12), respectively, for k_x. Both sets of curves are close to each other, especially for $2a/\lambda(n_1^2 - n_5^2)^{\frac{1}{2}} \geqq 1$.

The dashed-dotted lines are the couplings obtained by A. L. Jones[5] for two parallel cylinders of refractive index $n_1 = 1.8$ embedded in a medium $n_5 = 1.5$.[5] As expected, if the diameters of the round guides are equal to the widths of the rectangular guides, and if the separations are the same, the coupling between the round guides should be slightly smaller than that between the rectangular ones.

The normalized coupling equation (34) for the E^x_{1q} mode has been plotted in Fig. 11, using for k_x the exact solution of equation (20). For n_1/n_5 close to unity, the lines get close to the solid curves in Fig. 10 as the E^y_{1q} and E^x_{1q} modes approach degeneracy. The influence of the height b of the guides, the refractive indices n_2 and n_4, and the value of q in the coupling of either mode is not important since they only affect k_z.

To work some examples, assume

$$n_1 = 1.5, \qquad n_2 = n_3 = n_4 = n_5 = \frac{1.5}{1.01}, \quad \text{and} \quad a = 2b.$$

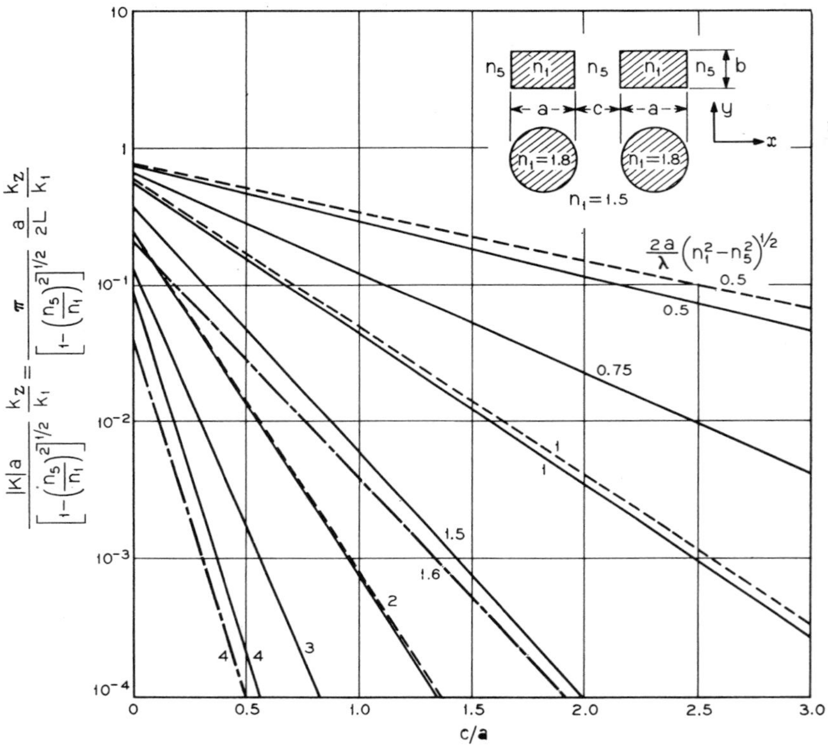

Fig. 10 — Coupling coefficient for $E_{1q}{}^y$ modes. ———— coupling calculated from trancendental equations; — — — — closed form approximations; —·—·— coupling between two cylindrical rods (A. L. Jones[5]).

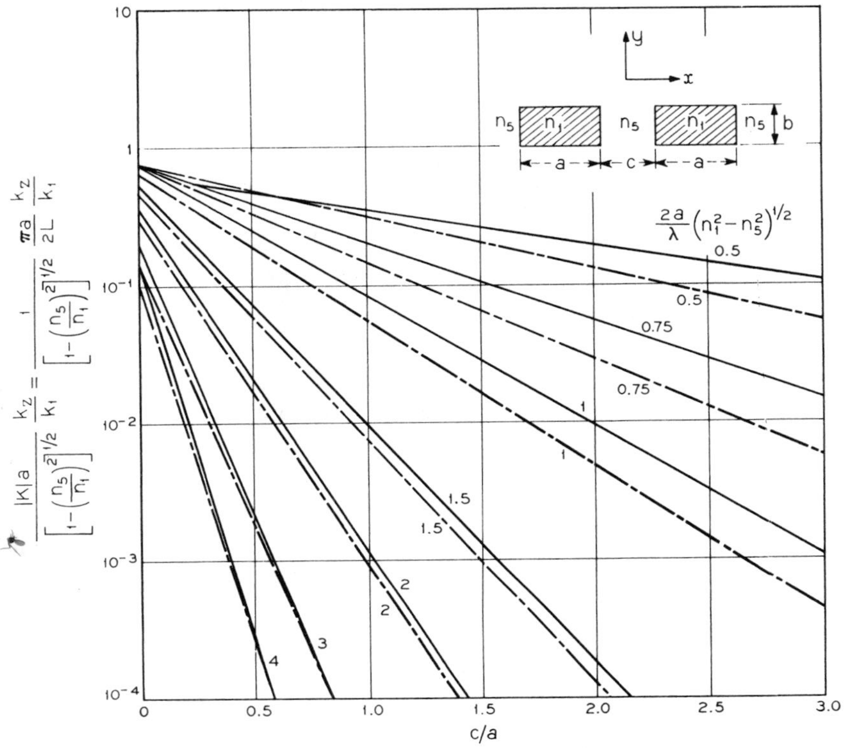

Fig. 11 — Coupling coefficient for E_{1q}^x modes. ———— E_{1q}^x coupling for $n_1/n_5 = 1.5$; —·—·— E_{1q}^x coupling for $n_1/n_5 = 1.1$.

To insure that each guide only supports the E_{11}^x and E_{11}^y modes, the normalized dimension b according to Fig. 6b must be chosen to be

$$\frac{2b}{\lambda}(n_1^2 - n_4^2)^{\frac{1}{2}} = 0.75.$$

Consequently

$$b = 1.77\lambda, \qquad a = 3.54\lambda, \quad \text{and} \quad \frac{k_z}{k_1} \cong 1.$$

From Fig. 10 we obtain the coupler length L for complete power transfer:

$$L = 6540\lambda \quad \text{for} \quad c = a \quad \text{and} \quad L = 262\lambda \quad \text{for} \quad c = \frac{a}{4}.$$

How far apart should two guides of length l be spaced to have small coupling? If the transfer coefficient $|T| = l|K| \ll 1$, from equation (33) we derive

$$c = \xi_5 \log\left[2 \frac{l}{|T|} \frac{k_x^2}{k_z} \frac{\xi_5}{a} \frac{1}{1 + k_x^2 \xi_5^2}\right]. \tag{35}$$

For the same guide dimensions of the previous example and for

$$l = 1 \text{ cm}, \lambda = 1\mu, \quad \text{and} \quad T = 0.01,$$

we derive, from either equation (35) or Fig. 10, that $c/a = 2.5$. Consequently, both guides 3.54μ wide and 1 cm long would couple -40dB if their separtion is 8.9μ.

Now we evaluate how a small change of the refractive index between the guides modifies their coupling. Such would be the case if the medium between the guides is, for example, an electrooptic material and we change the applied field to modulate or switch the output.

For E_{11}^x and E_{11}^y modes, assuming well-guided modes ($k_x A_5/\pi \ll 1$) and $n_1 - n_5/n_1 \ll 1$, the ratio between couplings for two values of refractive index in medium 5 (for example, n_5 and $n_5(1 + \delta)$), result from equations (34) and (12):

$$\frac{K_1}{K_2} = \frac{L_2}{L_1} = \exp\left\{-\pi\left(\frac{n_1^2}{n_5^2} - 1\right)^{-1} \frac{c\delta}{A_5}\left[1 - \left(\frac{2}{\pi} + \frac{a}{A_5}\right)^{-2}\right]^{\frac{1}{2}}\right\}. \tag{36}$$

That ratio is 1/2 if

$$\delta = 0.22\left(\frac{n_1^2}{n_5^2} - 1\right)\frac{A_5}{c}\left[1 - \left(\frac{2}{\pi} + \frac{a}{A_5}\right)^{-2}\right]^{-\frac{1}{2}}. \tag{37}$$

A directional coupler with coupling coefficient K_1 and length $L = \pi/|2K_1|$ would transfer all the power from one guide to the other. If the refractive index of the medium between the guides was changed from n_5 to $n_5(1+\delta)$ such that equation (37) is satisfied, the power would emerge at the end of the input guide. The larger the separation c of the guides, and the smaller the difference of refractive indexes $n_1 - n_5$, the smaller the change of refractive index required.

Following the example above, for

$$n_1 = 1.5, \qquad n_2 = n_3 = n_4 = n_5 = \frac{1.5}{1.01},$$

$$a = 1.5A_5 = 3.54\lambda, \quad \text{and} \quad c = a,$$

the percentage change of index required is only $\delta = 0.0033$.

V. DIRECTIONAL COUPLER MADE WITH SLIGHTLY DIFFERENT GUIDES

Consider the directional coupler of Fig. 12 in which the two guides have slightly different heights: one measures $b + h$ and the other $b - h$.

Let us qualitatively plot the coupling coefficient as a function of h, Fig. 13. Because of simple arguments of symmetry, the absolute value of coupling coefficient is stationary (first derivative zero) around $h = 0$. Therefore, the coupling coefficient between two guides of height b_1 and b_2 is the same as that of the coupling between two identical guides of height $1/2(b_1 + b_2)$, provided that $|b_1 - b_2|$ is small enough.

This reasoning applies to guides with different widths, heights, and refractive indices, provided that the differences are small enough. Unfortunately, as in most perturbation analysis, we don't know what "small enough" is unless we calculate the next higher order term.

VI. SUMMARY AND CONCLUSIONS

A dielectric rod (Fig. 4a) of rectangular cross section a by b surrounded by different dielectrics supports, through total internal reflection, two families of hybrid modes. They are essentially TEM modes polarized either in the x or the y direction; we call them E^x_{pq} and E^y_{pq}. The sub-indices state the number of extrema (p in the x direction and q in the y direction) of the magnetic or electric transverse field components.

Dispersion curves for guides of different proportions and different surrounding dielectric are plotted in Figs. 6a through k. Typical dimensions for several guides capable of supporting only the fundamental modes E^x_{11} and E^y_{11} are contained in Table I.

By picking dielectrics with similar indexes, the guide dimensions can be made large compared with λ, thus reducing the tolerance requirements. The dimensions a and b can be picked arbitrarily and still achieve

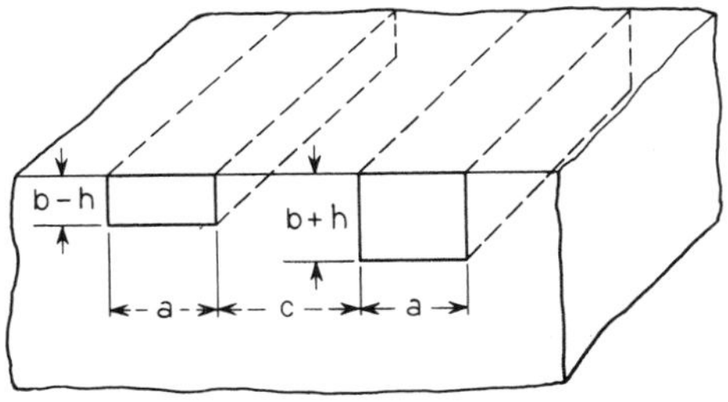

Fig. 12 — Directional coupler with guides of different heights.

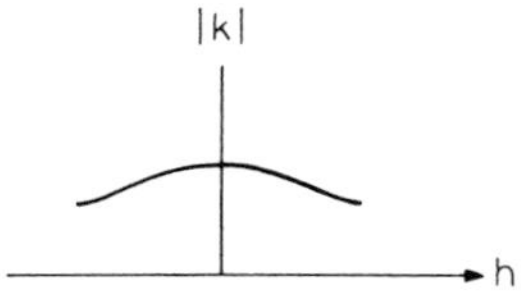

Fig. 13 — Qualitative behavior of the coupling coefficient as a function of h.

a guide which supports only the fundamental modes if one can choose the refractive indexes. The design is achieved with the help of either equation (14) or Fig. 7.

The penalty one pays with most of these guides is that the fundamental modes are almost degenerate; consequently, symmetry imperfections tend to couple them. A lossy layer added to the interface $y = b/2$ (Fig. 4a) should attenuate the E^x_{11} mode more than the E^y_{11}. As an alternative, the guide can be made to support only the E^y_{11} mode by metalizing the same interface. Dispersion curves are shown in Fig. 8.

Since the field is not confined, there is coupling between two of these guides (Fig. 3). Design curves for directional couplers are given in Figs. 10 and 11.

Typically, for $n_1 = 1.5$, $n_2 = n_3 = n_4 = n_5 = 1.5/1.01$, $a = 3.54\lambda$, $b = a/2 = 1.77\lambda$, and $c = a/4 = 0.88\lambda$, according to equation (33) the length necessary for 3dB coupling is $L/2 = 131\lambda$. This length increases exponentially with the separation between the guides.

Increasing the refractive index between the guides by a 3 per thousand doubles the coupling.

What is a reasonable separation to prevent coupling? Using the numbers of the previous example, two parallel guides 1 cm long separated by 2.5 times the width of each guide have a coupling of -40 dB.

The dielectric waveguides and the directional couplers described show great promise as basic elements for integrated optical circuitry because they:

(*i*) Can be made single mode even though their transverse dimensions can be large compared with the free space wavelength of operation. Consequently, the tolerance requirements can be relaxed.

(*ii*) Permit the building of compact optical components.

(*iii*) Are mechanically stable and alignment problems are minimized.

(*iv*) Are relatively simple structures and lend themselves to being fabricated with high precision integrated circuit techniques.

(*v*) Can include active devices of comparable small dimensions.

APPENDIX A

Field Analysis of the Directional Coupler

We solve Maxwell's equations for the directional coupler whose cross section is depicted in Fig. 3. The structure is symmetric with respect to the $x = 0$ plane; therefore, the modes have electric fields which are either symmetric or antisymmetric with respect to that plane. Consequently, the guide we have to study is simpler (Fig 14): if the plane $x = 0$ is an electric short circuit, the modes of the coupler propagating along z are antisymmetric; if the plane $x = 0$ is a magnetic short circuit, the modes are symmetric. As is known, it is the interaction of these symmetric and antisymmetric modes traveling with different phase velocities along z that represents the effect of coupling.

As discussed in Section II, by neglecting the power propagating through the shaded areas, the fields must be matched only along the sides of region 1. We find that two families of modes can satisfy the boundary conditions; we call them E^x_{pq} and E^y_{pq}. Each mode in the first family has most of its electric field polarized in the x direction, while each mode of the second family has the electric field almost completely polarized in the y direction. The subindexes p and q characterize the member of the family by the number of extrema that these transverse field components have along the x and y directions, respectively. For example, the E^x_{11} mode has its electric field virtually along x, its magnetic field along y; the amplitudes of the field have one maximum in each direction.

Each family of modes will be studied separately.

A.1 E^y_{pq} *Modes: Polarization Along y*

The field components in the νth of the five areas in Fig. 14 are:[13]

$$H_{x\nu} = \exp(-ik_z z + i\omega t)\begin{cases} M_1 \cos(k_x x + \alpha)\cos(k_y y + \beta) & \text{for } \nu = 1 \\ M_2 \cos(k_x x + \alpha)\exp(-ik_{y2}y) & \text{for } \nu = 2 \\ M_3 \cos(k_y y + \beta)\exp(-ik_{x3}x) & \text{for } \nu = 3 \\ M_4 \cos(k_x x + \alpha)\exp(ik_{y4}y) & \text{for } \nu = 4 \\ M_5 \cos(k_y y + \beta)\sin(k_{x5}x + \gamma) & \text{for } \nu = 5 \end{cases}$$

$$H_{y\nu} = 0$$

$$H_{z\nu} = -\frac{i}{k_z}\frac{\partial^2 H_{x\nu}}{\partial x\,\partial y} \tag{38}$$

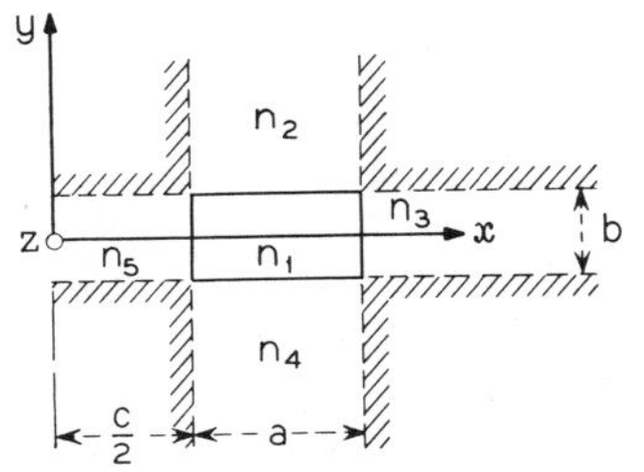

Fig. 14 — Coupler cross section with plane $x = 0$ either an electric or magnetic short circuit.

$$E_{x\nu} = -\frac{1}{\omega\epsilon n_\nu^2 k_z}\frac{\partial^2 H_{x\nu}}{\partial x\,\partial y}$$

$$E_{y\nu} = \frac{k^2 n_\nu^2 - k_{y\nu}^2}{\omega\epsilon n_\nu^2 k_z} H_{x\nu}$$

$$E_{z\nu} = \frac{i}{\omega\epsilon n_\nu^2}\frac{\partial H_{x\nu}}{\partial y}$$

in which M_ν determines the amplitude of the field in the νth medium; α and β locate the field maxima and minima in region 1; γ equal to 0° or 90° implies that the plane $x = 0$ is an electric (antisymmetric mode) or magnetic (symmetric mode) short circuit, respectively; ω is the angular frequency; ϵ and μ (appearing in $k^2 = \omega^2\epsilon\mu$) are the permittivity and permeability of free space.

In the νth medium the refractive index is n_ν, and the propagation constants $k_{x\nu}$, $k_{y\nu}$, and k_z are related by

$$k_{x\nu}^2 + k_{y\nu}^2 + k_z^2 = \omega^2\epsilon\mu n_\nu^2 = k_\nu^2 . \tag{39}$$

To match the fields at the boundaries between the region 1 and the regions 2 and 4, we have assumed in equation (38)

$$k_{x1} = k_{x2} = k_{x4} = k_x \tag{40}$$

and similarly to match the fields between media 1, 3, and 5,

$$k_{y1} = k_{y3} = k_{y5} = k_y . \tag{41}$$

Before finding the characteristic equations, let us assume the refractive index n_1 of the guide to be slightly larger than the others. That is

$$\frac{n_1}{n_{\substack{2\\3\\4\\5}}} - 1 \ll 1. \tag{42}$$

As a consequence only modes made of plane wavelets impinging at grazing angles on the surface of medium 1 are guided. Since this implies that

$$\begin{matrix} k_x \\ k_y \end{matrix} \ll k_z \,, \tag{43}$$

the field components E_x in equation (38) can be neglected.

Now we match the remaining tangential components along the edges of region 1 and from equation (38) we obtain

$$\tan\left(k_y \frac{b}{2} \pm \beta\right) = i \frac{n_1^2}{\begin{matrix} n_2^2 \\ 4 \end{matrix}} \frac{\begin{matrix} k_{y2} \\ 4 \end{matrix}}{k_y} . \tag{44}$$

$$\tan\left[k_x \begin{Bmatrix} \dfrac{c}{2} \\ a + \dfrac{c}{2} \end{Bmatrix} + \alpha\right] = i \frac{\begin{matrix} k_{x5} \\ 3 \end{matrix}}{k_x} \begin{bmatrix} ictn\left(k_{x5} \dfrac{c}{2} + \gamma\right) \\ 1 \end{bmatrix} . \tag{45}$$

Where there are two choices, the upper ones go together and the lower ones go together.

T. Li pointed out that each of these equations considered separately is the characteristic equation of a boundary value problem simpler than that of Fig. 14.[8, 9] Thus for a dielectric slab infinite in the x and z directions and refractive indexes as depicted in Fig. 15a, the characteristic equation for modes with no H_y component coincides with

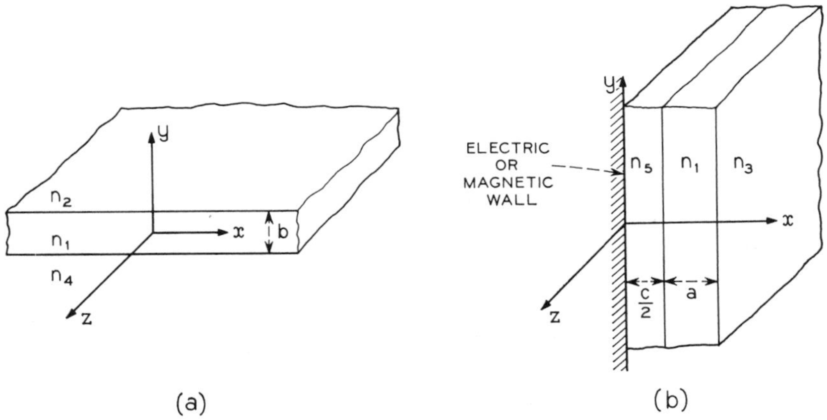

Fig. 15 — Dielectric slabs.

equation (44). Similarly, for two slabs infinite in the y and z directions and limited at $x = 0$ by an electric or magnetic short as in Fig. 15b, the characteristic equation for modes with $E_x = 0$ is equation (45).

A similar technique has been used by Schlosser and Unger to find the transmission properties of a rectangular dielectric guide immersed in another dielectric.[7] If the two guiding rods are so far apart that the coupling between them is a perturbation, then

$$| k_{x5}c | \gg 1 \tag{46}$$

and we can rewrite the characteristic equations (44) and (45) with the help of equations (39) and (46), making a and b explicit, as

$$k_y b = q\pi - \tan^{-1} \frac{n_2^2}{n_1^2} k_y \eta_2 - \tan^{-1} \frac{n_4^2}{n_1^2} k_y \eta_4 \tag{47}$$

$$k_x a = k_{x0} a \left[1 + \frac{2\xi_5}{a} \frac{\exp\left(-\dfrac{c}{\xi_5} - i2\gamma \right)}{1 + k_{x0}^2 \xi_5^2} \right] \tag{48}$$

where k_{x0} is the solution of

$$k_{x0} a = p\pi - \tan^{-1} k_{x0} \xi_3 - \tan^{-1} k_{x0} \xi_5 , \tag{49}$$

$$\eta_{\substack{2\\4}} = \frac{1}{\left| k_{y\substack{2\\4}} \right|} = \frac{1}{\left[\left(\dfrac{\pi}{A_{\substack{2\\4}}} \right)^2 - k_y^2 \right]^{\frac{1}{2}}} \tag{50}$$

$$\xi_{\substack{3\\5}} = \frac{1}{\left| k_{x\substack{3\\5}} \right|} = \frac{1}{\left[\left(\dfrac{\pi}{A_{\substack{3\\5}}} \right)^2 - k_{x0}^2 \right]^{\frac{1}{2}}} \tag{51}$$

and

$$A_{2,3,4,5} = \frac{\pi}{(k_1^2 - k_{2,3,4,5}^2)^{\frac{1}{2}}} = \frac{\lambda}{(n_1^2 - n_{2,3,4,5}^2)^{\frac{1}{2}}} . \tag{52}$$

In the transcendental equations (47) to (49), p and q are the arbitrary integers characterizing the order of the propagating mode, and the $\tan^{-1}$ functions are to be taken in the first quadrant. The angles $k_x a$ and $k_y b$ measure the phase shift of any field component across the guiding rod in the x and y directions respectively, or in other words, the electrical width and height of each guide of the coupler. On the other hand, $k_{x0} a$ is the electrical width of each guide assuming no interaction between the guides, that is assuming $c \to \infty$.

Let us find the physical significance of $\eta_{2,4}$ and $\xi_{3,5}$. The amplitude of each field component in medium 2 (Fig 14) decreases exponentially along y. It decays by $1/e$ in a distance η_2 given by equation (50). Similarly η_4, ξ_3, and ξ_5 measure the "penetration depth" of the field components in media 4, 3, and 5, respectively.

The propagation constant along z for each mode of the coupler is, according to equations (39), (40), and (41),

$$k_z = (k_1^2 - k_x^2 - k_y^2)^{\frac{1}{2}}. \tag{53}$$

With the help of equation (48), the slightly different propagation constants of the symmetric ($\gamma = 90°$) and antisymmetric modes ($\gamma = 0$) are

$$\left.\begin{matrix} k_{zs} \\ \\ k_{za} \end{matrix}\right\} = k_{z0}\left[1 \pm 2\frac{k_{x0}^2}{k_{z0}^2}\frac{\xi_5}{a}\frac{\exp(-c/\xi_5)}{1 + k_{x0}^2\xi_5^2}\right]. \tag{54}$$

In this expression

$$k_{z0} = (k_1^2 - k_{x0}^2 - k_y^2)^{\frac{1}{2}} \tag{55}$$

is the propagation constant of the E_{pq}^y mode of a single guide ($c \rightarrow \infty$).

The coupling coefficient K between the two guides and the length L necessary for complete transfer of power from one to the other are related to the propagation constants k_{zs} and k_{za} by[12]

$$\begin{aligned} -iK &= \frac{\pi}{2L} = \frac{k_{zs} - k_{za}}{2} = 2\frac{k_{x0}^2}{k_{z0}}\frac{\xi_5}{a}\frac{\exp(-c/\xi_5)}{1 + k_{x0}^2\xi_5^2} \\ &= \frac{2}{\pi}\frac{A_5 k_{x0}^2}{a k_{z0}}\left[1 - \left(\frac{k_{x0}A_5}{\pi}\right)^2\right]^{\frac{1}{2}} \exp\left\{-\frac{\pi c}{A_5}\left[1 - \left(\frac{k_{x0}A_5}{\pi}\right)^2\right]^{\frac{1}{2}}\right\}. \end{aligned} \tag{56}$$

As expected, the coupling increases exponentially both by decreasing c and by increasing the penetration depth ξ_5 in medium 5.

All these formulas contain either k_{x0} or k_y, which are solutions of the transcendental equations (47) and (49). For well-guided modes, most of the power travels within medium 1 and consequently

$$\left(\frac{k_{x0}A_{\substack{3\\5}}}{\pi}\right)^2 \ll 1 \tag{57}$$

and

$$\left(\frac{k_y A_{\substack{2\\4}}}{\pi}\right)^2 \ll 1. \tag{58}$$

It is possible then to solve those transcendental equations in a closed though approximate form by expanding the $\tan^{-1}$ functions in power of those small quantities and keeping the first two terms of the expansions. The explicit solutions of equations (47), (49), (50), (51), (55), and (56) are given in Section III.

A.2 E^x_{pq} *Modes: Polarization in the x Direction*

The field components and propagation constants can be derived from those in Section A.1 by changing E to H and μ to $-\epsilon$, and vice versa. Except for their polarizations, the E^x_{pq} and E^y_{pq} modes are very similar and have comparable propagation constants. Using boldface type to distinguish the symbols corresponding to E^x_{pq} modes, from equations (56), (55), (47), (49), (50), and (51), we obtain

$$-i\mathbf{K} = \frac{\pi}{2\mathbf{L}} = 2\frac{\mathbf{k}^2_{x0}}{\mathbf{k}_{z0}}\frac{\boldsymbol{\xi}_5}{a}\frac{\exp(-c/\boldsymbol{\xi}_5)}{1 + (\mathbf{k}_{x0}\boldsymbol{\xi}_5)^2} \tag{59}$$

where

$$\mathbf{k}_{z0} = (k_1^2 - \mathbf{k}^2_{x0} - \mathbf{k}^2_y)^{\frac{1}{2}} \tag{60}$$

and $\mathbf{k}_{x0}$ and $\mathbf{k}_y$ are solutions of the transcendental equations

$$\mathbf{k}_y b = q\pi - \tan^{-1}\mathbf{k}_y\boldsymbol{\eta}_2 - \tan^{-1}\mathbf{k}_y\boldsymbol{\eta}_4 \tag{61}$$

and

$$\mathbf{k}_{x0}a = p\pi - \tan^{-1}\frac{n_3^2}{n_1^2}\mathbf{k}_{x0}\boldsymbol{\xi}_3 - \tan^{-1}\frac{n_5^2}{n_1^2}\mathbf{k}_{x0}\boldsymbol{\xi}_5 \tag{62}$$

in which

$$\boldsymbol{\eta}_{\substack{2\\4}} = \frac{1}{\left[\left(\dfrac{\pi}{A_{\substack{2\\4}}}\right)^2 - \mathbf{k}^2_y\right]^{\frac{1}{2}}} \tag{63}$$

and

$$\boldsymbol{\xi}_{\substack{3\\5}} = \frac{1}{\left[\left(\dfrac{\pi}{A_{\substack{3\\5}}}\right)^2 - \mathbf{k}^2_{x0}\right]^{\frac{1}{2}}} \tag{64}$$

As in Section A.1, the transcendental equations (61) and (62) can be solved in closed, though approximate, form provided that

$$\left(\frac{\mathbf{k}_{x0}A_{\substack{3\\5}}}{\pi}\right)^2 \ll 1 \tag{65}$$

and

$$\left(\frac{\frac{\mathbf{k}_y A_2}{4}}{\pi}\right)^2 \ll 1. \tag{66}$$

The explicit results are given in Section III.

REFERENCES

1. Miller, S. E., "Integrated Optics: An Introduction," B.S.T.J., this issue, pp. 2059–2069.
2. Schineller, E. R., "Summary of the Development of Optical Waveguides and Components," Report 1471, Wheeler Laboratories, Inc., April 1967.
3. Kaplan, R. A., "Optical Waveguide of Macroscopic Dimension in Single-Mode Operation," Proc. IEEE, *51,* No. 8 (August 1963), pp. 1144–1145.
4. Goell, J. E., "A Circular-Harmonic Computer Analysis of Rectangular Dielectric Waveguides," B.S.T.J., this issue, pp. 2133–2160.
5. Jones, A. L., "Coupling of Optical Fibers and Scattering in Fibers," J. Opt. Soc. Amer., *55,* No. 3 (March 1965), pp. 261–271.
6. Bracey, M. F., Cullen, A. L., Gillespie, E. F. F., and Staniforth, J. A., "Surface-Wave Research in Sheffield," I.R.E. Trans. Antennas and Propagation, *AP-7,* Special Supplement (December 1959), pp. S219-S225.
7. Schlosser, W., and Unger, H. G., "Partially Filled Waveguides and Surface Waveguides of Rectangular Cross Section," *Advances in Microwaves,* New York: Academic Press, 1966, pp. 319–387.
8. Collin, R. E., *Field Theory of Guided Waves,* New York: McGraw-Hill, 1966, pp. 470–477.
9. Nelson, D. F., and McKenna, J., "Electromagnetic Modes of Anisotropic Dielectric Waveguides at p-n Junctions," J. Appl. Phys., *38,* No. 10 (September 1967), pp. 4057–4074.
10. Fox, A. G., and Li, T., "Resonant Modes in a Maser Interferometer," B.S.T.J., *40,* No. 2 (March 1961), pp. 453–488.
11. Marcatili, E. A. J., "Bends in Optical Dielectric Guides," B.S.T.J., this issue, pp. 2103–2132.
12. Miller, S. E., "Coupled Wave Theory and Waveguide Applications," B.S.T.J., *33,* No. 3 (May 1954), pp. 661–719.
13. Schelkunoff, S. A., *Electromagnetic Waves,* New York: D. van Nostrand, 1943, p. 94.

Wave Propagation in Thin-Film Optical Waveguides Using Gyrotropic and Anisotropic Materials as Substrates

SHYH WANG, MEMBER, IEEE, MANHAR L. SHAH, AND JOHN D. CROW, STUDENT MEMBER, IEEE

Abstract—**Wave propagation in optical waveguides on substrates of magnetic, optically active, or birefringent material is analyzed. The conditions for TE $\rightleftarrows$ TM mode conversion are derived and computer calculations showing the characteristic of a quartz-substrate mode converter is obtained. A mathematical analysis of a gyrator using two mode converters in series is given and the schemes for an isolator, modulator, optical switch, and optical read-out devices are mentioned.**

I. Introduction

In a thin-film optical waveguiding structure consisting of a thin film on a substrate material, the optical properties of the thin film and substrate may be used to control the propagation of the light in the film. The difficulty in controlling the optical properties of a thin film (often a single crystal material is required) makes the substrate an attractive medium to use to perform optical control functions. We have undertaken a study of wave propagation in an optical waveguide using a gyrotropic or anisotropic material as the substrate. Our theoretical study predicts that TE $\rightleftarrows$ TM mode conversion is possible in these structures. Once mode conversion is achieved, devices like the gyrator, isolator, and optical readout are possible. By utilizing the electrooptic and magnetooptic effects, the dielectric properties of the substrate can be externally controlled, and the modulator and optical switch are possible.

Manuscript received June 4, 1971; revised August 16, 1971. This research was sponsored by the U. S. Army Research Office, Durham, N. C., under Grant DA-ARO-D-31-124-71-G 36, the National Science Foundation under Grant GK-13197, and the Joint Services Electronics Program under Grant AF-AFOSR-68-1488.

The authors are with the Department of Electrical Engineering and Computer Sciences and the Electronics Research Laboratory, University of California, Berkeley, Calif. 94720.

II. Wave Propagation and Conditions for Complete Mode Conversion

We consider an optical waveguide structure as shown in Fig. 1, where all the materials are assumed lossless. The film is isotropic with an index of refraction $n_f = \sqrt{\epsilon_f/\epsilon_0}$. The substrate material is gyrotropic or anisotropic and, for generality, we take the top layer as anisotropic. We will

Reprinted from *IEEE J. Quantum Electron.*, vol. QE-8, part 2, pp. 212–216, Feb. 1972.

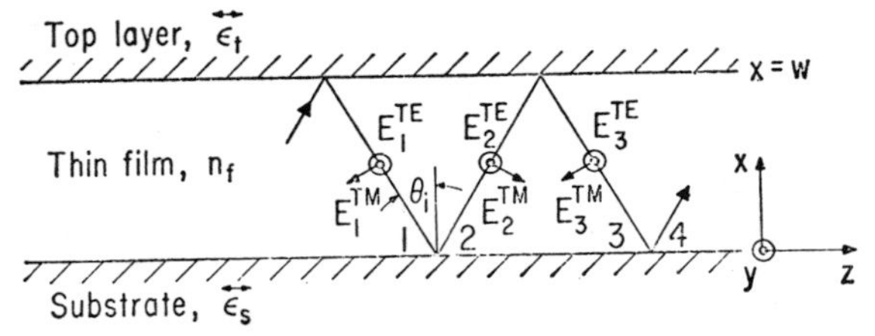

Fig. 1. Light-guiding structure showing the E-field components of the TE and TM modes of the light wave and the angle of incidence θ_i.

only consider mode converting mechanisms as taking place in the substrate so the top layer material will be assumed oriented such that its dielectric tensor $\bar{\epsilon}_t$ has only diagonal elements. The dielectric tensors for the substrate $\bar{\epsilon}_s$ and top layer are therefore

$$\bar{\epsilon}_s = \epsilon_0 \bar{\kappa}_s = \epsilon_0 \begin{bmatrix} K_{11} & K_{12} & K_{13} \\ K_{12}^* & K_{22} & K_{23} \\ K_{13}^* & K_{23}^* & K_{33} \end{bmatrix} \tag{1a}$$

$$\bar{\epsilon}_t = \epsilon_0 \bar{\kappa}_t = \epsilon_0 \begin{bmatrix} K_1 & 0 & 0 \\ 0 & K_2 & 0 \\ 0 & 0 & K_3 \end{bmatrix} \tag{1b}$$

where K_{12}, K_{13}, and K_{23} are real for an anisotropic substrate and they are imaginary for a gyrotropic substrate. All the diagonal elements are real. There are two cases for the form of $\bar{\epsilon}_s$ that are useful for mode conversion. Adopting the terminology of the magnetooptic Kerr effect they are: 1) $K_{13} = K_{23} = 0$, the longitudinal configuration; and 2) $K_{12} = K_{13} = 0$, the polar configuration.

A light wave is propagated in the film with the form $\exp[j\omega t - jk_0(\beta z \mp bx)]$ where $\beta = n_f \sin \theta_i$, $b = n_f \cos \theta_i$, k_0 = free space wavenumber, and θ_i = incidence angle as shown in Fig. 1. For light guiding, the wave will be totally reflected from the substrate and top layer, so the fields of the wave in these media will have the form $\exp[j\omega t - jk_0\beta z \pm k_0 p x]$ where $\beta = n_f \sin \theta_i$ as required by Snell's law, and $k_0 p$ is the decay constant of the evanescent field.

Guided waves in isotropic dielectrics [1], reflection and refraction from anisotropic and gyrotropic media [2], and modes in the anisotropic region of a p-n junction [3] have all been thoroughly discussed, so we will only briefly mention the approach as applied to light-guiding structures. The two eigenmodes of the substrate, obtained from the wave equation $k^2\boldsymbol{E} - \boldsymbol{k}(\boldsymbol{k}\cdot\boldsymbol{E}) = k_0^2\bar{\kappa}_s\boldsymbol{E}$, where $\boldsymbol{k} = k_0(-\hat{x}jp + \hat{z}\beta)$, are characterized by decay constants p_1 and p_2

$$p_{1,2}^2 = \frac{1}{2K_{11}} \{[K_{33}(\beta^2 - K_{11}) + K_{11}(\beta^2 - K_{22}) + |K_{12}|^2] \pm \sqrt{[K_{33}(\beta^2 - K_{11}) - K_{11}(\beta^2 - K_{22})]^2 + 2|K_{12}|^2[K_{33}(\beta^2 - K_{11}) + K_{11}(\beta^2 - K_{22})] + 4|K_{12}|^2 K_{11}K_{33} + |K_{12}|^4}\} \tag{2a}$$

for the longitudinal configuration, and

$$p_{1,2}^2 = \frac{1}{2K_{11}} \{[K_{33}(\beta^2 - K_{11}) + K_{11}(\beta^2 - K_{22})] \pm \sqrt{[K_{33}(\beta^2 - K_{11}) - K_{11}(\beta^2 - K_{22})]^2 - 4|K_{23}|^2 K_{11}(\beta^2 - K_{11})}\} \tag{2b}$$

for the polar configuration. These eigenmodes are not pure TE ($E \perp$ plane of incidence) or pure TM ($E \parallel$ plane of incidence) so that a pure TE or TM wave in the film cannot remain pure TE or TM upon total reflection from the substrate-film interface due to the boundary conditions. The reflected fields are therefore expressed by the scattering matrix

$$\begin{bmatrix} E_2^{TE} \\ E_2^{TM} \end{bmatrix} = \begin{bmatrix} r_{EE} & r_{EM} \\ r_{ME} & r_{MM} \end{bmatrix} \begin{bmatrix} E_1^{TE} \\ E_1^{TM} \end{bmatrix}. \tag{3}$$

For the longitudinal configuration

$$r_{EE} = F^{-1}(G_1 J_2^* L_1^{-1} - G_2 J_1^* L_2^{-1}) \tag{4a}$$

$$r_{EM} = j2n_f \cos\theta_i (p_1 - p_2)(F\beta K_{12})^{-1} \tag{4b}$$

$$r_{ME} = j2n_f \cos\theta_i (p_1 - p_2)(F\beta K_{12}^*)^{-1} \tag{4c}$$

$$r_{MM} = F^{-1}(G_1^* J_2 L_1^{-1} - G_2^* J_1 L_2^{-1}) \tag{4d}$$

where

$$G_{1,2} = K_{33}\cos\theta_i - jn_f p_{1,2} \tag{5a}$$

$$J_{1,2} = n_f \cos\theta_i - jp_{1,2} \tag{5b}$$

$$L_{1,2} = \beta^2 K_{33} - K_{11}K_{33} - K_{11}p_{1,2}^2 \tag{5c}$$

$$F = G_1 J_2 L_1^{-1} - G_2 J_1 L_2^{-1}. \tag{5d}$$

For the polar configuration r_{EE}, r_{MM}, J, L, and F remain unchanged, while r_{EM}, r_{ME}, and G become:

$$r_{EM} = -2n_f \cos\theta_i (p_1 - p_2)(FK_{23}^*)^{-1} \tag{6a}$$

$$r_{ME} = 2n_f \cos\theta_i (p_1 - p_2)(FK_{23})^{-1} \tag{6b}$$

$$G_{1,2} = K_{11}p_{1,2}\cos\theta_i - j(\beta^2 - K_{11})n_f. \tag{7}$$

The elements of the scattering matrix (3) for the substrate-film interface can be expressed in complex polar form as $r_{EE} = |r_{EE}| \exp(j\phi_{EE})$, $r_{EM} = |r_{EM}| \exp(j\phi_{EM})$, etc. It can be shown that $|r_{EE}| = |r_{MM}| \equiv r$, $|r_{EM}| = |r_{ME}| \equiv r'$; and $r^2 + r'^2 = 1$ as expected by conservation of energy. We note here that r', which is a measure of the converted mode amplitude, is determined primarily by two physical factors. A larger magnitude of the off-diagonal tensor element K_{ij}, or a greater penetration depth of the evanescent waves into the substrate will increase r'. An important feature of (3) is that $F^2 r_{EE} r_{MM}$ and $F^2 r_{EM} r_{ME}$ are real and have opposite signs. Physically, this means that after two mode-converting reflections the TE → TM → TE (for example) will be 180° out of phase with the original TE so that a complete depletion of the original mode, i.e., complete mode conversion, should be possible.

To further investigate the possibility of complete mode conversion we must study the phase terms of (3). These phase terms (suffered during total reflection at the substrate boundary) must be combined with the phase shifts of the waves suffered during propagation in the film and during total reflection at the top layer boundary in such a way as to ensure a constructive buildup of the converted mode power. Since the top layer is assumed to have no off-diagonal dielectric tensor elements, and hence no mode conversion, the scattering matrix for the film-top layer interface is

$$\begin{bmatrix} E_3^{\mathrm{TE}} \\ E_3^{\mathrm{TM}} \end{bmatrix} = \begin{bmatrix} \exp(j\phi_{\mathrm{E}}) & 0 \\ 0 & \exp(j\phi_{\mathrm{M}}) \end{bmatrix} \begin{bmatrix} E_2^{\mathrm{TE}} \\ E_2^{\mathrm{TM}} \end{bmatrix}. \tag{8}$$

For one complete zig-zag path (1-2-3 in Fig. 1) we can express the fields at *3* in terms of the fields at *1* by using (3) and (8), and defining some new parameters in order to bring out the symmetry of the scattering matrix. This scattering matrix is

$$\begin{bmatrix} E_3^{\mathrm{TE}} \\ E_3^{\mathrm{TM}} \end{bmatrix} = \exp(-j\psi) \cdot \begin{bmatrix} r\exp\{-j(\phi_a+\phi_b)\} & r'\exp\{j(\phi_a'-\phi_b)\} \\ -r'\exp\{-j(\phi_a'-\phi_b)\} & r\exp\{j(\phi_a+\phi_b)\} \end{bmatrix} \begin{bmatrix} E_1^{\mathrm{TE}} \\ E_1^{\mathrm{TM}} \end{bmatrix} \tag{9}$$

where $\psi = 2k_0bW - \psi_a - \psi_b$ with $\psi_a = (\phi_{\mathrm{MM}} + \phi_{\mathrm{EE}})/2$, $\psi_b = (\phi_{\mathrm{M}} + \phi_{\mathrm{E}})/2$, $2k_0bW$ is the phase shift suffered by the wave traveling from the substrate to the top layer and back, $\phi_a = (\phi_{\mathrm{MM}} - \phi_{\mathrm{EE}})/2$, $\phi_a' = \phi_{\mathrm{EM}} - \psi_a$, and $\phi_b = (\phi_{\mathrm{M}} - \phi_{\mathrm{E}})/2$. We notice from (4)–(7) that either $\phi_a' = 0$ or $\phi_a' = \pi/2$.

From (9), light propagation in the guide can be characterized by the matrix $\exp(-j\psi)\,[R]$ where

$$[R] = \begin{bmatrix} r\exp(-j\phi) & r'\exp(j\phi') \\ -r'\exp(-j\phi') & r\exp(j\phi) \end{bmatrix}, \tag{10}$$

$$\phi = \phi_a + \phi_b, \quad \phi' = \phi_a' - \phi_b.$$

That is, the fields at any point in the guide can be related to the fields at another point a distance $q2W \tan\theta_i$ (q = integer) down the guide by

$$\begin{bmatrix} E_q^{\mathrm{TE}} \\ E_q^{\mathrm{TM}} \end{bmatrix} = \exp(-jq\psi)[R]^q \begin{bmatrix} E_1^{\mathrm{TE}} \\ E_1^{\mathrm{TM}} \end{bmatrix} \tag{11}$$

where, by application of Sylvester's theorem [4],

$$[R]^q = \begin{bmatrix} \cos q\alpha - jr\sin\phi\dfrac{\sin q\alpha}{\sin\alpha} & r'\exp(j\phi')\dfrac{\sin q\alpha}{\sin\alpha} \\ -r'\exp(-j\phi')\dfrac{\sin q\alpha}{\sin\alpha} & \cos q\alpha + jr\sin\phi\dfrac{\sin q\alpha}{\sin\alpha} \end{bmatrix} \tag{12}$$

and $r\cos\phi = \cos\alpha$. Because $r\cos\phi < 1$ and usually $r' \ll 1$, significant mode conversion can result only when $\sin\alpha = \sqrt{1 - r^2\cos^2\phi}$ is small, i.e., $\cos\phi \approx 1$.

If we define $\delta\phi \equiv 2(\phi - m\pi)$ (m = integer), the condition for significant mode conversion becomes $\delta\phi = (\phi_{\mathrm{MM}} + \phi_{\mathrm{M}}) - (\phi_{\mathrm{EE}} + \phi_{\mathrm{E}}) - 2m\pi \approx 0$. Since ϕ_{EE}, ϕ_{MM}, ϕ_{E}, and ϕ_{M} are all less than π the only way we can satisfy $\delta\phi = 0$ is to take $m = 0$. Hence significant mode conversion can take place only among degenerate (or near degenerate) modes. Now, the above analysis implicitly assumes a single TE and TM mode light guide because we have considered only one value of θ_i; and a single TE and TM mode light guide can be realized experimentally by the proper choice of film thickness W. However our analysis will remain valid for the case of a low-mode-density light guide, because the mode conversion among the modes with different mode number or with different θ_i will be negligible due to the phase matching condition $\delta\phi = 0$. Under the condition $\delta\phi = 0$,

$$[R]^q = \begin{bmatrix} \cos q\alpha & \exp(j\phi')\sin q\alpha \\ -\exp(-j\phi')\sin q\alpha & \cos q\alpha \end{bmatrix}. \tag{13}$$

We can define a distance l_0 as the distance for complete mode conversion. From (13) we find that $l_0 = (\pi W \tan\theta_m)/\sqrt{G_0}$ where $\theta_m = \theta_i$, and $\sqrt{G_0} = r' = |r_{\mathrm{EM}}|$ when $\delta\phi = 0$.

III. Mode-Converting Structures

The condition $\delta\phi \approx 0$ cannot be met with an isotropic substrate and top layer since $\phi_{\mathrm{MM}} > \phi_{\mathrm{EE}}$ and $\phi_{\mathrm{M}} > \phi_{\mathrm{E}}$ [5] for this case. It is therefore necessary to have at least one anisotropic material for the three-layered light guide, with its diagonal dielectric tensor elements such that the difference in phase shift suffered by TE and TM modes in total reflection from one boundary is compensated at the other boundary. This type of structure is demonstrated elsewhere for isotropic substrates of magnetooptic materials with anisotropic top layers [6]. In this paper we will demonstrate the feasibility of a mode converting structure using an anisotropic substrate, an isotropic top layer, and an isotropic film. We have chosen quartz as a substrate material because it is a readily available birefringent material, obtainable in crystals of high optical quality. By orienting the c axis in the x–y plane of Fig. 1, we can obtain the off-diagonal dielectric tensor elements for the longitudinal configuration as well as matching the phases of the TE and TM modes for certain incident angles θ_i. The indices of refraction used for quartz are $n_0 = 1.5431$ and $n_e = 1.5522$ [7]; the thin film was taken as glass with an index, $n_f = 1.6$; and the top layer was taken as air, $n_t = 1.0$ or glass, $n_t = 1.5$. These values are for a free-space light wavelength of $\lambda = 0.6278\ \mu$.

Fig. 2 shows the amplitude of the converted field (as a fraction of the input field amplitude) C versus the normalized distance down the guide, as found by multiplying $[R]$ by itself q times on a computer. It is thus an independent check of the analytic expression for $[R]^q$ (13) and verifies that complete mode conversion is obtained when $\theta_i = \theta_m$ and C follows the expected $\sin(az)$ behavior of coupled mode problems [8]. The maximum converted

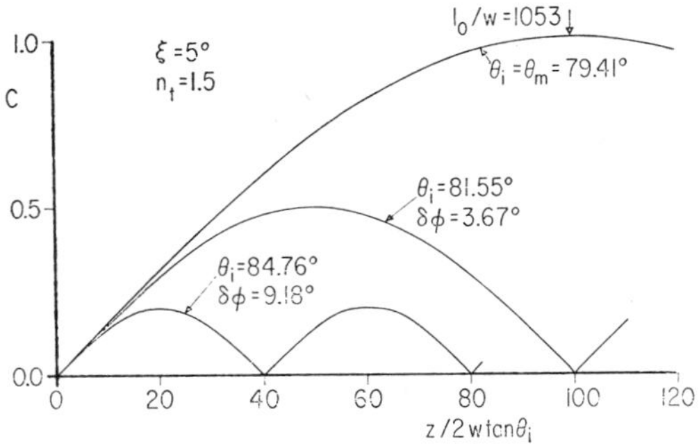

Fig. 2. Amplitude of the converted mode (as a fraction of the incident field amplitude) versus normalized distance, $z/2W \tan \theta_i$ (equals the number of mode-converting bounces) for quartz-substrate mode converter.

field amplitude achievable when $\delta\phi \neq 0$ is also shown in Fig. 2. This leads to a definition of an incident-angle bandwidth $\Delta\theta_i$ as the difference between θ_i and θ_m when the maximum possible power conversion is 50 percent.

Fig. 3 shows the one-reflection power conversion factor G_0 and $\Delta\theta_i$ as a function of ξ, the angle between the c axis of the quartz and the x axis of Fig. 1. It is seen from our calculations that, with an air top layer, the mode converting structure can operate "phase matched" for $0 < \xi < 10°$, while with a glass top layer, ξ can be as large as 30° with $\theta_i = \theta_m$, thus greatly increasing the off-diagonal dielectric elements and therefore the power conversion factor G_0.

Fig. 4 shows l_0/W and the angle θ_m as a function of the substrate orientation ξ. If we consider a typical case of $\xi = 5°$, $n_t = 1.5$, and a thin-film thickness $W = 1\ \mu$, the length of waveguide needed for complete mode conversion is ≈ 1 mm. The bandwidth on the incidence angle $\Delta\theta_i$ for this example is $\approx 3°$, which is a greater tolerance than the tolerance on coupling light into a waveguide mode [9], [10]. For many configurations of mode-converting structures this is true, implying that alignment necessary for mode converting is no more difficult than for light guiding.

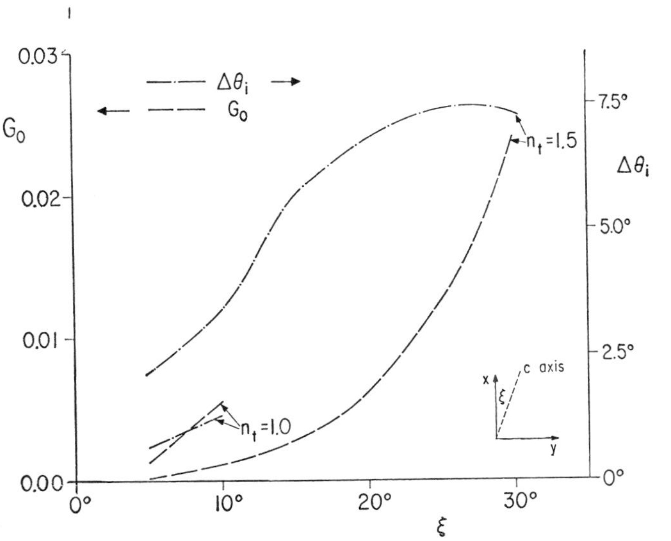

Fig. 3. Initial power conversion rate G_0 and incident-angle bandwidth $\Delta\theta_i$ as a function of angle ξ (orientation of c axis of quartz).

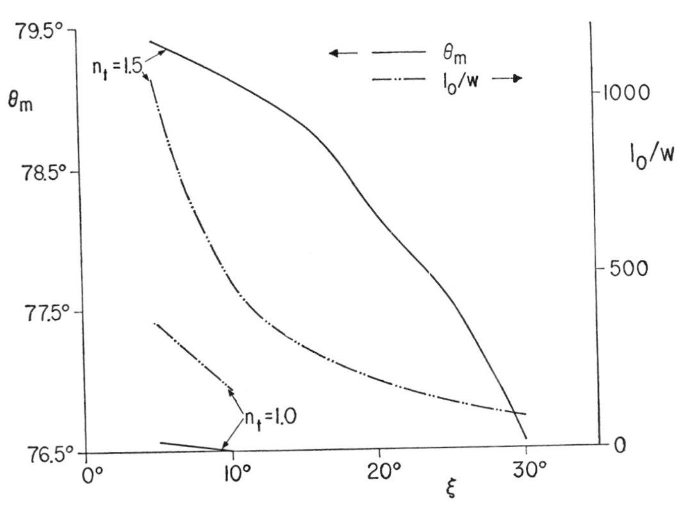

Fig. 4. Normalized complete mode-conversion length l_0/W and phase-matching angle θ_m as a function of ξ.

IV. Applications

Two mode converters can be utilized to construct an optical gyrator. The gyrator and a polarization selective power absorber can be combined to yield an optical isolator. We will give a brief discussion of the design of these devices and conclude by mentioning other devices that are equally feasible.

The essential property of a gyrator is its nonreciprocal nature. Equations (2), (3), and (9) indicate that the anisotropic and gyrotropic mode converters are nonreciprocal in the longitudinal configuration and reciprocal in the polar configuration. We have found that the proper combinations of anisotropic and gyrotropic mode converters for a gyrator are

1) gyrotropic polar + anisotropic longitudinal;
2) gyrotropic longitudinal + anisotropic polar;
3) gyrotropic polar + anisotropic polar + phase shifter;
4) gyrotropic longitudinal + anisotropic longitudinal + phase shifter;

where the phase shifter is just some specified length of waveguide (nonmode converting). Consider case 3) as an example, where mode converter I uses a magnetooptic substrate and mode converter III uses an anisotropic dielectric substrate (see Fig. 5). From (4) and (6) we find that $\phi_a' = \pi/2$ for mode converter I and $\phi_a' = 0$ for mode converter III. Then for a waveguide length such that the magnitude of the diagonal elements of $[R]^a$ are equal to $1/\sqrt{2}$ we have the following relations between the input and output fields for forward ($+z$) propagation:

$$\underset{\longrightarrow}{\begin{bmatrix} E_{\text{out}}^{\text{TE}} \\ E_{\text{out}}^{\text{TM}} \end{bmatrix}} = \frac{1}{2}\begin{bmatrix} 1 & \exp(-j\phi_b) \\ -\exp(j\phi_b) & 1 \end{bmatrix}\begin{bmatrix} 1 & 0 \\ 0 & \exp(-j\gamma) \end{bmatrix} \times \begin{bmatrix} 1 & j\exp(-j\phi_b') \\ j\exp(j\phi_b') & 1 \end{bmatrix}\underset{\longrightarrow}{\begin{bmatrix} E_{\text{in}}^{\text{TE}} \\ E_{\text{in}}^{\text{TM}} \end{bmatrix}} \quad (14)$$

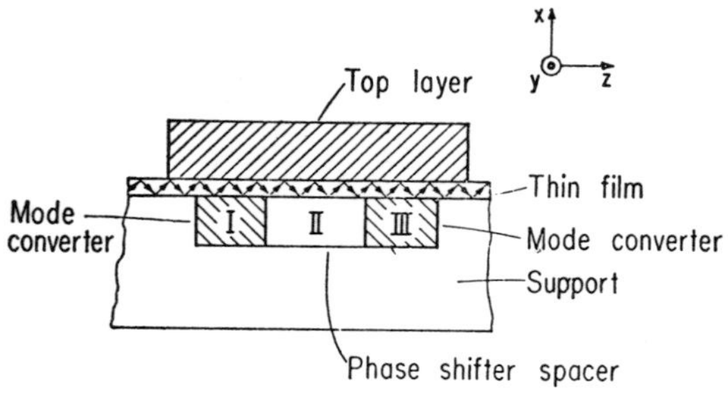

Fig. 5. Schematic showing the use of mode converters in series to obtain a gyrator. Mode converters I and III represent a gyrotropic and an anisotropic mode converter. The light wave propagates in the z direction.

and for backward ($-z$) propagation:

$$\underleftarrow{\begin{bmatrix} E_{out}^{TE} \\ E_{out}^{TM} \end{bmatrix}} = \frac{1}{2}\begin{bmatrix} 1 & j\exp(-j\phi_b') \\ j\exp(j\phi_b') & 1 \end{bmatrix}\begin{bmatrix} 1 & 0 \\ 0 & \exp(-j\gamma) \end{bmatrix} \times \begin{bmatrix} 1 & \exp(-j\phi_b) \\ -\exp(j\phi_b) & 1 \end{bmatrix}\underleftarrow{\begin{bmatrix} E_{in}^{TE} \\ E_{in}^{TM} \end{bmatrix}}. \qquad (15)$$

If we choose the diagonal dielectric properties of mode converters I and III such that $\phi_b = \phi_b'$ (this can be done by crystal orientations or proper top-layer indices of refraction) and take $\gamma = \pi/2$ (by choosing the proper length for the phase shifter) (14) and (15) reduce to

$$\underrightarrow{\begin{bmatrix} E_{out}^{TE} \\ E_{out}^{TM} \end{bmatrix}} = \begin{bmatrix} 1 & 0 \\ 0 & -j \end{bmatrix}\underrightarrow{\begin{bmatrix} E_{in}^{TE} \\ E_{in}^{TM} \end{bmatrix}} \qquad (16a)$$

$$\underleftarrow{\begin{bmatrix} E_{out}^{TE} \\ E_{out}^{TM} \end{bmatrix}} = \begin{bmatrix} 0 & \exp(j\phi_b) \\ j\exp(-j\phi_b) & 0 \end{bmatrix}\underleftarrow{\begin{bmatrix} E_{in}^{TE} \\ E_{in}^{TM} \end{bmatrix}}. \qquad (16b)$$

This demonstrates the nonreciprocal nature of the gyrator, i.e., a forward TE or TM propagates unchanged while a backward TE or TM is completely mode converted.

The other gyrator structures are also described by (16). The selection of the best structure will depend primarily on the dielectric tensors of the materials involved and the substrate configuration desired. For example, we have found that for a given material the longitudinal mode converter is a superior (larger G_0 and $\Delta\theta_i$) configuration to the polar case. And the large birefringence of naturally anisotropic crystals can make them more efficient mode converters than either of the field-induced mode converters. Therefore, if we wish to utilize electrooptic and magnetooptic effects, structure 4 is the best; while if naturally anisotropic crystals can be used then structure 2 is better.

The amount of mode conversion can be controlled externally by using an electrooptic or magnetooptic material as the substrate. The application of a dc electric or magnetic field could then be used to switch the modes; ac fields could be used to modulate the modes; and finally, digital information could be read by the light-guide modes.

An example of this last function is discussed in [6] where a magnetooptic mode converting structure is used to nondestructively read digital information stored as magnetic bubbles.

V. Conclusions

Our computer studies have demonstrated that gyrotropic and anisotropic materials can be used to make feasible mode-converting structures. The fabrication of these devices can be expected to reveal practical limitations to device performance. For example, reflections and mode mismatch caused by substrate discontinuities, and scattering and absorption losses due to imperfect dielectrics will result in a nonzero insertion loss. A more critical practical tolerance may be the control on the film thickness. If the film thickness varies, the modes will no longer be degenerate and mode conversion will be reduced. We have studied perturbations to the film thickness and have found that thickness tolerances necessary can generally be met by current polishing techniques. (For example, the above-mentioned quartz mode converter can have a variation in W of up to 2000 Å.)

Preliminary experiments performed at our laboratory using an anisotropic quartz crystal as substrate ($\xi = 5°$), a photoresist film (KOR), and a lead glass prism to couple light into and out of the film have demonstrated approximately 50 percent mode conversion and this work will be reported elsewhere when completed.

References

[1] R. E. Collin, *Field Theory of Guided Waves.* New York: McGraw-Hill, 1960.
[2] G. N. Ramachandran and S. Ramaseshan, "Crystal optics," in *Encyclopedia Phys.* (*Handbuch der Physik*), vol. 25/1, 1961.
[3] D. F. Nelson and J. McKenna, "Electromagnetic modes of anisotropic dielectric waveguides at p-n junctions," *J. Appl. Phys.*, vol. 38, no. 10, p. 4057, 1967.
[4] R. A. Frazer, W. J. Duncan, and A. R. Collar, *Elementary Matrices.* Cambridge, England: University Press, 1952, p. 78.
[5] M. Born and E. Wolf, *Principles of Optics.* New York: Pergamon, 1959, p. 46.
[6] S. Wang, J. D. Crow, and M. Shah, "Studies of magnetooptic effects for thin-film optical-waveguide applications," *IEEE Trans. Magn.*, vol. MAG-7, pp. 385–387, Sept. 1971.
[7] *American Institute of Physics Handbook.* New York: McGraw-Hill, 1957, pp. 6–23.
[8] W. H. Louisell, *Coupled Mode and Parametric Electronics.* New York: Wiley, ch. 1.
[9] M. L. Dakss, L. Kuhn, P. F. Heidrich, and B. A. Scott, "Grating coupler for efficient excitation of optical guided waves in thin films," *Appl. Phys. Lett.*, vol. 16, p. 523, 1970.
[10] P. K. Tein, R. Ulrich, and R. J. Martin, "Modes of propagating light waves in thin deposited semiconductor films," *Appl. Phys. Lett.*, vol. 14, p. 291, 1969.

Part 4
Mode Launching

MODES OF PROPAGATING LIGHT WAVES IN THIN DEPOSITED SEMICONDUCTOR FILMS

P. K. Tien, R. Ulrich, and *R. J. Martin*
Bell Telephone Laboratories, Inc.
Holmdel, New Jersey 07733
(Received 12 March 1969)

We report theory and experiment on modes of propagating light waves in deposited semiconductor films. The modes are excited by a novel prism-film coupler which is also used for the measurement of their phase velocities. Up to 50% of the incident laser energy has been fed into a single mode of propagation. The positions and linewidths of the modes, the wave intensity inside the film, and a dramatic view of the mode spectrum displayed by the scattered light are discussed in detail.

Guidance of light waves in semiconductor films may lead to numerous new applications in the laser and electro-optical field. In particular, one might explore nonlinear optics of guided waves which offer new possibilities of phase-matching at high concentrations of light energy. In many of these applications a distance of propagation of the order of 1 cm would be sufficient. Such a length does not require materials of excessively high optical quality, and standard techniques of film deposition can be employed. In our experiments, we used sputtered films of ZnO and films of ZnS evaporated by electron bombardment. The films used range from 800 to 30 000 Å thick. We report here: (1) the use of a novel prism-film coupler for exciting in those films any selected mode of propagating light wave, (2) a method of probing the modes by measuring their phase velocities, (3) a dramatic view of the mode spectrum of a multimode film as a series of bright lines displayed by the scattered light.

In earlier experiments,[1-4] light guidance was studied in *pn* junctions when the light was focussed onto the edge of the junction. These experiments were usually handicapped by simultaneous excitation of many modes and by excessive scattering at the edge of the junction. We have avoided these difficulties by the use of the prism-film coupler shown in Fig. 1. A laser beam enters a prism of sufficiently high refractive index n_3. It reaches the base of the prism at an angle of incidence θ_3 and is totally reflected. The film is placed at a close distance S parallel to the prism base. Normally, the incident power is totally reflected. Under certain conditions, however, the light energy can be transferred into the film by "optical tunneling." This coupling is effected by the evanescent fields that are excited in the gap S by the total reflection. The

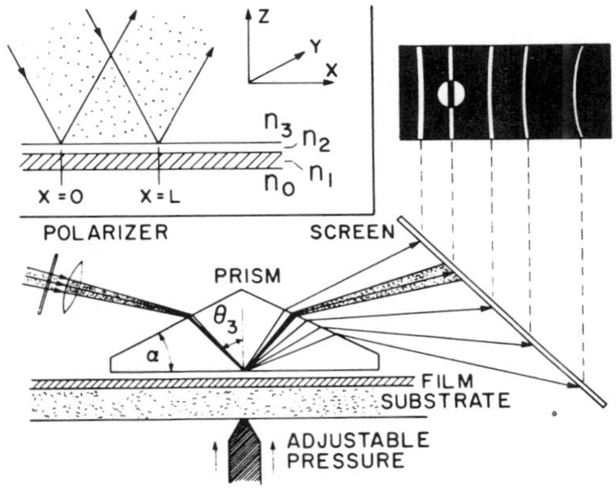

Fig. 1. Experimental arrangement for observation of coupling and intermode scattering. This setup was also used for the determination of the propagation constants β.

conditions for coupling are (1) the incident beam must have the proper angle of incidence so that the evanescent fields in the gap S travel with the same phase velocity as the mode to be excited in the film. This direction θ_3 will be called a synchronous direction. (2) As the modes of the film have a distinct polarization (TM or TE) the incident beam must have the same polarization as the mode to be excited. (3) The film must be placed close enough to the prism base. Typically, S is in the order of half a wavelength.

In order to state the condition (1) more precisely, let the propagation constant of the mode under consideration be β. For the incident beam the component of the propagation vector parallel to the film is $kn_3 \sin \theta_3$, where $k = \omega/c$, and ω and c are, re-

Reprinted with permission from *Appl. Phys. Lett.*, vol. 14, pp. 291–294, May 1, 1969.

spectively, the angular frequency and the velocity of light in vacuum. Then a direction θ_3 is a synchronous one if $\beta = kn_3 \sin\theta_3$. Since different modes in the film generally have different β's, we can thus selectively excite any one of them by adjusting θ_3. Alternatively, the propagation constant β or the phase velocity ω/β of a mode can be determined by measuring the synchronous direction of θ_3 of that mode.

Theory. For simplicity, we will neglect losses in the various media, restrict ourselves to a one-dimensional analysis in which all quantities are independent of y, and assume an arrangement infinitely extended in x and y directions. For TM modes the nonvanishing field components are E_x, E_z, H_y. The film thickness is W. Let us use the subscripts 0, 1, 2, 3 to denote quantities in the substrate, film, gap and prism, respectively. In order to support propagating modes, the film refractive index n_1 must be $n_1 > n_0$ and $n_1 > n_2$. For the prism $n_3 > \beta/k$ is required. We have one set of waves in each of the four media. The waves in the prism and in the film are ordinary plane waves. Let the incident and reflected waves in the prism have amplitudes A_3 and B_3. They vary as $\exp(-ib_3z)$ and $\exp(ib_3z)$, respectively, and their wave vectors form the angles $\pi - \theta_3$ and θ_3 with the z axis. Similarly, the waves A_1, B_1 in the film vary as $\exp(\pm ib_1z)$ and travel at angles $\pi - \theta_1$ and θ_1. In the coupling gap S and in the substrate, the fields are evanescent waves. These vary as $\exp(\pm p_2z)$ and $\exp(p_0z)$, respectively. Here all b's and p's are real positive quantities. We do not lose the generality of the problem, if we let all waves vary as $\exp(i\beta x - i\omega t)$ and allow the amplitudes to be complex functions of x. As the waves described must satisfy the wave equations we have for $i = 1$ or 3 and $j = 0$ or 2

$$\beta^2 = (kn_i)^2 - b_i^2 = (kn_j)^2 + p_j^2 \quad , \quad \beta = kn_1 \sin\theta_i \ . \tag{1}$$

To facilitate our later discussion, we introduce the following angles ψ_{ij} (half the reflection phase at the interface between the media i and j) and the quantities ψ_{12}' and t_1^2:

$$\tan\psi_{ij} = (n_i/n_j)^2 p_j/b_i \quad \text{for TM modes}, \tag{2a}$$

$$\tan\psi_{ij} = p_j/b_i \quad \text{for TE modes}, \tag{2b}$$

$$\psi_{12}' = \psi_{12} + \sin 2\psi_{12} \cos 2\psi_{32} \exp(-2p_2S), \tag{3}$$

$$t_1^2 = 4 \sin 2\psi_{12} \sin 2\psi_{32} \exp(-2p_2S) \ . \tag{4}$$

By matching the wave amplitudes at the boundaries between the media, the intensity of the waves inside the film is obtained. This intensity exhibits sharp maxima at a finite number of discrete values of β's. These maxima represent the modes of propagation of light in the film when the prism is present. The propagation constants β of the modes are the solutions of

$$2b_1(\beta)W - 2\psi_{12}'(\beta) - 2\psi_{10}(\beta) = 2m\pi \ , \tag{5}$$

where the ψ_{ij} satisfies (2) and $0 \le \psi_{ij} \le \pi/2$. The integer $m = 0, 1, 2, 3, \ldots$ is called the order of the mode. For weak coupling ($p_2S > 1$), ψ_{12}' is given by (3). From (3) and (5), the positions of the modes in β are seen to vary slightly with the coupling. For a given thickness W there exists a finite number of discrete β's, their number increasing with W. From the symmetry of the problem it follows that to each mode or order m there exists an infinite number of degenerate modes, whose propagation vectors all lie in the xy plane. To evaluate the linewidths of the modes, we replace $2m\pi$ in (5) by δ. We find the half intensity points at $\delta = 2m\pi \pm t_1^2$, and the linewidth $2t_1^2$. The finesse[5] is then π/t_1^2 in the δ scale. As $S \to \infty$, $\psi_{12}' \to \psi_{12}$ and (5) is reduced to the usual equation of modes for the film alone.[4] The linewidths in this case vanish since losses have been neglected.

Next we show how the light intensity in the film builds up in the direction of propagation due to its coupling to the input laser beam. Let this beam be incident in a synchronous direction and illuminate the prism base uniformly between the limits $x = 0$ and $x = L$ (Fig. 1). The amplitudes of the various waves are normalized so that $|A|^2$ is the power flow through a unit area parallel to the xy plane. We now ask for the x dependence of the intensities of the A_1 and B_1 waves constituting the field in the film. For $0 < x < L$ we find

$$|A_1(x)|^2 = |B_1(x)|^2 = 4t_1^{-2}|A_3|^2\{1 - \exp[-xt_1^2/(4W\tan\theta_1)]\}^2 \tag{6}$$

and for $x > L$

$$|A_1(x)|^2 = |B_1(x)|^2 = |A_1(L)|^2 \exp[-(x-L)t_1^2/(2W\tan\theta_1)] \ , \tag{7}$$

where for weak coupling, t_1^2 is given by (4). The power density (6) inside the film builds up gradually from $x = 0$ and then approaches an asymptotic value $4t_1^{-2}|A_3|^2$. A typical value of t_1 being 0.1, the power density inside the film can theoretically be several orders of magnitude higher than that of the incident beam, a feature important to nonlinear optics. The total power carried by the film is $2W|A_1(x)|^2 \tan\theta_1$. Thus it follows from (6) that by proper choice of S or L a maximum of about 81% of the incident power can be transferred into a film under the prescribed uniform illumination. For $x > L$, the light intensity (7) in the film decreases exponentially since here the prism serves as an output coupler. In order to retain the power $|A_1(L)|^2$ inside the film beyond $x = L$, it is therefore necessary to decouple film and prism in the region $x > L$. This can be done by increasing there the gap S or by cutting away the prism beyond $x = L$, as will be discussed later in the experiment.

Experiment. Both ZnO and ZnS films, deposited on glass, have been used for the experiments. The

ZnO films were sputtered in an argon–oxygen atmosphere from sintered ZnO powder. The c axes of the crystallites in the film are oriented within a cone of 5° from the film normal. These films were polished after growth to reduce their surface roughness. The ZnS films, evaporated by electron bombardment, are not oriented. The coupling prism was of rutile, α = 26°, optical axis parallel to the y axis of Fig. 1. Thus, light coupling to TM modes of the film propagates as ordinary ray in the prism, and that coupling to TE modes as extradordinary ray. The film is pressed with adjustable pressure against the prism base (Fig. 1). Residual dust particles, serving as spacers, and the elasticity of the substrate permit adjustment of S by variation of the pressure. The linear polarized laser beam is directed on the coupling spot at the prism base. As shown in Fig. 1, the reflected light is observed on a screen. The whole prism-film assembly, including the screen but not the laser, is mounted on a turntable so that the direction θ_3 can be varied at will.

In the experiment, a bright spot produced by the reflected wave B_3 is always observed on the screen. When the film is turned into any one of the synchronous directions, three phenomena can be observed: first, a reduction in the intensity of the bright spot. This dip, occasionally down to 50%, means that part of the incident light has been fed into the film and was then lost by absorption or scattering before it could be coupled back into the prism again. Thus, the dip indicates that a mode was launched into the film. Second, a small area of the film near the coupling spot appears very bright. Third, we observe one or more (depending on the film thickness) bright lines on the screen (Fig. 1). One of these lines always intersects the bright B_3 spot. These lines will be called m lines because they display the modes of different orders m, as will be explained later. With increasing coupling the lines become brighter first, and then become broader and shift. All these phenomena disappear if any of the three coupling conditions mentioned in the beginning is violated. These observations are explained as follows: The laser beam, incident in the xy plane at a synchronous angle, excites in the film a mode of order m. This mode propagates in the positive x direction. By scattering, one part of its light is directed into the substrate and causes the observed bright appearance of the film in the coupling area. Another part of the light of the main mode is scattered into other modes that can propagate in the film. Two cases are to be distinguished here. First, scattering into those modes that are azimuthally degenerate with the main mode of excitation. The light of these modes, being coupled out of the film by the very same prism, produces the m line that intersects the bright B_3 spot. Second, intermode scattering into modes of order numbers m different from that of the main mode, including their degenerates. When coupled out on the screen, the light of these modes produces the other m lines. The m lines are therefore a direct display

Table I. Comparison between observed and theoretically expected propagation constants of a ZnO film.ª Film thickness W = 15 881 ± 60 Å, refractive index of glass substrate n_0 = 1.5127 at 6328 Å and n_0 = 1.5206 at 4880 Å.

λ	Polaris	n_1	m	β/k Observed	β/k Theory	Diff.
6328 Å	TE	1.9732	0	a	1.9647	a
			1	1.9383	1.9389	−0.0006
			2	1.8961	1.8954	0.0007
			3	1.8329	1.8332	−0.0003
			4	1.7518	1.7510	0.0008
			5	1.6469	1.6473	−0.0004
			6	1.5248	1.5249	−0.0001
	TM	1.9779	0	a	1.9686	a
			1	a	1.9404	a
			2	1.8933	1.8929	0.0004
			3	1.8251	1.8249	0.0002
			4	1.7353	1.7355	−0.0002
			5	1.6242	1.6246	−0.0004
4880 Å	TE	2.0428	0	2.0360	2.0377	−0.0017
			1	2.0215	2.0223	−0.0008
			2	1.9973	1.9964	0.0009
			3	1.9603	1.9596	0.0007
			4	1.9117	1.9115	0.0002
			5	1.8511	1.8513	−0.0002
			6	1.7786	1.7781	0.0005
			7	1.6907	1.6910	−0.0003
			8	1.5898	1.5892	0.0006
	TM	2.0485	0	a	2.0430	a
			1	a	2.0265	a
			2	a	1.9987	a
			3	1.9598	1.9593	0.0005
			4	1.9089	1.9078	0.0011
			5	1.8432	1.8433	−0.0001
			6	1.7649	1.7652	−0.0003
			7	1.6727	1.6729	−0.0002
			8	1.5677	1.5687	−0.0010

ªThese modes could not be observed with the 26° rutile prism.

of the spectrum of the film modes.

An interesting detail of Fig. 1 is the dark line visible in the bright B_3 spot. The diameter of this spot corresponds to the aperture of the incident beam which is about 1° (inside the prism) in this case. At the chosen coupling, however, the linewidth of the mode is so narrow, that only a small fraction of this θ_3 range is acceptable to the film. The ratio of the separation between adjacent m lines to the width of the dark line is roughly equal to the finesse F discussed in the theory. In one case, we have approximately F = 50, indicating t_1^2 = 0.06.

At weak coupling the synchronous directions become very sharply defined. Then the synchronous angles θ_3 and thus the β's can be determined quite accurately from a measurement of these directions. The refraction of the incident beam at the first prism face must be taken into account. Table I shows the results of such measurements on a ZnO film. Also given are the theoretical β's computed from (5), the lineshifts caused by the difference

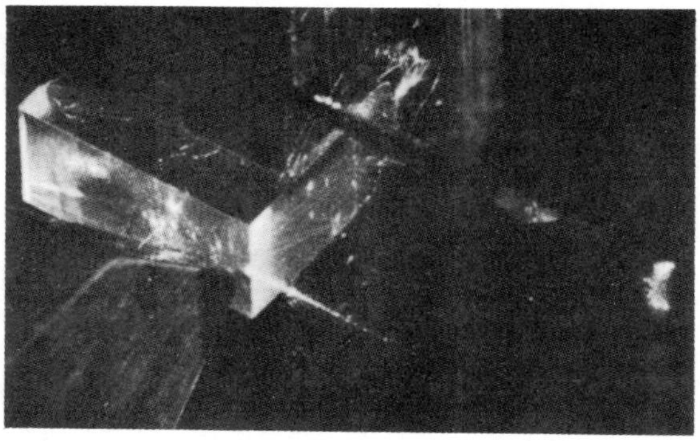

Fig. 2. The path of a light wave propagating in a ZnS film is visible as a bright streak. Film thickness 1500 Å, wavelength 6328 Å, height of the prism 0.5 cm, viewed through the substrate.

between ψ_{12} and ψ_{12} being neglected. In arriving at these theoretical β's both the film thickness W and its refractive index n_1 were treated as free parameters. They were adjusted for best agreement between the observed and computed β's of the 6 lowest TE modes at 6328 Å. The thickness thus determined, $W = 15881 \pm 60$ Å was subsequently used for the computation of all theoretical β's, while the refractive index n_1 was fitted separately for each group. It is seen that good agreement is obtained between the theoretical and measured β's. Because of the crystalline orientation of the ZnO film the TE modes propagate as ordinary ray in the ZnO. The value $n_1 = 1.973 \pm 0.001$ at 6328 Å determined by the present experiment is only slightly less than the value of 1.988 interpolated from Bond's measurement[6] of bulk ZnO. The agreement is equally good for the other groups of Table I. This supports electrical and x-ray measurements in indicating the good crystalline perfection of our ZnO films.

In another experiment we show that a mode, once it is launched into the film, will propagate freely provided that coupling back into the prism is prevented. For this purpose we cut the prism of Fig. 1 along its plane of symmetry and removed the right half-prism. The laser beam is adjusted so that the 90° corner plays the role of the point $x = L$ discussed in the theory. Figure 2 is a photograph of the arrangement. The film (ZnS, 1500 Å thick) is pressed against the prism base by a pointed tool. The laser beam, not visible in this photograph, enters the prism through the hypotenuse face from the upper left. It is set to the synchronous direction of this single-mode film and a bright streak of light is seen extending from the coupling spot toward the lower right. The streak is about 1 cm long and consists of light scattered from the propagating mode, much like a laser beam is visible in dusty air. When the continuity of the film is interrupted by scratching across the streak with a fine point, the streak ends sharply at the scratch and the scratch itself radiates brightly. We have also used two such 90° prisms, one as an input coupler and the other for output, 3 mm apart. In addition to the desired bright spot of directly transmitted light, again an m line is observed on the output screen. Details of the theory and experiments will be described elsewhere.

[1] D. F. Nelson and K. F. Reinhart, Appl. Phys. Letters 5, 148 (1964).

[2] A. Ashkin and M. Gershenzon, J. Appl. Phys. 34, 2116 (1963).

[3] R. F. Kazarinov, O. V. Konstantinov, V. I. Perel, and A. L. Efros, Fiz. Tverd. Tela 7, 1506 (1965) [English transl.: Soviet Phys.—Solid State 7, 1210 (1965)].

[4] D. F. Nelson and J. McKenna, J. Appl. Phys. 38, 4057 (1967).

[5] M. Born and E. Wolf, *Principles of Optics* (Pergamon Press, Inc., New York, 1959, pp. 322–343.

[6] W. L. Bond, J. Appl. Phys. 36, 1674 (1965).

Evanescent Field Coupling into a Thin-Film Waveguide

J. E. MIDWINTER

***Abstract*—We have derived expressions for the field induced in a thin-film waveguide by plane waves incident upon an adjacent frustrated total reflecting interface. From these, we have deduced an expression for the coupling coefficient for a Gaussian TEM_{00} beam into such a guide as a function of the guide and beam parameters.**

Introduction

TIEN *et al.* [1] have recently described an experiment in which they successfully coupled a large proportion of the energy from a gas-laser beam into a thin-film deposited on a suitable plane substrate. They achieved this by bringing a high-refractive-index prism close to the guiding structure and allowing the laser beam to be "totally reflected" from the inside surface of the prism. Coupling into the guide then occurred through the evanescent field of the prism. Efficient overall coupling was achieved by sharply terminating the prism at the point of maximum energy density in the guide. We present here a detailed analysis of the coupling produced by such a combination in a way that demonstrates the physics behind the coupling process and the dependence of that coupling on the physical parameters of the film–prism combination. The analysis follows closely that employed previously for evanescent coupling between two prisms or other optical elements [2]. However, it differs insofar as we emphasize the presence of sideways propagation in the waveguide and we note the importance of the condition for waveguide mode propagation in the coupling process.

General Discussion

There appear to be two basic methods of analyzing the coupler situation. One technique would be to derive an expression for the field coupling through the evanescent field gap for an infinitesimal section of a finite incident beam. Using this coupling, the field components in the guide would be integrated along the waveguide direction across the incident beam section, making due allowance for energy leakage in the reverse direction out of the guide, to derive an expression for the growth of the waveguide mode. This is apparently the approach followed by Tien *et al.* [1]. It appears from their paper that it was necessary in their analysis to assume weak coupling.

We present here a different approach to the problem, in which we work with infinite plane waves and subsequently produce the spatial location of the beam through the summation of an infinite number of plane waves, that is to say, by Fourier transformation. The approach has several attractive features. The first is that it demonstrates very simply the fundamental cause of the coupling process. It also yields exact expressions for the phase distribution of the waves involved and it predicts effects that are not apparent from the analysis of Tien *et al.* [1]. These effects have been observed experimentally and will be reported elsewhere [3].

Our analysis proceeds in three stages. Initially the "prism" of the coupler is allowed to extend to infinity so that it becomes a half-space. We consider a structure composed of a high-index half-space, a low-index layer that is equivalent to the air gap of Tien's analysis, a high-index layer (the waveguide) finally capped by another low-index half-space that is the substrate. We allow a single plane wave of infinite extent to impinge on this structure from the high-index half-space side at an angle of incident θ to the guide plane normal. Since the wave must be totally reflected, if not at the high-index half-space interface then at the waveguide substrate interface, we find that for the infinite plane-wave case a reflected wave of equal amplitude at an angle of reflection θ to the waveguide plane normal results. However, this wave undergoes subtle phase changes when θ is such that a waveguide mode is excited. The second part of the analysis uses the phase-change results of stage 1 to predict the form of the reflected spot from a Gaussian TEM_{00} mode beam incident upon the guide plane from the high-index half-space side but with the structure still considered infinite in extent. This is done by Fourier decomposition of the Gaussian beam into the plane-wave components in the angular spectrum, convolving these plane-wave components with the appropriate phase shifts on reflection and then resumming them. This analysis shows that as a waveguide mode is excited the reflected spot appears to shift away from the incident spot, as if energy were being transported down the guide before leaking out again. In addition, we obtain detailed results for the spot shape and phase structure as a function of evanescent coupling, guide mode, beam size, and angle of incidence. Finally, the third part of the analysis considers the effects of truncating the high-index half-space in a plane perpendicular to the plane of incidence and the guide plane. This results in a structure with the sharp cut-off characteristic of the prism coupler of Tien [1] and we show how optimum coupling is achieved in this situation.

Manuscript received October 29, 1969; revised February 16, 1970.
The author was with the Perkin-Elmer Corporation, Norwalk, Conn. 06852. He is now with the Materials Research Center, Allied Chemical Corporation, Morristown, N. J. 07960.

Reprinted from *IEEE J. Quantum Electron.*, vol. QE-6, pp. 583–590, Oct. 1970.

Plane-Wave Formulation

The infinite plane structure that we have just described is illustrated in Fig. 1, in which we also define the coordinates for the problem. The z axis is normal to the guide plane and x lies in the guide plane along the direction of waveguide-mode propagation.

It is easy to show that in this geometry, since $\partial/\partial y = 0$, two independent solutions to Maxwell's equation exist [4], one involving the waveguide TE mode fields E_y, B_x, and B_z coupled to one another, and the other involving a TM mode with fields B_y, E_x, and E_z coupled together. We will focus our attention on the TE mode solution, which is algebraically the easier. Formally, the TM mode solution is identical apart from the addition of more complex coefficients.

In each of the layers, we postulate plane-wave electric fields of the form indicated below and in Fig. 1.

$$E_y(1) = A \exp(-i\gamma_1 z) + A' \exp(i\gamma_1 z) \quad (1)$$

$$E_y(2) = B \exp(\beta_2 z) + B' \exp(-\beta_2 z) \quad (2)$$

$$E_y(3) = C \exp(-i\gamma_3 z) + C' \exp(i\gamma_3 z) \quad (3)$$

$$E_y(4) = D \exp(\beta_4 z). \quad (4)$$

Quite generally, Maxwell's equations are written in Gaussian units as

$$\bar{\nabla} \times \bar{E} = -\frac{1}{c}\frac{\partial \bar{B}}{\partial t} \quad (5)$$

$$\bar{\nabla} \times \bar{H} = \frac{4\pi}{c}\bar{J} + \frac{1}{c}\frac{\partial \bar{D}}{\partial t} \quad (6)$$

$$\bar{\nabla}\cdot\bar{D} = 4\pi\rho \quad (7)$$

$$\bar{\nabla}\cdot\bar{B} = 0. \quad (8)$$

We have assumed that μ the magnetic permeability is unity, that $\bar{D} = \epsilon\bar{E}$ where ϵ is the scalar dielectric constant, and we will also assume that ρ and $\bar{J}$ are zero, i.e., no free charges or electric conduction currents.

Now we assume that the fields $\bar{E}$ and $\bar{H}$ both have sinusoidal time dependence of the form $\exp(-i\omega t)$, to immediately obtain

$$\Delta^2\bar{E} + \frac{\omega^2\epsilon}{c}\bar{E} = 0 \quad (9)$$

$$\Delta^2\bar{B} + \frac{\omega^2\epsilon}{c}\bar{B} = 0. \quad (10)$$

If we now assign an x dependence for $\bar{B}$ and $\bar{E}$ of the form $\exp(ik_x x)$ and we set $k_0 = \omega/c$, then we can solve for the TE and TM modes separately. Using (5) and (6) we obtain the following results for the TE mode:

$$-\frac{\partial E_y}{\partial z} = ik_0 B_x \quad (11)$$

$$k_x E_y = k_0 B_z \quad (12)$$

$$\frac{\partial B_x}{\partial z} = ik_x B_z - ik_0\epsilon E_y \quad (13)$$

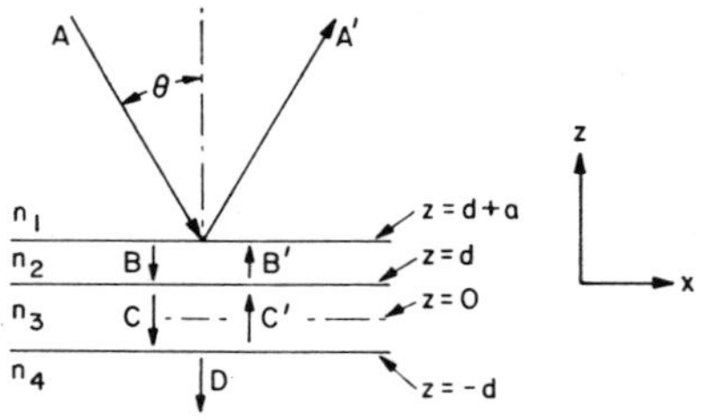

Fig. 1. Definitions of geometry and waves.

and a similar set for the TM mode. For the TE mode, only E_y has to be found, because B_x and B_z follow by simple differentiation. For the TM mode, we would seek a solution for B_y only. Now, if we recall that the film is infinite in y so that $(\partial^2 B_y)/(\partial y^2) = 0$ and $(\partial^2 B_y)/\partial y^2) = 0$, then from (9) we obtain for the TE mode:

$$\frac{\partial^2 E_y}{\partial z^2} = (k_x^2 - \epsilon k_0^2)E_y. \quad (14)$$

Finally, we note that the fields in the three media are related by the following boundary conditions that stem from the assumptions that ρ and $\bar{J}$ are both zero.

$$\Delta B_{\text{nor}} = \Delta D_{\text{nor}} = 0$$
$$\Delta E_{\text{tan}} = \Delta H_{\text{tan}} = 0 \quad (15)$$

where subscript nor signifies normal to the interface and subscript tan signifies tangential to it.

Using (11) and (1)–(4), it follows directly that

$$B_x(1) = \frac{\gamma_1}{k_0}[A \exp(-i\gamma_1 z) - A' \exp(i\gamma_1 z)] \quad (16)$$

$$B_x(2) = \frac{1}{ik_0}[\beta_2 B \exp(\beta_2 z) - \beta_2 B' \exp(-\beta_2 z)] \quad (17)$$

$$B_x(3) = \frac{\gamma_3}{k_0}[C \exp(-i\gamma_3 z) - C' \exp(i\gamma_3 z)] \quad (18)$$

$$B_x(4) = -\frac{1}{ik_0}[\beta_4 D \exp(\beta_4 z)] \quad (19)$$

and using relations (1)–(4) and (14) we obtain the results:

$$\begin{aligned} \gamma_1^2 &= n_1^2k_0^2 - k_x^2 = n_1^2k_0^2\cos^2\theta \\ \beta_2^2 &= k_x^2 - n_2^2k_0^2 \\ \gamma_3^2 &= n_3^2k_0^2 - k_x^2 \\ \beta_4^2 &= k_x^2 - n_4^2k_0^2 \\ k_x &= n_1k_0\sin\theta. \end{aligned} \quad (20)$$

In order to discover the relationship between the various waves in the structure, we must apply the appropriate continuity conditions at the interfaces of the structure. Thus for the TE mode we equate E_y and B_x, respectively, at each of the three interfaces to obtain six simultaneous equations in the amplitudes A, A', B, B', etc. We write these equations in the form $a_1A = b_1A' + c_1B + d_1B' + e_1C + f_1C' + g_1D$; $a_2A = b_2A'$, $\cdots$, etc. We also give the matrix of the coefficients (see Table I). For ease of reading, we have extracted common factors from each

TABLE I
MATRIX OF COEFFICIENTS FOR COUPLER-GUIDE EQUATIONS*

		a	b	c	d	e	f	g
$Z = d + a$	E_y	-1	1	$\exp(\beta_2 a)$	$\exp(-\beta_2 a)$	0	0	0
	B_x	$(-\gamma_1)/k_0$	$-(\gamma_1)/k_0$	$(-\beta_2)/(ik_0)\exp(\beta_2 a)$	$(\beta_2)/(ik_0)\exp(-\beta_2 a)$	0	0	0
$Z = d$	E_y	0	0	1	1	1	1	0
	B_x	0	0	$(\beta_2)/ik_0$	$(\beta_2)/(ik_0)$	$(\gamma_3)/(k_0)$	$-(\gamma_3)/(k_0)$	0
$Z = -d$	E_y	0	0	0	0	$\exp(2i\gamma_3 d)$	$\exp(-2i\gamma_3 d)$	1
	B_x	0	0	0	0	$(\gamma_3)/(k_0)$ $\cdot\exp(2i\gamma_3 d)$	$(\gamma_3)/(k_0)$ $\cdot\exp(2i\gamma_3 d)$	$(\beta_4)/(ik_0)$

*Common factors have been extracted from each column.

Column	Common Factor
a	$\exp(-i\gamma_1(d+a))$
b	$\exp(i\gamma_1(d+a))$
c	$\exp(\beta_2 d)$
d	$\exp(-\beta_2 d)$
e	$\exp(-i\gamma_3 d)$
f	$\exp(i\gamma_3 d)$
g	$\exp(-\beta_4 d)$

column and placed them at the base of the column and we indicate at the left side the interface and the field being equated. The general form of the plane-wave solution is now given by an expression of the form

$$\frac{A}{|bcdefg|} = \frac{A'}{|acdefg|} = \frac{B}{|badefg|} = \text{etc.} \qquad (21)$$

where $|bcdefg|$ represents the determinant formed by the columns b, c, d, e, f, g of the matrix of coefficients above. We note that, in general, each of these determinants will be complex, signifying that a phase change will occur as well as a possible amplitude change. Since we have seven unknowns (A, A', B, etc.) and only six equations, it is necessary to specify one variable and evaluate the remainder in terms of it. We choose the wave A to be specified for obvious reasons.

incident

PROPERTIES OF PLANE-WAVE SOLUTIONS

Before we examine the expressions for a localized Gaussian beam incident on the infinite structure, we will study the properties of the plane-wave response derived above. The conclusions are strictly valid only for an infinite structure.

We first note that the determinants $|bcdefg|$ and $|acdefg|$ are of the general form $J(G + iH)$ and $J^*(G - iH)$, respectively. Hence, we have immediately the result that the amplitude of A' is equal to the amplitude of A. There is a phase shift, however, so that

$$A' = A\left[\frac{J^*(G - iH)}{J(G + iH)}\right] = A\exp(2i\psi) \qquad (22)$$

where

$$J = \frac{-2}{k_0^3}[\exp(i\gamma_1(d+a))\exp(-\beta_4 d)] \qquad (23)$$

and J^* is its complex conjugate.

$$G = \gamma_1[\exp(\beta_2 a)X - \exp(-\beta_2 a)Y] \qquad (24)$$

$$H = \beta_2[\exp(\beta_2 a)X + \exp(-\beta_2 a)Y] \qquad (25)$$

$$X = (\gamma_3^2 - \beta_2\beta_4)\sin(2\gamma_3 d) - \gamma_3(\beta_2 + \beta_4)\cos(2\gamma_3 d) \qquad (26)$$

$$Y = (\gamma_3^2 + \beta_2\beta_4)\sin(2\gamma_3 d) + \gamma_3(\beta_2 - \beta_4)\cos(2\gamma_3 d). \qquad (27)$$

If we set $\psi = \phi + \phi'$, we can directly obtain an expression for the phase shift ϕ due solely to the guiding structure. This is the component arising out of the terms G and H. The term ϕ' is due primarily to the choice of axes for the problem and is contained in J and J^*. Setting $\tan(\phi) = H/G$ we have the result that

$$\tan\phi = \frac{\beta_2}{\gamma_1}\left[\frac{\exp(\beta_2 a)X + \exp(-\beta_2 a)Y}{\exp(\beta_2 a)X - \exp(-\beta_2 a)Y}\right]. \qquad (28)$$

We now note that the condition $X = 0$ corresponds to the condition for a freely propagating mode of the guide in the absence of any "coupler" (high-index half-space).

This is equivalent to setting the subdeterminant

$$\begin{vmatrix} d_3 & e_3 & f_3 & g_3 \\ d_4 & \cdot & \cdot & \cdot \\ d_5 & \cdot & \cdot & \cdot \\ d_6 & \cdot & \cdot & \cdot \end{vmatrix} = 0, \qquad (29)$$

which yields the formally equivalent conditions

$$\tan(2\gamma_3 d) = \frac{\gamma_3(\beta_4 + \beta_2)}{\gamma_3^2 - \beta_2\beta_4} \qquad (30)$$

$$(2\gamma_3 d - m\pi) = \tan^{-1}(\beta_4/\gamma_3) + \tan^{-1}(\beta_2/\gamma_3) \qquad (31)$$

the familiar conditions for the TE$_m$ mode. We note with interest that φ only becomes rapidly varying in the region defined as

$$|\exp(\beta_2 a)X| \leq |\exp(-\beta_2 a)Y|. \quad (32)$$

Since X, Y, γ_1, β_2, γ_3, and β_4 are all functions of θ the condition, $|X| = 0$ is, in turn, a condition upon the angle θ of plane-wave incidence. It corresponds to the angle of propagation in the "prism" of the coupler that excites the TE_m mode of the guide. We will call the situation defined by $|X| = 0$, "waveguide-mode resonance" and the value of θ so defined is signified by θ_M. We shall see that when $|X| \approx 0$, resonant coupling of energy from the incident wave into the guide occurs. Well away from incidence we have the results that $|\exp(\beta_2 a)X| \gg |\exp(-\beta_2 a)Y|$ and $\phi = \tan^{-1}(\beta_2/\gamma_1)$. As resonance is approached, ϕ goes first to $\pi/2$ (when the denominator is zero), it takes the value $\tan^{-1}(-\beta_2/\gamma_1)$ when $X = 0$ and finally returns to $\tan^{-1}(\beta_2/\gamma_1)$. Thus φ changes by a total amount π as θ tracks through the waveguide resonance and the phase of the reflected wave changes by 2π. This change is plotted in Fig. 2.

The sharpness of the change depends critically upon the thickness of the low-index layer (as $\exp(-2\beta_2 a)$. As the thickness increases, the change ($\Delta\theta$) of the incident angle θ over which ϕ shifts by π becomes very small. In the central region we may approximate this response by a straight line characterized by $((\partial\phi)/(\partial\theta))_{\theta=\theta_M}$ where θ_M is the angle of incidence that sets $X \equiv 0$. Then, we can use a small-angle expansion to express $(\partial\phi)/(\partial\theta)$ in the form

$$\left(\frac{\partial\phi}{\partial\theta}\right)_{\theta_M} = \cos^2(\phi)_{\theta_M}\cdot(\beta_2/\gamma_1)\cdot 2\cdot\left(\frac{-\partial X}{\partial\theta}\right)_{\theta_M}(\exp(-2\beta_2 a)Y)^{-1} \quad (33)$$

where

$$(\phi)_{\theta_M} = \tan^{-1}(-\beta_2/\gamma_1).$$

Hence,

$$\left(\frac{\partial\phi}{\partial\theta}\right)_{\theta_M} = \left[\left[\frac{\gamma_1^2+\beta_2^2}{\gamma_1^2}\right]^{-1}\frac{\beta_2}{\gamma_1}\cdot\left(\frac{\partial X}{\partial\theta}\right)\exp(2\beta_2 a)Y^{-1}\right]_{\theta_M}. \quad (34)$$

We will see that the value of $((\partial\phi)/(\partial\theta))_{\theta_M}$ is critically important in assessing the performance of the coupler. Suffice it to say now that it is clearly a rapidly varying function of the thickness of the low-index layer.

The function X is easily differentiated to form $(\partial X)/(\partial\theta)$ since

$$\frac{\partial(\gamma_3)^2}{\partial\theta} = -\frac{\partial(\beta_2)^2}{\partial\theta} = -\frac{\partial(\beta_4)^2}{\partial\theta}.$$

Then we find that

$$\left(\frac{\partial X}{\partial\theta}\right) = -n_1^2 k_0^2 \sin(2\theta)\cdot(\beta_2+\beta_4)$$
$$\cdot\left[\left(\frac{\beta_2+\beta_4}{2\beta_2\beta_4}\right)+d\right)\sin(2\gamma_3 d)$$
$$+\left(d\cos(2\gamma_3 d) - \frac{1}{2\gamma_3} + \frac{\gamma_3}{2\beta_2\beta_4}\right)\cos(2\gamma_3 d)\Bigg]. \quad (35)$$

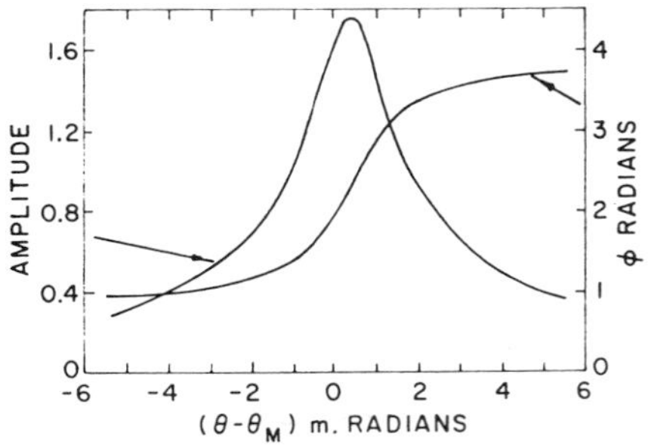

Fig. 2. Amplitude of guide-wave and reflected-wave phase shift against θ.

We have now derived essentially all the relations necessary to find a solution to our problem. However, before we do this we note in Fig. 2 a plot of the amplitude of the wave C, in the guide, for constant amplitude incident as the angle of incidence varies. Not unexpectedly, we see that as waveguide resonance is approached the amplitude increases and reaches a maximum at resonance.

In Appendix I we give the full forms of these determinants after they have been multiplied out and terms have been collected.

Coupling a Gaussian Beam

We now examine a case of particular interest, namely that of coupling a Gaussian beam (TEM_{00}) into the guide structure. We start by examining the formal solution for the approximate case in order to obtain a feel for the problem. We consider a Gaussian beam incident upon the coupler-guide plane at an angle θ to the normal and we wish to examine its form after reflection at the surface. Thus we define coordinates u, v, and w as shown in Fig. 3 for the reflected waves where the w axis lies at θ to the plane normal, nominally along the center of the reflected beam, and u and v are the two transverse directions with v identically equal to y. We ignore the change in coordinates due to the reflection per se and concentrate only on the effect of the phase shift φ upon the Gaussian beam components. Hence, our incident beam in spatial terms is given by

$$E(r) = E_0 \exp(-r/R^2) \quad (36)$$

where

$$r^2 = u^2 + v^2.$$

This can be expressed as a sum of plane waves as

$$E(r) = \int_{-\infty}^{\infty}\int_{-\infty}^{\infty} E(k_u, k_v)\exp(-i(\boldsymbol{k}\cdot\boldsymbol{r}))\,dk_u\cdot dk_v \quad (37)$$

where

$$|\boldsymbol{k}|^2 = k_u^2 + k_v^2 + k_w^2 = k_T^2 + k_w^2$$

$$E(k_u, k_v) = (1/4\pi)E_0 R^2 \exp(-k_T^2 R^2/4). \quad (38)$$

We can now define a small angular deviation $\Delta\theta$ from the center of the incident beam that lies at angle θ to the normal where $\Delta\theta = k_T/k_w$. Since θ is defined to lie in the x–z plane, this reduces to $\Delta\theta = (k_u/k_w)$.

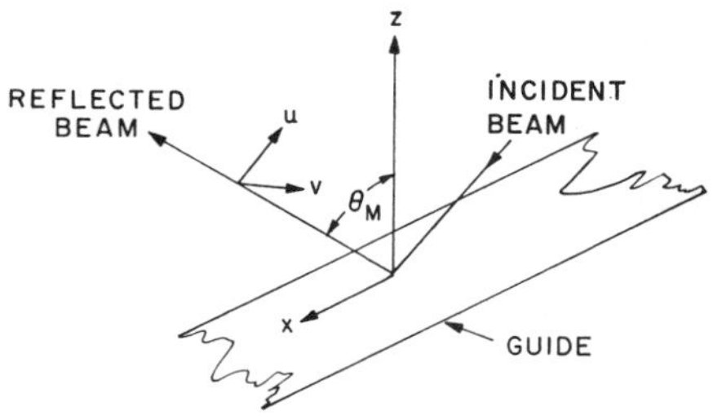

Fig. 3. Definition of Gaussian beam coordinates ($u = x \sin \theta$, $v = y$).

Now let us examine the effect of a $\Delta\theta$ dependent phase shift on the plane-wave components of the Gaussian beam. Suppose we assume that $\phi = ((\partial\phi)/(\partial\theta))_{\theta_M} \cdot \Delta\theta$, that is, we approximate the curve of ϕ versus θ of Fig. 2 to a straight line of slope $((\partial\phi)/(\partial\theta))_{\theta_M}$. Then (37) has to be multiplied by a phase factor $\exp(-i(k_u((\partial\phi)/(\partial\theta))_{\theta_M}/k_w)$. The reflected and phase-shifted beam is now identical to the reflected beam in the absence of a phase shift apart from a coordinate change of $u + \Delta u$ in the place of u where Δu is given by

$$\Delta u = (1/k_w)\left(\frac{\partial\phi}{\partial\theta}\right)_{\theta_M}. \qquad (39)$$

Physically the energy has entered the waveguide and traveled sideways in it before leaking back again into the coupling prism or the high-index half-space. If we can arrange for the linear shift Δu to be large enough so that the exit beam is totally displaced from the position that a simply reflected beam would take, then for a small distance along the waveguide, essentially all the Gaussian beam energy will be contained in the guide. Termination of the high-index half-space in this region would cause that energy to remain trapped in the guide and 100 percent coupling would have been achieved.

Unfortunately this is not possible. When we examine the case more carefully we find that a total phase shift from one wing of the Gaussian beam angular spectrum to the other of more than 2π is required. Thus, our approximation to a straight line ϕ versus θ response is not valid. We must make a more accurate approximation. Since the π phase flip in ϕ occurs over a small range of θ, typically $10^{-2} \rightarrow 10^{-3}$ radians, it is reasonable to recast (28) in the form

$$\tan\phi = \frac{\beta_2}{\gamma_1}\left[\frac{\left(\frac{\partial X}{\partial\theta}\right)_{\theta_M}\cdot\Delta\theta + \exp(-2\beta_2 a)Y}{\left(\frac{\partial X}{\partial\theta}\right)_{\theta_M}\cdot\Delta\theta - \exp(-2\beta_2 a)Y}\right]. \qquad (40)$$

Now, (β_2/γ_1), $((dX)/(d\theta))_{\theta_M}$, $\exp(-2\beta_2 a)$, and Y are essentially constant over the range of interest for $\Delta\theta$. We can, therefore, replace (40) with the particularly simple result

$$\tan\phi = \left[\frac{K\cdot k_u + 1}{K\cdot k_u - 1}\right] \quad \text{or} \quad \left(\phi + \frac{\pi}{4}\right) = \tan^{-1}(-K\cdot k_u) \qquad (41)$$

where

$$K = \left(\frac{\partial X}{\partial\theta}\right)_{\theta_M} \exp(2\beta_2 a)(k_w\cdot Y)^{-1}.$$

This result is exact if $\beta_2 = \gamma_1$ and it is an excellent approximation over at least the range $2\gamma_1 > \beta_2 > 0.5\gamma_1$. The prime effect of the (β_2/γ_1) factor is to increase or decrease slightly the length of the linear slope region of the ϕ versus θ curve, with a corresponding sharpening or smoothing of the curve at either end. The total phase change and the slope at the center remain essentially unaltered. Using this result, we have numerically integrated (37) in the form

$$E'(k, u) = (1/4\pi)E_0R^2 \cdot \int_{-\infty}^{\infty} \exp(-k_u^2R^2/4)\exp[-i(k_u + 2\phi(k_u))]\,dk_u \qquad (42)$$

where

$$\phi(k_u) = \tan^{-1}(-K\cdot k_u).$$

$E'(K, u)$ is now an expression for the electric field of the reflected beam at the waveguide-coupler interface. It includes all the effects due to the phase changes caused by the waveguide, namely, a changed amplitude with position coordinate relative to the input beam and also data on the phase distribution within the exit beam itself since $E'(K, u)$ is complex. We have evaluated the modulus $|E'(K, u)|^2$, which gives the output-spot spatial power distribution and also the relative spatial phase distribution defined as

$$\eta(u) = \tan^{-1}[\mathscr{I}(E'(K, u))/\mathscr{R}(E'(K, u))]. \qquad (43)$$

Equation (40) contains the variables d, E_0, u, K (and k_u, which is integrated out). For the purposes of evaluation, we arbitrarily set the factors $(1/4\pi)E_0R^2 = 1 = R^2/4$, which, in turn, defines u to be in units of half Gaussian-spot radii, i.e., $u = 2$ defines the $1/e$ point in the electric field of the incident beam. The variable K has been set successively to 0, 1, 2, 3, and 6, to produce the set of curves shown in Fig. 4 ($K = 0$ effectively gives the input beam shape for reference).

From Fig. 4, we see clearly how the reflected spot is shifted laterally with respect to the input spot. We also see that the shift[1] is not simple, but that as we increase the coupling parameter K (decrease the low-index film thickness), the spot not only shifts but also splits. Ultimately for very large K, the first split-off spot, which is centered at $u = -1$ for $K = 6$ in the figure, becomes the dominant one, the shifted spot decays to nothing, and the curve of power versus u is identical to that for $K = 0$. This situation corresponds to no evanescent gap.

An alternative method of presenting these data is to

[1] *Note added in proof:* A linear shift in a reflected beam due to phase changes in its plane wave components has previously been observed in studies of the Goos–Haenchen effect. For further reading on this and numerous references, see H. K. V. Lotsch, "Reflection and refraction of a beam of light at a plane interface," *J. Opt. Soc. Am.*, vol. 58, pp. 551–561, April 1968.

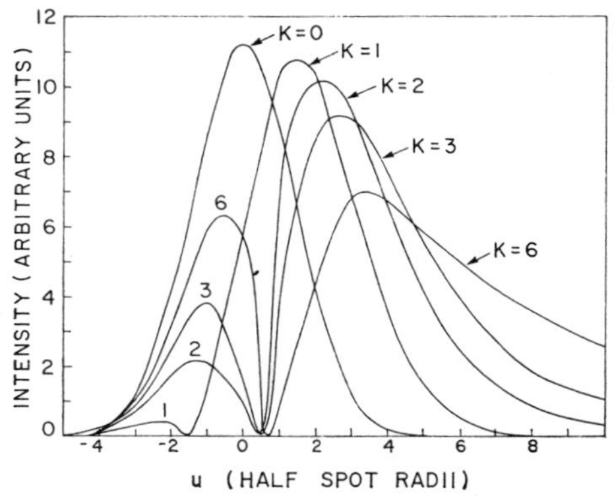

Fig. 4. Intensity profiles of incident ($K = 0$) and reflected spots for various values of coupling parameter K against spatial coordinate u.

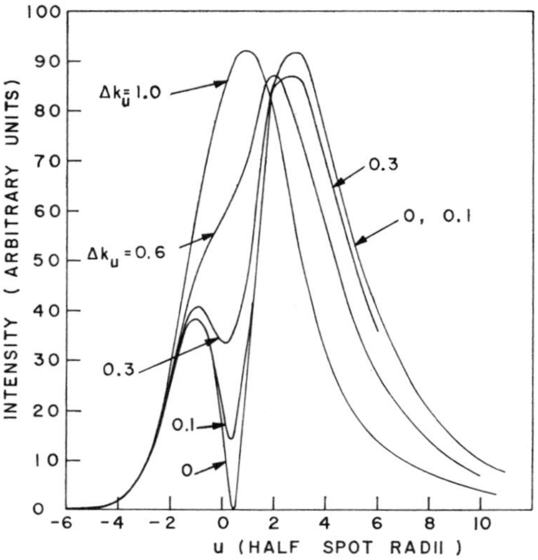

Fig. 5. Change in reflected spot shape as the input Gaussian beam is angularly tuned through θ_M. ($\Delta k_u = 1$ corresponds to angular mismatch of ½ beam divergence). Values are for $K = 3$.

examine how the reflected spot behaves as in the input beam direction θ is varied relative to the mode-coupling angle θ_M. This is readily done by numerically weighting the function $\phi(k_u)$,

$$\phi(k_u) = \tan^{-1} (K \cdot (k_u - \Delta k_u)), \tag{44}$$

whence the beam offset $(\theta - \theta_M)$ is given by $(\Delta k_u / k_w)$. Note that the half-beam divergence to the $1/e$ point corresponds to $\Delta k_u = (2/R)$ (numerically $R = 2$ in this paper). In Fig. 5 we show the results of such an evaluation for $K = 3$ and values of Δk_u of 0, 0.1, 0.3, 0.6, and 1.0. The spot format is independent of the sign of Δk_u. This sequence of curves illustrates the spot shapes seen experimentally [3] as a laser beam is angularly tuned through the angle of incidence for maximum waveguide coupling. The curves for other values of K show essentially the same features. We will see that the value $K = 3$ corresponds to optimum coupling.

In Fig. 6 we show the corresponding phase distribution, $\eta(u)$, in the reflected spot as a function of u for values of $\Delta k_u = 0; 0.2; 0.4; 0.6$. We note that perfect angular match-

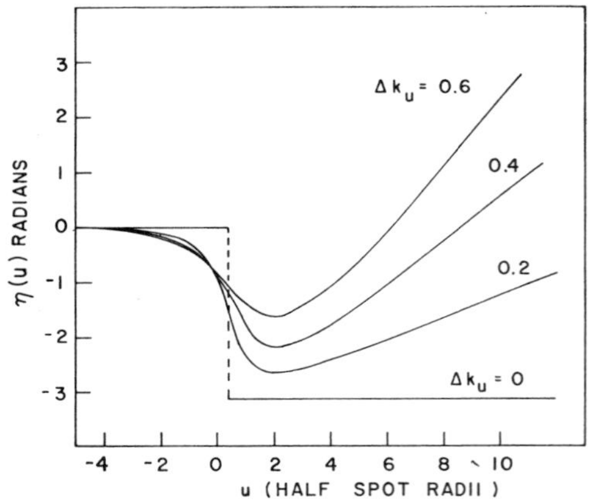

Fig. 6. Phase distribution $\eta(u)$ of reflected spot for various value. of angular mismatch as a function of u.

ing of the input beam to the waveguide mode results in a uniform phase distribution for each of the two spots, the shifted and the split off, these being in antiphase to each other. A small angular mismatch produces a nearly linearly varying phase across the shifted spot tail. This can be seen experimentally as the presence of interference fringes in the intermediate field region [3], the near field being the waveguide coupler interface, the far field being the Fourier transform of the near field or the pattern at infinity.

Study of Optimum Coupling

Our analysis so far has been concerned only with infinite structures. However, we have shown that when a localized beam is incident upon the structure, the reflected beam is partially shifted sideways when coupling to a waveguide mode occurs. We will now calculate the energy contained within the waveguide as a function of position coordinate u and seek the conditions for it to be a maximum. We numerically evaluate the integral

$$I(K, u) = \int_{-\infty}^{u} E'(K, u')\, du' \tag{45}$$

for a series of values of K and u. The value of $I(0, u)$ represents the integrated energy that has arrived from the incident beam at the guide structure in the space preceding the point u. The value of $I(K, u)$ is the energy that has left the guide structure up to that point. Hence, the value of $I(0, u) - I(K, u)$ represents the energy contained in the guide at the point u. We, therefore, define the coupling efficiency factor as follows:

$$\mathcal{E}(K, u) = [I(0, u) - I(K, u)]/I(0, \infty). \tag{46}$$

This is the fraction of the total incident beam energy that is in the guide as a function of coupling parameter K and the position coordinate u. Note that when $u = 0$, $\mathcal{E}$ must be zero and also for $u = \infty$. At some point in between it will take on its maximum value. In Fig. 7 we will plot the values of $\mathcal{E}(K, u)$ for a sequence of K values, against u. From these curves, we extract maximum values

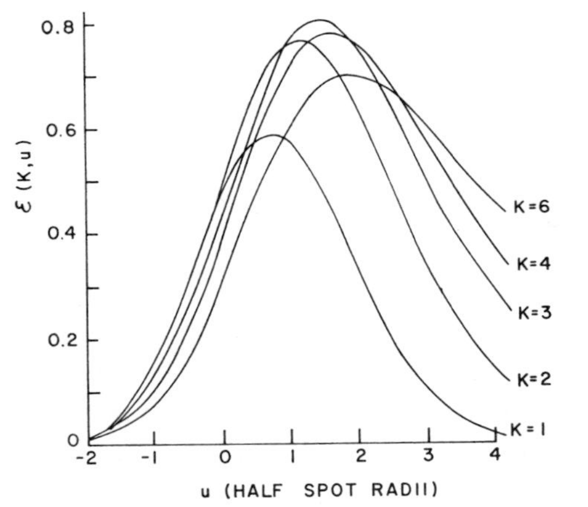

Fig. 7. Values of coupling $\mathcal{E}(K, u)$ versus u for various K values.

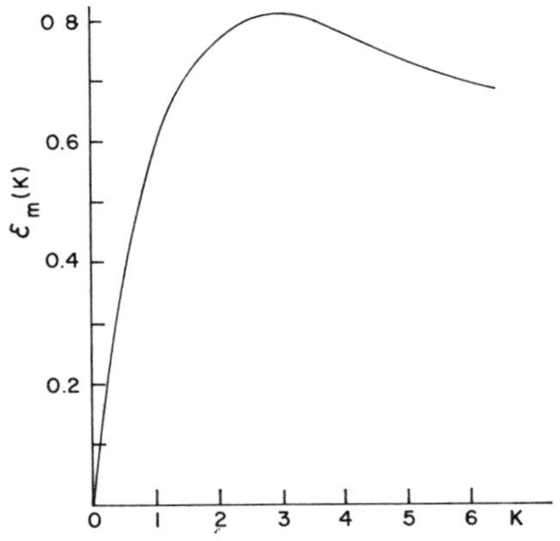

Fig. 8. $\mathcal{E}_m(K)$, the maximum values of $\mathcal{E}(K, u)$ versus K.

($\mathcal{E}_m(K)$) of $\mathcal{E}(K, u)$ for each value of K and in Fig. 8 we plot $\mathcal{E}_m(K)$ versus K.

We see that $\mathcal{E}_m(K)$ itself has a maximum value of about 81 percent at a K value of about 3. This is indicative of the coupling efficiency possible from the Gaussian TEM_{00} beam to the thin-film waveguide. Actually, this value still applies only to the infinite structure. We see from Fig. 7 that although at the point $u = 1.56$ and for $K = 3$, 81 percent of the incident energy is in the waveguide; this energy immediately starts to leak out again and when $u = 4$ only 28 percent remains. Thus to obtain efficient coupling in practice, the high-index half-space must be truncated in the y–z plane at the point corresponding to $u = 1.56$. Strictly speaking the act of truncation invalidates our plane-wave analysis. However, we note that at the point $u = 1.56$, 94 percent of the incident beam energy has arrived and thus we would not expect our plane-wave description to be seriously disturbed by the truncation operation.

Conclusion

We have shown how the coupling through the evanescent field of the TIR prism to a thin-film waveguide comes about, and how it depends critically upon the beam parameters and the film parameters. We have studied the intensity and phase distribution in the reflected spot on the coupler-waveguide interface and we will report elsewhere experimental verification of the effects predicted [3]. For a Gaussian beam incident on this interface we have calculated an optimum coupling of energy from the incident beam into the thin-film waveguide of 81 percent. This agrees with the previously published result and it is not subject to any restriction of weak coupling. The theory presented may be readily generalized to cover the case of any other form of incident beam.

Appendix I

We list below the form of the subdeterminants of the matrix 21 that are of prime interest to us. We use the relations for J, J^*, G, H, X, and Y to simplify the expressions [see (23)–(27)]

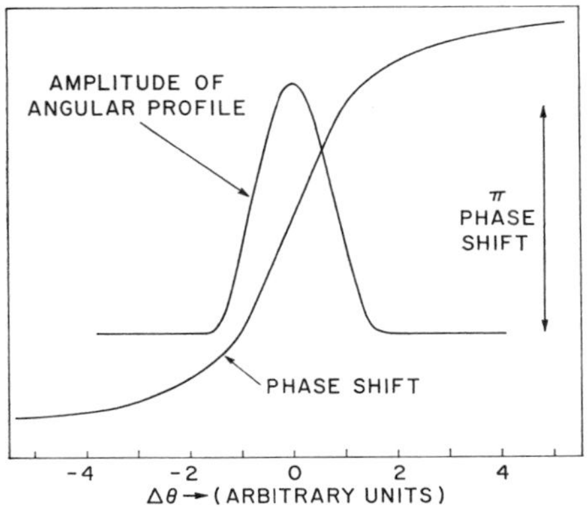

Fig. 9. The relative disposition of the phase term ϕ and the Gaussian angular spectrum for most efficient coupling.

$$|bcdefg| = J(G + iH)$$

$$|acdefg| = J^*(G - iH)$$

$$|badefg| = 4i\gamma_1 \exp\left(-(\beta_4 + \beta_2)d\right) X k_0^{-3}$$

$$|bcaefg| = 4i\gamma_1 \exp\left(-(\beta_4 - \beta_2)d\right) Y k_0^{-3}$$

$$|bcdafg| = 4i\gamma_1\beta_2 \exp\left(-(i\gamma_3 + \beta_4)d\right)(\gamma_3 + i\beta_4) k_0^{-3}$$

$$|bcdeag| = 4i\gamma_1\beta_2 \exp\left((i\gamma_3 - \beta_4)d\right)(\gamma_3 - i\beta_4) k_0^{-3}$$

$$|bcdefa| = 8i\gamma_1\gamma_3\beta_2.$$

Appendix II

We note that (40) can be recast in an analytically simpler form by the use of some standard Fourier transforms [5]. The function $\exp(-2i\phi(k_u))$ is identical apart from a sign change and absolute phase shift of $\pi/4$ to

$$f\cdot(k_u) = (1 + iK\cdot k_u) / (1 - iK\cdot k_u)$$
$$= (2/K)(1 - iK\cdot k_u)^{-1} - 1.$$

The Fourier transform of $f(k_u)$ is given by the relation

$$F(u) = (2/K)\exp(-u/K)U(u) - \delta(u)$$

where

$$U(u) = 1 \qquad u > 0$$
$$= 0 \qquad u \leq 0.$$

Equation (42) can now be written as

$$E'(K, u) = E_0 \int_{-\infty}^{\infty} \exp(-l^2/d^2)$$
$$\cdot [(2/K) \exp(-(l - u)/K)U(l - u) - \delta(l - u)]\, dl.$$

ACKNOWLEDGMENT

The author wishes to thank R. E. Hufnagel, E. L. Kerr, J. L. Kreuzer, and F. Zernike for many helpful discussions. Further, he would like to thank R. E. Hufnagel and F. Zernike together with the referees of this paper for their constructive comments on the script.

References

[1] P. K. Tien, R. Ulrich, and R. J. Martin, "Modes of propagating light-wave in thin deposited semiconductor films," *Appl. Phys. Letters*, vol. 14, pp. 291–294, May 1969.
[2] See, for example and further references, L. Bergstein and C. Shulman, "The frustrated total reflection filter, spectral analysis," *Appl. Opt.*, vol. 5, pp. 9–21, January 1966.
[3] J. E. Midwinter and F. Zernike, "Experimental studies of evanescent wave coupling into a thin-film waveguide," *Appl. Phys. Letters*, vol. 16, pp. 198–200, March 1, 1970.
[4] See, for example, C. H. Walter, *Travelling Wave Antennas*. New York: McGraw-Hill, 1965.
[5] A. Papoulis, *The Fourier Integral and Its Applications*. New York: McGraw-Hill, 1962.

Variable Tunneling Excitation of Optical Surface Waves

JAY H. HARRIS, MEMBER, IEEE, AND RICHARD SHUBERT, STUDENT MEMBER, IEEE

Abstract—**A method for coupling an optical beam into thin films utilizing shaped tunneling regions is described. It is shown that in principle all of the power can be coupled into the film. For ease of fabrication, a uniform gradient structure is preferable, however, and yields coupling of over 90 percent for incident coherent Gaussian beams. Experimental coupling values with gradient tunneling regions are in excess of 50 percent. Good mode isolation and elimination of prism edge effects are major features of the coupler. The experimental coupling values are obtained from a transmission measurement utilizing two prisms.**

INTRODUCTION

THE MECHANISM of tunneling has been shown to be an effective method for coupling power from an optical beam into a dielectric waveguide [1]–[3], [9]. Fig. 1(a) illustrates the basic features of the technique. A beam that is large compared to both the optical wavelength and the thickness of the film is incident from a high index region at an angle greater than the critical angle of a low index tunneling layer atop the high index waveguide. The angle of incidence is also greater than the critical angle of the substrate so that no power passes through the substrate. Under appropriate specification of the thickness of the tunneling layer, the major portion of the power in the incident beam may be caused to pass into the waveguide. When the region in which the beam is incident is terminated, the energy in the film is converted into a propagating surface wave. If scattering at the edge of the high index region is neglected, maximum coupling into the surface wave from a Gaussian beam has been shown to be 75 percent and measured values well over 50 percent have been reported [3].

In this paper we extend the discussion of the coupling problem to include tunneling layers that vary in thickness along the direction of incidence of the beam. In principle an appropriately designed tunneling structure can provide nearly complete coupling of the beam into the waveguide. For practical purposes involving ease of fabrication, simple tunneling structures appear preferable. The uniform gradient structure is the simplest of these and is discussed in detail. It is shown that theoretical coupling values of over 90 percent can be achieved. Furthermore, such a structure has the advantage of removing the edge of the high index region from interaction with the light beam. This removal eliminates both the theoretical question of scattering at the edge of the high index region and the practical difficulties of obtaining a square corner at the edge of the high index region. Experimental results bear out these advantages.

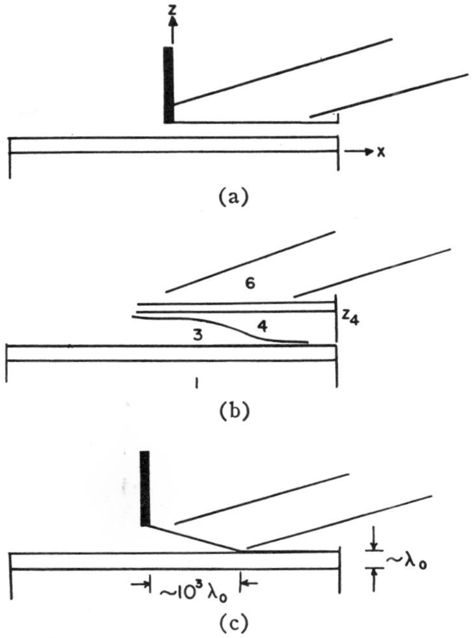

Fig. 1. Beam incidence for surface wave coupling by means of the tunneling mechanism. (a) Uniform tunneling layer below prism with forward face painted black. (b) General variable thickness tunneling layer and dielectric phase matching regions. (c) Uniform gradient tunneling region.

Coupling efficiency is computed for the general structure shown in Fig. 1(b). The structure consists of a substrate of low refractive index (region 1), a dielectric waveguide of high refractive index (region 2), and a tunneling region of low index (region 3) whose thickness may vary. Above the tunneling region are high index regions 4 and 5 in which the permittivity may vary with x. Finally, there is the high index region 6 through which the beam enters. Regions 4 and 5 are included in Fig. 1(b) to indicate how coupling efficiency can be maximized for arbitrary tunneling region thickness variations, as will be discussed later. For practical purposes, regions 4, 5, and 6 consist of a single homogeneous material and represent a shaped prism.

Manuscript received May 20, 1970; revised August 28, 1970. The work reported here was supported under NSF Grant GK10319.
The authors are with the Department of Electrical Engineering, University of Washington, Seattle, Wash. 98105.

Reprinted from *IEEE Trans. Microwave Theory Tech.*, vol. MTT-19, pp. 269–276, Mar. 1971.

The efficiency of coupling through the tunneling region is obtained from the reciprocity theorem which yields the field excited in the film E' and which we write [3]

$$\int E' \cdot J dv = -(2\pi)^2(2/\omega\mu) \int u_6 E^i(k) \cdot E(-k) dk \quad (1)$$

where J is a fictitious current distribution that has the transverse shape of a surface wave propagating in and near the guiding film and is the source of a field $E(r)$. Here $E(k)$ is the transform of the field $E(r)$ in the plane that forms the lower boundary of region 6, $E^i(k)$ is the transform of the incident polarized beam in the same plane, and $u_6 = (k_6^2 - k^2)^{1/2}$ is the z component of the plane wavenumbers. It is assumed that the incident electric field is polarized in the y direction and is a collimated beam of width much greater than the wavelength. The exact integral (1) can then be closely approximated by the following integral over the fields at the lower boundary of region 6:

$$\int E' \cdot J dv = -(2u_6^i/\omega\mu) \int E_y^i(\varrho) E_y(\varrho) d\varrho \quad (2)$$

where $u_6^i = (k_6^2 - k_x^{i2})^{1/2}$ and k_x^i is the component of the propagation constant of the beam that lies parallel to the film. The power coupled into a surface wave beam of the same form as $J(E \propto J)$ may be shown by direct waveguide considerations to be

$$P = (k^{(1)}/2\omega\mu)\left(\int E' \cdot J dv\right)^2 \Big/ \left(\int J^2 dv\right) \quad (3)$$

where $k^{(1)}$ is the propagation constant of the surface wave. Equation (3) combined with (2) yields the coupling efficiency when the incident beam (E_y^i) is of unit power and will be used to determine optimum coupling conditions.

The Field Produced by a Current Source in the Waveguide

To compute coupling efficiency it is necessary to determine the field produced by the current source J. This is accomplished to the Wentzel–Kramer–Brillouin (WKB) approximation [4] as follows. Assume that J is located in a portion of the structure in which the surface wave propagates without loss, i.e., to the left of the tunneling region in Fig. 1(b). J itself has the form of the transverse behavior of a surface wave and by reason of the orthogonality conditions with respect to the continuous spectrum excites only the surface wave. The wave that is launched in the positive x direction enters a transition region whose linear dimension is on the order of the beamwidth. In this region the surface wave is converted to a leaky wave.

In the transition region it is assumed that the leaky wave propagates as the Wentzel–Kramer–Brillouin type of solution in the particular sense that reflected surface wave components are neglected. Reflections may be treated using the methods of Marcuse [5] and Snyder [6], for example, but they will be suppressed in the present analysis. The principal justification for this suppression is the fact that the transition region length is typically on the order of 10^2 to 10^3 λ so that the transition region constitutes a gentle match from the surface wave to the portion of the continuous spectrum that is of interest.

To obtain the Wentzel–Kramer–Brillouin type of wave for a spatially varying tunneling region, the dual stipulations that there are no reflected waves and that power is conserved are imposed. At the same time the requirement that the wave above the tunneling region satisfy the wave equation to first-order derivatives in the amplitude of the wave is relaxed. Specifically, we derive the field solution from the leaky waves that propagate in the presence of a uniform tunneling region. The electric field above such a tunneling region at the lower boundary of region 5 is of the form [3]

$$E_y = A \exp[ik^0 x - y^2/2\sigma^2 + iu_4 t_4] \quad (4)$$

where k^0 is the complex propagation constant of the leaky wave and t_4 is the thickness of region 4. k^0 will be in the form $k^0 = k^{(1)} + k^{(2)} + i\alpha$, where $k^{(1)}$ is the propagation constant of the surface wave in the presence of an infinitely thick tunneling region and $k^{(2)}$ and α are small corrections to k^0 due to the leakage of energy. We also assume that there are no reflections from region 5 (implying region 5 matches region 4 to region 6 at the angle of interest. The Gaussian in (4) represents the transverse spatial variation. Radiation from an aperture distribution of the form (4) occurs over a narrow angular spectrum centered about $k_x = k^{(1)} + k^{(2)}$, $k_y = 0$, so long as α is small compared with $k^{(1)}$. The longitudinal magnetic field associated with (4) may thus be approximated by the plane wave field

$$H_x = (Au_4^{(2)}/\omega\mu) \exp[ik^0 x - y^2/2\sigma^2 + iu_4 t_4] \quad (5)$$

where $u_4^{(2)} = [k_4^2 - (k^{(1)} + k^{(2)})^2]^{1/2}$.

The power radiated by the leaky wave expressed in (4) and (5) over a region $x_a < x < x_b$, $z = z_4$, may be computed by integrating $E_y H_x^*$ over the region with the result

$$P = |A|^2 (u_4^{(2)}/2\omega\mu)(\sqrt{\pi}\sigma/2\alpha)[e^{-2\alpha x_a} - e^{-2\alpha x_b}]. \quad (6)$$

Equation (6) may be used to compute $|A|$ when the total power in the leaky wave is known. Our interest, however, lies in determining $|A|$ when the tunneling region thickness and therefore α varies with x. A simple starting point is to investigate the amplitude of A when the tunneling region has one thickness in region a $(0 < x < x_a)$ and another thickness in region b $(x > x_a)$. Labeling parameters with a subscript a or b corresponding to the first or second region and assuming that P_0 is the total available power then under the assumption of no reflected waves, we find A_a is independent of region b and may be expressed

$$|A_a| = (2P_0\omega\mu/\pi^{1/2}\sigma)^{1/2}(\alpha/u_4^{(2)})_a^{1/2}. \tag{7}$$

To conserve power, we then have

$$P_0 = P_0(1 - e^{-2\alpha_a x_a}) + |A_b|^2(\pi^{1/2}\sigma/4\omega\mu)(u_4^{(2)}/\alpha)_b e^{-2\alpha_b x_b} \tag{8}$$

implying

$$|A_b| = 2(P_0\omega\mu/\pi^{1/2}\sigma)^{1/2}(\alpha/u_4^{(2)})_b^{1/2}. \tag{9}$$

The continuous function that incorporates the behavior (9), as well as the appropriate exponential behavior when we pass to the limit in which the tunneling region varies continuously, is

$$E_y = 2(P_0\omega\mu/\pi^{1/2}\sigma)^{1/2}(\alpha(x)/u_4^{(2)}(x))^{1/2} \cdot \exp\left(i\int_0^x k^0 dx - y^2/2\sigma^2 + iu_4^{(2)}(x)t_4(x)\right). \tag{10}$$

We assume that (10) is the appropriate solution for the varying tunneling region. It is possible to prove that (10), combined with the continuous equivalent of (5), leads to conservation of power so long as $\alpha > 0$. This is accomplished by considering the integral

$$P = \frac{1}{2}\int_0^\infty dx \int_{-\infty}^\infty dy E_y H_x^* = P_0 \int_0^\infty 2\alpha \exp\left(-2\int_0^x \alpha dx\right)dx = P_0\int_0^\infty e^{-u}du = P_0. \tag{11}$$

Equations (11) are valid so long as $\alpha > 0$.

The solution (10) is also in the form that is obtained from the full wave solution for the fields when the tunneling layer is uniform [3]. A factor $\alpha^{1/2}$ results from the residue evaluation of the expression for E_y at the complex pole that gives rise to the leaky wave. On the other hand, (10) is not a solution of the approximate wave equation that customarily gives rise to the Wentzel–Kramer–Brillouin solution, viz.,

$$k^{0\prime}A + 2k^0A' = 0. \tag{12}$$

This result is to be expected, however, since the leaky wave above the tunneling layer is basically a source excited field, and should be found as the solution to an inhomogeneous equation rather than (12). Within the guiding film, the wave satisfies an equation of the form (12) which results in a field whose amplitude does not approach zero when α approaches zero.

Phase Matching Considerations

To match a field of the form (10) to a collimated beam, a linear phase distribution across the upper bounding plane of region 5 is required. Elimination of nonlinear phase terms in (10) is effected through cancellation of the appropriate portion of $\mathrm{Re}\int_0^x k^0dx + u_4^{(2)}(x)t_4(x)$. Parameter t_4 is fixed through specification of the tunneling layer thickness t_3, which in turn is employed to obtain a prescribed $\alpha(x)$. Cancellation of those terms that arise from $\mathrm{Re}\int_0^x k^0 dx$ therefore rests on specification of $u_4^{(2)}(x)$, which implies that the refractive index n_4 is varied along the structure to provide desired phase properties. Such index control is possible in principle although obtaining a specific index distribution offers considerable fabrication difficulties and is unnecessary for the uniform gradient profile. In this section we wish to indicate the nature of the phase term $\mathrm{Re}\int_0^x k^0dx$ and show how the tunneling structure can be designed in principle to provide a linear phase distribution across the plane z_4.

The phase $\mathrm{Re}\int_0^x k^0 dx$ an integral over the eigenvalues of the structure, i.e., the zero of the system function $D_{1,5}$ [3]. Since there are no reflections from the interface at z_4 when a matching region (region 5) is provided, the eigenvalue is also a solution of $D_{1,4}=0$ which reduces to the eigenvalue equation

$$\sin\left(\tan^{-1}\frac{K_1}{iK_2} + \tan^{-1}\frac{K_3}{iK_2} - u_2t_2\right) = \sin\left(\tan^{-1}\frac{K_1}{iK_2} - \tan^{-1}\frac{K_3}{iK_2} - u_2t_2\right) \cdot \exp\left[-2i\tan^{-1}\frac{K_3}{iK_4} - 2|u_3|t_3\right] \tag{13}$$

where K_j is the wave impedance of layer j and is u_j for the TE waves under discussion or u_j/ϵ_j for TM waves. For a tunneling layer of infinite thickness ($t_3 \to \infty$), the right side of (13) is zero and the surface wave eigenvalue $k^{(1)}$ is obtained. Second-order terms obtained by differentiating the left side of (13) and suppressing the arctangent derivatives yield

$$k^{(2)}, \alpha = \left\{\frac{u_2}{t_2k}\left(\frac{2K_2|K_3|}{K_2^2 + |K_3|^2}\right) \cdot e^{-2|u_3|t_3}\left[\frac{|K_3|^2 - K_4^2}{|K_3|^2 + K_4^2} + \frac{2|K_3|K_4}{|K_3|^2 + K_4^2}\right]\right\}_{k^{(1)}}. \tag{14}$$

The exponential in (14) is real since u_3 is imaginary at $k = k^{(1)}$. Of importance in (14) is the fact that $K^{(2)}$ is zero when $|K_3| = K_4$ so that the second-order phase term introduced when the surface wave is converted to a leaky wave may be made zero by selecting either the wave propagation constant (which can be specified through selection of the film thickness), or the tunneling region or region 4 indices to satisfy the relation

$$k^{(1)^2} = (k_3^2 + k_4^2)/2. \tag{15}$$

That condition (15) results in a zero value of $k^{(2)}$ is evident in (13) where the phase $2\tan^{-1} K_3/iK_4$ becomes $\pi/2$ so that the right side of (13) is purely imaginary. This zero remains whether or not the derivatives of the arctangent functions on the left of (13) are included in (14).

A third-order term (second-order correction) to the real part of the eigenvalue may be found by iteration. Employing the same linear expansion of the left of (13) as was used to obtain (14), we find

$$k^{(3)} = \mathrm{Re}\left\{\frac{u_2}{t_2 k}\sin\left(\tan^{-1}\frac{K_1}{iK_2} - \tan^{-1}\frac{K_3}{iK_2} - u_2 t_2\right)\right.$$
$$\left.\cdot\exp\left[-2i\tan^{-1}\frac{K_1}{iK_4} - 2\,|\,u_3\,|\,t_3\right]\right\}_{k^{(1)}+i\alpha}. \quad (16)$$

Under the condition (15), (16) is of order α^2, i.e., $0(10^{-6}/\lambda)$ and is entirely negligible.

What we have shown is that it is possible to reduce $\mathrm{Re}\int_0^x k^0 dx$ to a linear function of x. When the tunneling layer thickness varies in an arbitrary manner, which is the way in which α is controlled, there remains a phase distribution $u_4^{(2)}(x)t_4(x)$ along the plane z_4. Since $t_4(x)$ may be taken to be nonzero, $u_4^{(2)}t_4$ can in principle be linearized by choosing $k_4(x)$ to satisfy

$$(k_4^2 - k^{(1)2})^{1/2} = (a_0 + a_1 x)/t_4(x). \quad (17)$$

When the phase is linearized, an effective match to the incident beam can be achieved. When the tunneling layer thickness varies linearly, there is no need to vary the refractive index of region 4.

Coupling Efficiency for an Arbitrary Structure

Coupling efficiency can be computed from (2) and (3) using the Wentzel–Kramer–Brillouin field (10) once the parameter P_0 of (10) is related to J of (3). P_0 is the power in the surface wave launched in the x direction. Since the current J launches a magnetic field of amplitude $J/2$, the power P_0 in the leaky wave that is of width $\sigma(\sigma \gg \lambda)$ is

$$P_0 = \left[\int J^2 dv\right]\omega\mu/2^3 k^{(1)} \quad (18)$$

and the power coupled into the surface wave from (2) and (3) is

$$P = (u_4{}^i/\omega\mu)^2\left[\int E_y{}^i E_y d\varrho\right]^2/4P_0. \quad (19)$$

Equation (19) yields the coupling efficiency when $E_y{}^i$ is an incident beam of unit power, i.e.,

$$E_y{}^i = (\mu/\epsilon_4)^{1/4}(2/\pi\sigma^2)^{1/2}$$
$$\cdot\exp\left[-\frac{(x-x_0)^2}{2\sigma k_4/u_4{}^i} - y^2/2\sigma^2 + ik_x{}^i x\right] \quad (20)$$

where $\theta = \sin^{-1} k_x{}^i/u_4{}^i$ is the angle of incidence of the beam and x_0 is the center of the beam on z_4. Placing (20) and (10) into (19) and carrying out the integration over y yields for the coupling efficiency

$$P = \left\{\int_0^\infty \left[\left(\frac{2d}{u_4{}^{(2)}/u_4{}^i}\right)^{1/2}\right.\right.$$
$$\cdot\exp i\left(\int_0^x k^0 dx + u_4{}^{(2)}t_4\right)\Bigg]\Bigg[(\pi^{1/2}\sigma')^{-1/2}$$
$$\left.\left.\cdot\exp\left(-\frac{(x-x_0)^2}{2\sigma'^2} - ik_x{}^i x\right)\right]dx\right\}^2 \quad (21)$$

where $\sigma' = \sigma k_4/u_4{}^i$ is introduced as the effective beamwidth of the incident beam.

Expression (21) contains terms α, k^0, $u_4{}^{(2)}$, and t_4 that may vary with x and are used to achieve high coupling. The term within each bracket when multiplied with its own complex conjugate and integrated over the indicated domain yields essentially unity provided the point of incidence of the beam x_0 is sufficiently removed from $x=0$, the beginning of the tunneling region, and provided $u_4{}^{(2)}$ is essentially constant. In the succeeding material $u_4{}^{(2)}$ will be taken to be a constant. Since both terms of the integrand yield unity, optimization of coupling consists of providing a tunneling region that makes the first bracket in (21) look as much like the complex conjugate of the second bracket as possible. In principle amplitude and phase matching can be achieved to a high degree of accuracy, and nearly 100 percent efficiency is possible.

The Linear Tunneling Region

When the tunneling layer thickness varies in a linear fashion, a significant improvement in coupling efficiency over the uniform tunneling region is achieved without the necessity for providing dielectric phase matching. We take k_3 and k_4 to be constant in x and to satisfy the relation (15). The tunneling layer thickness is defined

$$t_3(x) = \begin{cases} t_0(1 - x/L), & 0 < x < L - \Delta \\ \delta, & x > L - \Delta \end{cases} \quad (22)$$

where Δ and δ are introduced as small quantities only to satisfy analytic properties discussed previously. Fig. 1(c) illustrates the tunneling region. The phase of (21) is then

$$\text{phase }(21) = k^{(1)}x + u_2{}^{(2)}t_0 x/L + k_x{}^i x = 0. \quad (23)$$

The zero in (23) is set in order to maximize the coupling. The negative value of $k_x{}^i$ obtained from (23) is appropriate since the excitation beam arrives from positive x values toward negative x. In practice the appropriate value of $k_x{}^i$ is obtained by rotating the incident beam relative to the film until the phase match is obtained.

When the phase is matched, the coupling efficiency is from (21):

$$P = \left\{ \int_0^\infty (2\alpha)^{1/2} \cdot \exp\left[-2\int_0^x \alpha dx\right]\left[(\pi^{1/2}\sigma')^{1/2} \cdot \exp\left[-(x-x_0)^2/2\sigma'^2\right]\right] dx \right\}^2. \quad (24)$$

Attenuation constant α is expressed in (15). The integral in the exponent of (24) may be computed to be

$$\int_0^x \alpha dx = \frac{u_2}{t_2 k^{(1)}} \frac{2|K_3|K_2}{K_2^2 + |K_3|^2} \cdot \frac{\exp[-2|u_3|t_0]}{2|u_3|t_0/L} (\exp[-2|u_3|t_0 x/L] - 1) \quad (25)$$

where α depends on two parameters t_0 and L which define the slope and the intersection point of the tunneling region. To consider the integral (24) further it is useful to define two different parameters, b and v, and to normalize the integration variable x through the definitions

$$b = 2^{1/2}|u_3|t_0\sigma'/L$$
$$v = x/2^{1/2}\sigma'$$
$$\exp - v_1 = 2^{1/4}\sigma'^{1/2}\left[\frac{u_2}{t_2 k^{(1)}} \frac{2|K_3|K_2}{K_2^2 + |K_3|^2}\right]^{1/2} \cdot \exp[-|u_3|t_0]. \quad (26)$$

The resulting expression for the efficiency is

$$P = \left\{ \int_0^{L/2^{1/2}\sigma'} (2^{3/4}/\pi^{1/4}) \cdot \exp\left[bv - v_1 - \frac{e^{-2v_1}}{2b}(e^{2bv} - 1) - (v - v_0)^2\right] dy \right\}^2. \quad (27)$$

Our objective is to maximize (27). The integral is a three-parameter function involving b, v_1, and $v_0 = x_0/\sigma'\sqrt{2}$. These parameters define the slope of the tunneling region, the initial thickness of the tunneling region, and the point of incidence of the center of the beam, respectively. The limit of the integral is chosen at a point at which the integrand is small and does not represent an important parameter. Attempts have been made to maximize (27) in analytic fashion, but without success because of the presence of the three parameters. As a consequence, we have resorted to computational methods. To simplify these computations we have imposed what seems a reasonable condition for the maximization, which is that the peaks in the incident Gaussian and in the radiating surface wave be coincident. Interpreted in terms of the parameters in (27) this imposes the restriction

$$v_1 = bv_0 - \ln b^{1/2}. \quad (28)$$

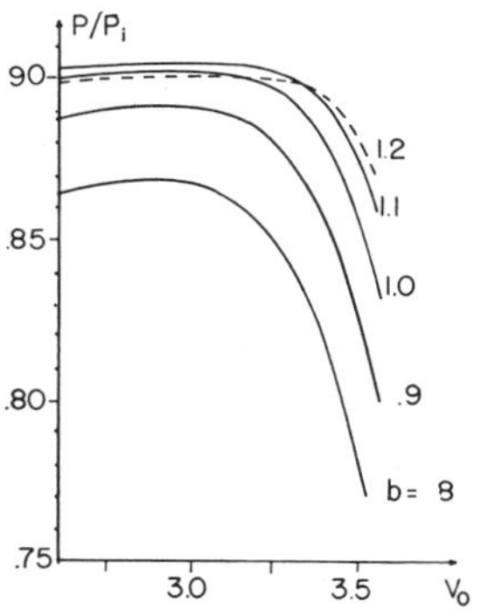

Fig. 2. Surface wave coupling efficiency with uniform gradient tunneling region as a function of incident beam position, with tunneling region slope as parameter.

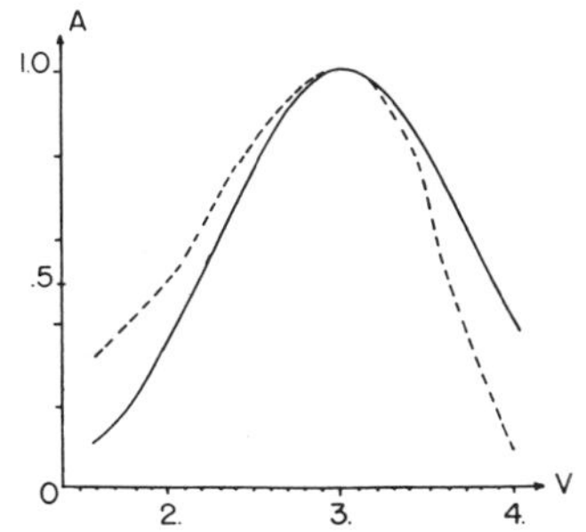

Fig. 3. Transverse leaky wave electric field (dotted) above a uniform gradient tunneling layer produced by a source in the waveguide under conditions that yield 90.5 percent coupling efficiency. The incident Gaussian (solid) is shown for comparison.

The computations have been further simplified by ranging v_0 about a point one unit from the effective end of the tunneling region which corresponds to a distance $x_0 = 2^{3/2}\sigma'$ from the end and b about the value unity. At these values of x_0 and b, the Gaussian and leaky wave amplitudes agree to quadratic terms in their exponents. Results of these computations for an arbitrary L of $6\sigma'$ are shown in Figs. 2 and 3.

Fig. 2 shows the coupling efficiency as a function of v_0, the point of incidence of the beam, with the tunneling layer slope b as parameter. The curves rise with increasing b to a value $b = 1.10$ and then decrease for $b > 1.10$. A maximum efficiency of 90.5 percent has been noted at $b = 1.10$, $v_0 = 3.05$. Additional curves for finer increments of b and for considerably larger values of b have been obtained but are not shown in Fig. 2. Fig. 3 illustrates the degree to which the amplitude distribution of the leaky wave and the incident Gaussian coincide under the conditions that yield 90.5 percent efficiency. The leaky wave field typically rises more slowly than the Gaussian and falls more rapidly. The integral of the product of the leaky wave and Gaussian fields actually yields the number 0.95 whose square is the indicated efficiency. If the tunneling region were shaped so that its slope was slightly larger than the indicated optimum linear slope to the left of the point of incidence of the beam and slightly less to the right, the dashed curve in Fig. 3 could be made to match the Gaussian closely and nearly 100 percent coupling could be achieved.

Utilizing (26) and (28), the optimum coupling conditions may be interpreted in terms of the physical parameters. The inverse of the slope is found to be

$$L/t_0 = 8.08(|u_3|/u_4)\sigma/\lambda. \qquad (29)$$

The factor $|u_3|/u_4$ is unity when the condition for phase matching (15) is utilized with TE waves and is ϵ_3/ϵ_4 for TM waves. Examination of (29) reveals that the required slope is quite gentle. For a 1-mm-diam beam ($\sigma = 10^{-3}/2$) at a wavelength of a half micron the slope is on the order of 1/10 000. Under somewhat relaxed conditions of phase matching which still does not introduce serious coupling loss, the slope may be increased to the order of 1/1000. The point of incidence of the center of the beam from the end of the tunneling region under optimum conditions is found to be

$$L - x_0 = 1.68\sigma' = 1.68(k_4/u_4)\sigma. \qquad (30)$$

The interpretation of (30) is that the edge of the beam is placed close to the point where the tunneling region ends, i.e., where the prism touches the film. The total length of the tunneling region is arbitrary, but should be at least twice the length expressed in (30).

Experimental Results

Gradient tunneling structures have been fabricated from flint glass ($n = 1.648$) rectangular prisms by polishing techniques. The prisms were inserted into a rotary random grinder at an appropriate angle and ground until the wedge region reached approximately halfway across the surface of the prism. This method was convenient and resulted in prisms with substantially improved characteristics over the uniform tunneling region, but the overall optical quality of the prisms was insufficient to provide theoretically predicted coupling efficiency.

Fig. 4 shows the subjective appearance of the prism when it is in place. The prism is clamped tightly to a thin-film waveguide. Because of the presence of both a clamp and black paint on the front surface of the prism, light enters the prism only through the glass substrate and at shallow angles. As a consequence, areas in which the prism is in contact with the film have a bright appearance while regions in which the tunneling layer is greater than about a half wavelength are dark in appearance. The sharp color contrast permits subjective evaluation of the quality of the film and prism surfaces. Our prisms were found to have substantial irregularities.

Measurements have been made with sputtered glass films and whirl-coated polymeric films in the 1–3 μ thickness range that are deposited on glass ($n = 1.520$) substrates. Our films have typically 1–2 dB/cm scattering losses which are comparable to figures reported by others [7]. Although overall system power transmission measurements may be made without difficulty, interpretation of the measurements in terms of coupling efficiency is difficult, particularly with our gradient tunneling prisms because the manner of their fabrication causes the input and output beam positions to be separated by 1 cm. As a case in point, on a 2.5-μ (three-mode) polymeric film we have measured an actual power transfer of 35.5 percent of the power in the incident beam. To relate this figure to coupling efficiency it is necessary to compensate for both beam reflections at the air–prism interfaces and for transmission scattering losses. The interface losses can be eliminated with quarter-wave matching films, but we have not troubled to do this nor have we attempted to match indices in order to satisfy (15). At normal incidence, reflections account for 6 percent losses at each interface. In practice the beam makes an angle on the order of 10° with respect to our prism faces so that reflection losses are somewhat greater than 6 percent at each interface.

Fig. 4. Experimental prism with gradient tunneling region atop a thin-film waveguide on a glass substrate. The beam enters the prism side facing the viewer and strikes the boundary between dark and light regions at the bottom of the prism. Viewed light passes through the substrate and does not penetrate the regions in which the tunneling region is thick thus leaving them dark. Surface imperfections and curvature in the onset of the tunneling region may be noted.

Scattering losses are found to be somewhat inhomogeneously distributed through the films. These have been measured by moving the output prism which in itself introduces reproducibility errors. Scattering losses are deduced from least-square-error fit of the data. Scattering losses of 1–2 dB amount to 21–37 percent of

Fig. 5. Mode patterns of the exit beams of the uniform gradient coupler as photographed on a screen 18 in from the exit prism of a sputtered glass waveguide. The transmission path in the multimode guide was sufficient to provide scattering into all five modes of the guide.

the power. Our prisms introduce an approximately 1-cm minimum transmission path in the film. If the indicated compensation terms are introduced, the 35.5 percent power transmission is interpreted as 51–64 percent coupling efficiency.

The principal mechanism for loss of coupling efficiency relates to the quality of the surface of the gradient tunneling region. Using simple techniques we have produced prisms that provide good isolation between modes in the output beams. Fig. 5 gives a typical example of mode isolation. On the other hand, the phase and amplitude errors introduced by our prisms are sufficient to prohibit achievement of theoretically predicted values of coupling efficiency. A measure of the coherence properties of the prism may be obtained from consideration of the spread in the output beam. In a typical case the output beamwidth has been measured to be about 1.8°. The coherence length of an aperture field that produces such a beamwidth is on the order of tens of wavelengths. At the input prism, therefore, comparable phase distortions are produced and coupling efficiency is significantly degraded.

Conclusions

A substantial improvement in efficiency and other performance criteria in Gaussian beam to thin-film surface wave couplers can be achieved by varying the thickness of the tunneling layer used to effect the coupling. Simple uniform gradient structures provide over 90 percent theoretical efficiency, and nearly 100 percent efficiency is achievable with variable index structures. Definitive experimental verification is dependent on the availability of good quality prisms. With prisms of detectably limited surface flatness, efficiencies in excess of 50 percent have been measured. The coupler described here should prove of substantial aid in the development of techniques for microoptical instrumentation and integrated processing devices [8].

References

[1] J. H. Harris and R. Shubert, "Optimum power transfer from a beam to a surface wave," in *Conf. Abstracts, URSI Spring Meeting*, 1969.
[2] P. K. Tien, R. Ulrich, and R. J. Martin, "Modes of propagating light waves in thin deposited semiconductor films," *Appl. Phys. Lett.*, vol. 14, 1969, p. 291.
[3] J. H. Harris, R. Shubert, and J. N. Polky, "Beam coupling to films," *J. Opt. Soc. Am.*, Aug. 1970.
[4] H. Bremmer, "The WKB approximation as the first term of a geometrical-optical series," *Commun. Pure Appl. Math.*, vol. 4, 1951, p. 105.
[5] D. Marcuse, "Mode conversion caused by surface imperfections of a dielectric slab waveguide," *Bell Syst. Tech. J.*, vol. 48, 1969, p. 3187.
[6] A. W. Snyder, "Excitation and scattering of modes on a dielectric or optical fiber," *IEEE Trans. Microwave Theory Tech.*, vol. MTT-17, Dec. 1969, pp. 1138–1144.
[7] J. E. Goell and R. D. Standly, "Sputtered glass waveguide for integrated optical circuits," *Bell Syst. Tech. J.*, vol. 48, 1969, p. 3445.
[8] R. Shubert and J. H. Harris, "Optical surface waves on thin films and their application to integrated data processors," *IEEE Trans. Microwave Theory Tech.*, vol. MTT-16, Dec. 1968, pp. 1048–1054.
[9] J. E. Midwinter, "Evanescent field coupling into a thin-film waveguide," *IEEE J. Quantum Electron.*, vol. QE-6, Oct. 1970, pp. 583–590.

Holographic Thin Film Couplers

By H. KOGELNIK and T. P. SOSNOWSKI

(Manuscript received June 17, 1970)

Recently P. K. Tien and his co-workers have described a prism coupler as a convenient means to feed light into a single mode of a guiding thin optical film.[1] Distributed couplers of this kind are of great interest for integrated optical devices. In this brief we describe thin film coupling with a thick dielectric grating. To produce these grating couplers, we have used the materials and techniques of holography where dielectric gratings are known to yield high ($\gtrsim$ 90%) diffraction efficiencies.[2] Independently A. Ashkin and E. Ippen[3] suggested that a grating could be used as a thin film coupling device, and very recently M. L. Dakss, *et al.*,[4] have reported successful light coupling into thin films by means of a (thin) phase grating made of photoresist.

Figure 1 shows a grating coupler. A diffraction grating, placed in the vicinity of the guiding film, diffracts the incident light. If the diffracted wave is phase-matched to a mode of the film, then coupling occurs and light is fed into the film. The grating coupler is a distributed coupler just as the prism coupler, and much of the concepts and the theory developed for the latter[1] can be applied.

In order to make a good and efficient grating coupler one has to (*i*) use lossless and scatterfree materials, (*ii*) suppress unwanted grating orders, and (*iii*) provide for a sufficiently deep spatial modulation of the optical phase shift to achieve strong coupling and the associated short coupling lengths. Point (*i*) restricts us to phase or dielectric gratings. There are two possibilities to satisfy point (*ii*). The first is to use a thick grating and light incident near the Bragg angle. Then, Bragg effects will suppress all but one diffraction order. We shall call this coupler type a "Bragg coupler". The second possibility is to use a grating with a very large number of lines (or fringes) per millimeter. This results in such a large diffraction angle that only one diffraction order can propagate while all the others are beyond cutoff. This large diffraction angle leads to a thin film mode which travels in a direction reverse to that of the incident light, which is why we shall call this coupler a "reverse coupler." In many of the holographic materials which are available today, the achievable refractive index changes are relatively small (10^{-5} to 10^{-2}). It is, therefore, easier to satisfy point (*iii*) by using a Bragg coupler, where the phase shift accumulates throughout the thickness of the grating. The resulting strong coupling leads to short coupling lengths. These are desirable for miniaturization, and they make it easier to maintain the tolerances which are needed for phase-matching.

Figure 1 shows the geometry of a grating coupler, the choice of the coordinate system, and several parameters of interest. In our case the

Reprinted with permission from *Bell Syst. Tech. J.*, vol. 49, pp. 1602–1608, Sept. 1970.

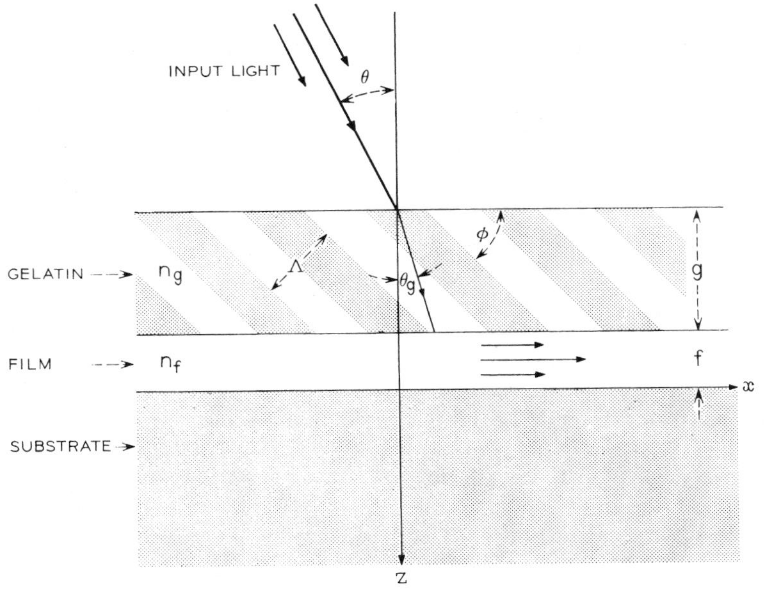

Fig. 1—Cross section of a grating coupler. Λ is the fringe spacing, φ the slant angle, and θ the angle of incidence. A transmission grating is shown. For a reflection grating the light is incident from the substrate side.

dielectric grating is formed in a layer of gelatin of refractive index n_g which is deposited directly onto the guiding film of index $n_f > n_g$. The grating is characterized by the grating vector **K** which is oriented perpendicular to the fringe planes and has a magnitude

$$K = 2\pi/\Lambda \tag{1}$$

where Λ is the fringe spacing. The fringes are slanted with respect to the grating surface by an angle ϕ. The propagation vector of the incident light is $\mathbf{k}_{in}$ and has a magnitude equal to the free space propagation constant $k_0 = 2\pi/\lambda$. In the gelatin the k-vector of this light is $\mathbf{k}_g$ which is of magnitude $n_g k_0$.

The diffracted wave has a k-vector equal to $(\mathbf{k}_g + \mathbf{K})$. It is phase-matched to a film mode when this k-vector has a tangential ($x-$) component equal to the propagation constant β of the film mode, i.e., when

$$\beta = (\mathbf{k}_g + \mathbf{K})_x = (\mathbf{k}_{in} + \mathbf{K})_x \,. \tag{2}$$

Figure 2 shows the k-vector diagrams for the Bragg coupler (a) and the reverse coupler (b). The latter is shown for reasons of comparison. Note that in this latter case the grating vector **K** is parallel to the x-axis (which is typical for a thin grating). The cross marks the k-vector of the -1 order which is generated beyond cutoff. The circles of radius

$n_g k_0$ and k_0 respectively indicate the locus of the k-vectors of the input light in the gelatin and in air for a variable angle of incidence. The line spaced a distance $\beta > n_g k_0$ away from the z-axis is the matching line. Phase-matching occurs when the vector sum $(\mathbf{k}_g + \mathbf{K})$ terminates on this line, as shown in the figure. The Bragg condition is obeyed when

$$\cos(\phi - \theta_g) = K/2n_g k_0 \tag{3}$$

where θ_g is the angle of incidence in the gelatin. Geometrically this implies that the vector sum $(\mathbf{k}_g + \mathbf{K})$ would terminate on the $n_g k_0$-circle.

It is clear from Fig. 2a that the Bragg condition and the phase-match condition cannot be met for the same angle of incidence. One can show that there is a minimum possible difference $\Delta\theta_{g\,\min}$ between the Bragg angle and the matching angle which is approximately given by

$$\Delta\theta_{g\,\min} \approx (n_f - n_g)/n_g \tag{4}$$

where we have assumed that this "detuning angle" is small and that $\beta \approx n_f k_0$. The detuning angle is typically a few degrees of arc. The angular width $2\Delta\theta_{g\,\mathrm{Bragg}}$ of the Bragg response between half-power points is[2]

$$2\Delta\theta_{g\,\mathrm{Bragg}} = \Lambda/g \tag{5a}$$

for transmission gratings, and

$$2\Delta\theta_{g\,\mathrm{Bragg}} = (\Lambda/g) \cot\theta_g \tag{5b}$$

for reflection gratings, where g is the grating thickness. These formulas

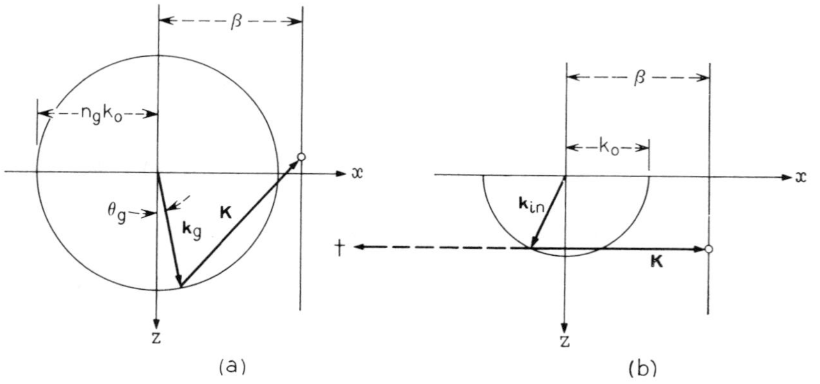

Fig. 2—k-vector diagrams for a Bragg coupler (a) and a reverse coupler (b). $\mathbf{K}$ is the grating vector; β, the propagation constant of a film mode.

assume unslanted gratings. Formulas are available also for slanted gratings[2], and one expects widths of the order of Λ/g, which is typically a few degrees. To take advantage of Bragg effects, we have to bring the matching angle as close as possible to the Bragg angle, and at least onto the wings of the Bragg response. Then we can expect suppression of unwanted orders and sufficiently strong coupling, which is, indeed, what we have observed experimentally.

Our experiments were done with Bragg couplers. We made the dielectric gratings in dichromated gelatin using the preparation and development techniques described in Ref. 5. Gelatin layers of about 4 μm thickness were deposited on the guiding films using a dipcoating technique. The films were low-loss sputtered films of Corning 7059 glass which were kindly supplied by J. E. Goell and D. R. Standley.[6] They were about 0.3 μm thick and had a refractive index of $n_f = 1.62$.

We designed the couplers for light of wavelength $\lambda = 0.6328$ μm, incident perpendicular to the film plane. This prescribes gratings with a slant angle ϕ of about 45° and a fringe spacing of $\Lambda = 0.25$ μm (i.e., 4000 lines/mm). To obtain the right grating parameters, some experimentation is necessary because the gelatin shrinks during development. We produced the desired fringe pattern holographically by exposing the sensitized gelatin at the shorter wavelength of $\lambda' = 0.4416$ μm, which is a line of the cadmium laser. The k-vector diagram of Fig. 3 indicates how the use of this shorter wavelength gives us fringes with the wanted k-vector at convenient angles of incidence. To get the proper interference angles in the gelatin, the two collimated light beams

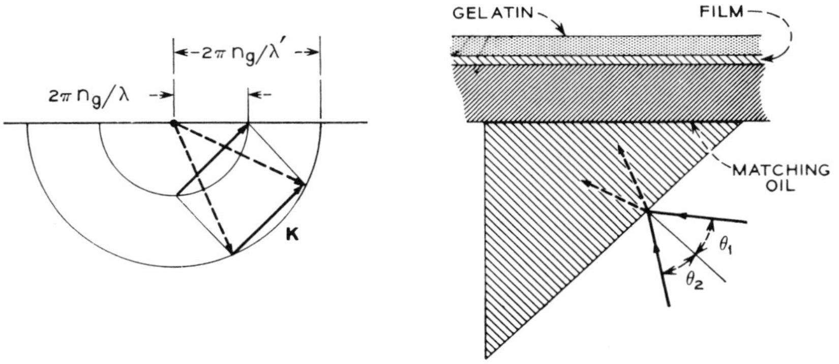

Fig. 3—Arrangement for holographic exposure used to make the grating coupler. The k-vector diagram shows how the grating vector **K** is conveniently produced by the interference of blue light (λ').

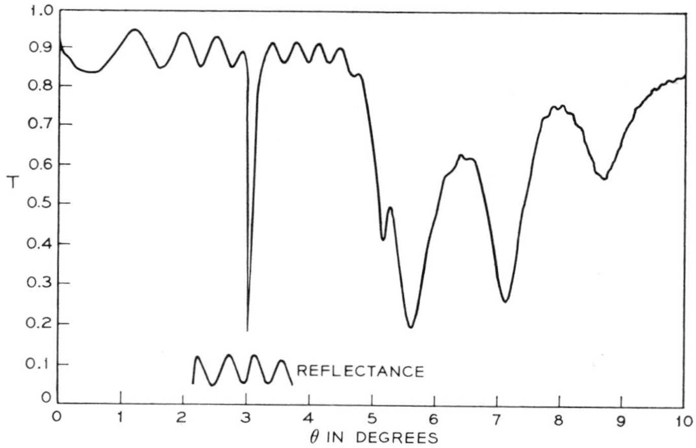

Fig. 4—Angular transmission spectrum for $2w_0 = 0.3$ mm. T is the transmittance of the coupler and θ the angle of incidence (in air). Light is incident from the substrate side.

were entered through a 45° prism which was joined to the film substrate with matching oil as shown in the figure. The angles of incidence on the prism were approximately $\theta_1 = 55°$ and $\theta_2 = 50°$. The edge of the coupler was produced by the shadow of a knife edge in one of the exposing beams about 5 cm away from the gelatin. The size of the resulting gratings was approximately 5 mm $\times$ 5 mm.

We made several Bragg couplers and studied their characteristics at 0.6328 μm. We obtained coupling both on transmission and on reflection from the grating, but the better results (which we report below) were obtained with reflecting couplers, i.e., with light incident from the substrate side. We used a gaussian laser beam with its waist positioned near the grating. The beam was positioned at the edge of the coupler. To optimize the coupling further, we varied the waist diameter $2w_0$ and the spacing between the coupler and the waist. For our best coupler we found optimum coupling in a diverging beam with a waist diameter of about $2w_0 = 0.3$ mm and a waist-coupler spacing of 16 cm. For this case the beam diameter at the coupler was $2w = 0.6$ mm. By analyzing the m-lines[1] we found a value for the optimum coupling length which was of the same magnitude. The m-lines also indicated that our film guide supported only one TE and one TM mode.

Figure 4 shows the angular transmission spectrum obtained for TE-coupling with the optimum beam parameters. This plot is obtained by monitoring the transmittance T through the substrate-film-gelatin sandwich for a variable angle of incidence θ. In the figure we notice a small modulation ($\Delta T \approx 0.1$) superimposed on the spectrum which is due to substrate resonances. The large dip near $\theta = 3°$ is caused by coupling into the film. The angular width of this dip is about 0.1°. The Bragg condition is obeyed near $\theta = 6°$. The Bragg response is strongly modulated by gelatin and film resonances. It has an overall width of about 3°. The measured reflectance of the coupler is also shown in the figure. From these data we arrived at the following power balance for optimum coupling: 3 percent of the incident light is absorbed in the substrate, film and gelatin, 8 percent reflected, 18 percent transmitted, and 71 percent coupled into the film.

Coupling to the TM film mode is relatively weaker for the above coupler because it was designed for normal incidence ($\theta \approx 0$) where TM diffraction is at a minimum.[2] The same beam parameters which produced optimum TE-coupling yielded TM-coupling of only 15 percent.

We would like to acknowledge fruitful discussions with J. E. Goell, R. D. Standley and R. Ulrich, and the technical assistance of M. Madden.

REFERENCES

1. Tien, P. K., Ulrich, R., and Martin, R. J., "Modes of Propagating Light Waves in Thin Deposited Semiconductor Films", Applied Physics Letters, *14*, No. 9 (May 1969), pp. 291–294.
2. Kogelnik, H., "Coupled Wave Theory for Thick Hologram Gratings", BSTJ, *48*, No. 9 (November 1969), pp. 2909–2947.
3. Ashkin, A., and Ippen, E., unpublished work.
4. Dakss, M. L., Kuhn, L., Heidrich, P. F., and Scott, B. A., "A Grating Coupler for Efficient Excitation of Optical Guided Waves in Thin Films", Appl. Phys. Letter, *16*, No. 12 (June 1970), pp. 523–525.
5. Brandes, R. G., Francois, E. E., and Shankoff, T. A., "Preparation of Dichromated Gelatin Films for Holography", Applied Optics, *8*, No. 11 (November 1969), pp. 2346–2348.
6. Goell, J. E., and Standley, R. D., "Sputtered Glass Waveguide for Integrated Optical Circuits", BSTJ, *48*, No. 10 (December 1969), pp. 3445–3448.

GRATING COUPLER FOR EFFICIENT EXCITATION OF OPTICAL GUIDED WAVES IN THIN FILMS

M.L. Dakss, L. Kuhn, P.F. Heidrich, and B.A. Scott

IBM Thomas J. Watson Research Center, Yorktown Heights, New York 10598

(Received 6 April 1970)

We report a new method of coupling a laser beam to thin-film optical guided waves which utilizes an optical grating that is made from photoresist and fabricated directly on the film. High efficiency coupling (~ 40 %) into a single mode in a glass film is observed.

The propagation of optical guided waves in thin semiconductor and dielectric films has received much interest lately.[1-3] The potential for producing miniaturized optical integrated circuits on a single wafer, resistant to vibrations and thermal effects, has raised the need for a simple, efficient means of coupling light into and out of thin-film waveguides. Previously described coupling techniques include edge illumination[1,2] and evanescent field coupling with a prism.[3] In this letter we describe an experiment demonstrating the efficient (~40 %) coupling of a laser beam into single propagating modes within a thin-film waveguide using diffraction by a grating that is fabricated on the film. A similar technique has been used to study optical plasma surface waves in metals.[4] The advantage of the grating is that it is a simple, reproducible, and permanent coupler which is compatible with planar device technology.

The grating coupling mechanism is explained with reference to Fig. 1. A laser beam incident on the phase grating at an angle θ has a phase variation in the x direction given by $\exp\{i(2\pi/\lambda_0)\times(\sin\theta)x\}$, where λ_0 is the vacuum wavelength. As the beam passes through the grating it undergoes a phase retardation $\Delta\phi \sin(2\pi x/s)$, where $\Delta\phi$ is the phase depth of the grating (which we have assumed for convenience to be sinusoidal), and s is the grating periodicity. A polarization wave varying as $\exp\{i[\Delta\phi \sin(2\pi x/s) + (2\pi/\lambda_0)(\sin\theta)x]\}$ is thus established on the surface of the film. This polarization wave can be treated[5] as a superposition of many waves with individual phase variations $\exp\{i[m(2\pi/s)x + (2\pi/\lambda_0)(\sin\theta)x]\}$, where m is any integer. The polarization wave will couple most strongly to a film guided wave of the form $\exp\{i(2\pi/\lambda_g)x\}$, where λ_g is the wavelength of the guided wave, if one of the waves in the superposition is phase matched to the guided wave. This condition implies:

$$c/v = \sin\theta + m\lambda_0/s\,, \qquad (1)$$

where v is the phase velocity of the guided wave, c is the velocity of light in vacuum, and where we have used the relation $\lambda_g = (v/c)\lambda_0$.

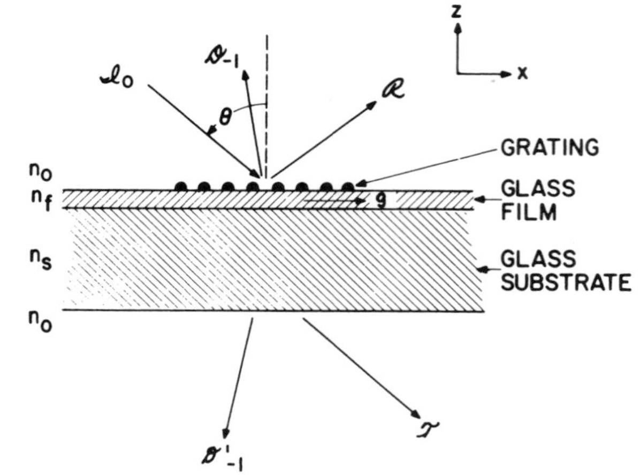

FIG. 1. Schematic drawing of the experiment. The incident beam ($\mathcal{I}_0$), transmitted beam ($\mathcal{T}$), reflected beam ($\mathcal{R}$), first-order diffracted beam in transmission ($\mathcal{D}'_{-1}$), and guided optical beams ($\mathcal{G}$) are shown.

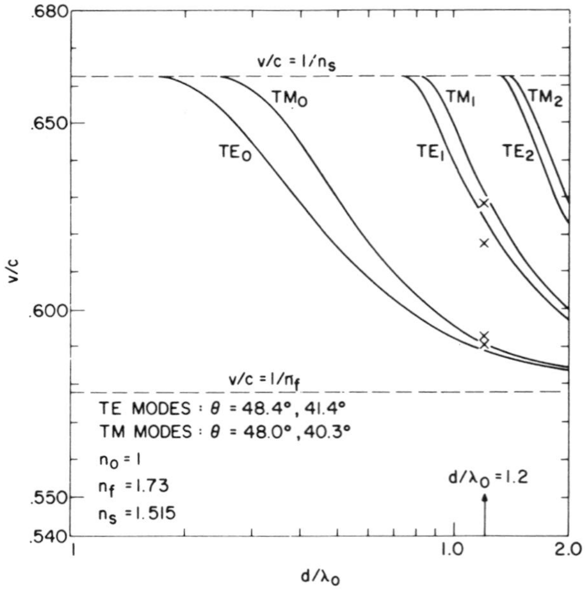

FIG. 2. Dispersion characteristics. Theoretical curves for the phase velocity of the six lowest modes of the film are plotted. The experimental points are plotted for $d/\lambda_0 = 1.2$.

Reprinted with permission from *Appl. Phys. Lett.*, vol. 16, pp. 523–525, June 15, 1970.

The experiment is shown schematically in Fig. 1. The waveguiding medium was a Corning 8390 dense flint glass film[6] (with a thickness $d \simeq 0.76\ \mu$ and refractive index $n_f \simeq 1.73$) that was rf sputtered onto a Corning 8370 glass substrate (with a refractive index $n_s = 1.515$). The grating, formed from Shipley AZ-1350 photoresist by exposure to a 4880-Å laser interferometer fringe pattern and by subsequent development, had a periodicity $s \simeq 0.665\ \mu$. The experiment was performed by directing a 6328-Å laser beam (~2 mm diam) at the grating near its edge and by varying the incident angle θ until coupling, as evidenced by the appearance of a bright streak in the film, was observed. TE and TM modes[7] could be selectively excited by choosing the correct beam polarization. The coupling angles for the two TE and two TM modes thus observed are listed in Fig. 2.

In order to identify the various modes observed, we calculated the dispersion characteristics of the film for several different values of n_f using the analysis of Ref. 7. In addition, by means of Eq. (1), we computed experimental values for the velocities of the observed modes from the measured values of the coupling angles. The best fit of the experimental velocity points to the theoretical dispersion curves occurred for the combination of parameters $n_f = 1.73$ and $d = 1.2\lambda_0 = 0.76\ \mu$ and is presented in Fig. 2. The slight discrepancy between theory and experiment may be related to the loading of the waveguide by the grating. The values of n_f and d determined by this fitting procedure were independently confirmed as follows. The film index was inferred from *in situ* microprobe analyses of the film composition and found to be 1.72 ± 0.03. The optical thickness of the film was determined by spectral transmission measurements. Using this optical thickness and $n_f = 1.72$, the thickness was found to be $d = 0.81 \pm 0.06\ \mu$.

The coupling efficiency was determined by monitoring the intensity of all of the beams leaving the grating except the guided wave. The decrease in the sum of these beams as θ was varied through the coupling condition is attributed to energy being carried off by the guided wave. The angular dependence of the various beam intensities and their sum, normalized to the incident beam intensity, is shown in Fig. 3 for the cases of coupling to the TE_1 and TE_0 modes. In both cases, the sum was found to vary from about 100% far from coupling to about 60% at the coupling angle. This implies an efficiency of coupling to film guided modes of about 40%. Systematic errors relating to drift in the input beam intensity and reproducibility of photomultiplier readings were about ± 5%. The $m = -2$ diffracted beam was not measured because it was trapped in the substrate. It was therefore not included in the sum. This beam had an intensity much smaller than the $m = -1$ orders which themselves produce only a small effect on the sum.

We believe that the increase in the reflected signal at the coupling angle corresponds to energy

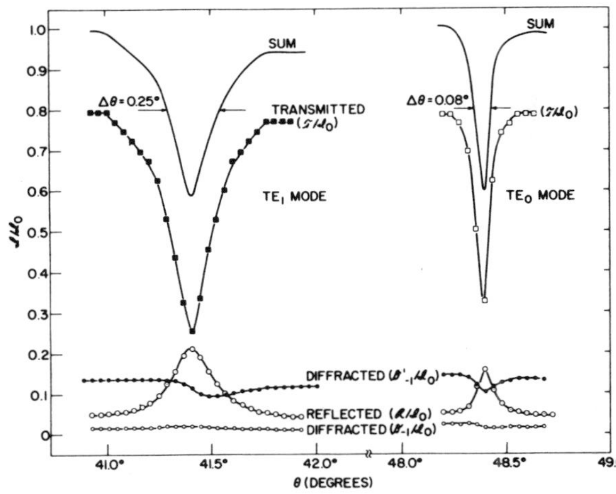

FIG. 3. Coupling characteristics. The figure shows the variation in intensity of the beams $\mathcal{T}$, $\mathcal{R}$, $\mathcal{D}_{-1}$, and $\mathcal{D}'_{-1}$ (normalized to $\mathcal{I}_0$) as the angle of incidence θ passes through the coupling condition for the TE_1 and TE_0 modes.

coupled into the film and then out. This increase would be minimized, and the coupling efficiency maximized, by matching the incident beam diameter to the optimum coupling length[8] and by varying the grating efficiency along the direction of propagation.[9] Further improvements might also be obtained by using a blazed grating fabricated, for example, using the techniques of Ref. 10.

Using the grating technique, we have also demonstrated coupling (~10%) into an epoxy film sandwiched between two glass flats.[2] We have also observed coupling out of the epoxy film by means of a second grating.

In conclusion, we have demonstrated an efficient method of coupling light into a thin-film optical waveguide using a grating coupler. This approach is simple and inexpensive, and appears to be compatible with the concept of mass-produced planar optical devices.

[1]E.R. Schineller, Microwaves 7, 77 (1968).
[2]R. Shubert and J.H. Harris, IEEE Trans. Microwave Theo. Tech. MIT-16, 1048 (1968).
[3]P.K. Tien, R. Ulrich, and R.J. Martin, Appl. Phys. Letters 14, 291 (1969).
[4]Y.Y. Teng and E.A. Stern, Phys. Rev. Letters 19, 511 (1967).
[5]J.W. Goodman, *Introduction to Fourier Optics* (McGraw-Hill, New York, 1968), Sec. 4.2.
[6]Sputtered glass waveguides (of a different glass composition) have also been reported by J.E. Goell and R.D. Standley, Bell System Tech. J. (Brief) 48, 3445 (1969).
[7]W.W. Anderson, IEEE J. Quantum Electron. QE-1, 228 (1965).
[8]C.C. Johnson, *Field and Wave Electrodynamics* (McGraw-Hill, New York, 1965), Sec. 9.4.
[9]R.E. Collin, *Foundations for Microwave Engineering* (McGraw-Hill, New York, 1966), Sec. 6.4.
[10]N.K. Sheridon, Appl. Phys. Letters 12, 16 (1968).

EXPERIMENTS ON LIGHT WAVES IN A THIN TAPERED FILM AND A NEW LIGHT-WAVE COUPLER

P.K. Tien and R.J. Martin
Bell Telephone Laboratories, Incorporated, Holmdel, New Jersey 07733
(Received 18 February 1971)

We discuss experiments in an asymmetric optical waveguide which is simply a dielectric film deposited on a substrate. The film has a tapered edge. Near this edge and depending on the angle of incidence, a guided light wave in the film may be totally reflected or it can enter into the substrate as radiation modes. In the latter case, the tapered film edge is used here to couple light energy into or out of the film.

Recently, the development of the prism film and grating couplers[1-3] has attracted much attention and opened new possibilities for electro-optical and nonlinear devices by guiding a light wave into a thin-film structure. We report here a new and simpler light-wave coupler which utilizes the cutoff property of an asymmetric optical waveguide. Consider in Fig. 1(a) a thin film deposited on a substrate. The film here is used as an optical waveguide. It is tapered to nothing in a distance between X_a and X_b, typically of 10—100 optical wavelengths. In the tapered region, a waveguide mode of the film is coupled to the radiation modes[4] of the substrate. Because of this coupling, a laser beam entering into the substrate can be coupled into the film and vice versa. The new coupler is simply a tapered film edge.

The physics of this coupler cannot be understood fully without discussing some interesting total reflection phenomena observed in a tapered film, which have not been discussed before. Figures 2(a)—2(d) are photographs of a ZnS film taken by a camera facing the top surface of the film. The film has a tapered edge and the tapered section [$X_a - X_b$ in Fig. 1(a)] is about 70 μm wide as measured by a Varian Å scope interferometer. Using the coordinates of Fig. 1, the photographs show the X-Y plane which is parallel to the surface of the film, and the film edge is parallel to the Y axis. Although the dark line in the photographs indicates the position of the edge, the width of the taper is about three times that of the dark line. We have, therefore, in the photographs, except for a small tapered section, a uniform film at the left of the dark line and no film at the right of it. A light beam from a He-Ne 6328-Å laser was fed into the uniform part of the film by a prism-film coupler (not shown) and propagates there as a $m=0$ waveguide mode. In Fig. 2(a), the guided wave was traveling toward the film edge at an incident angle $\theta_i = 75°$. Here the incident angle is in the plane of the film and it is measured with respect to the normal of the edge. The light wave was seen totally reflected at the film edge and the reflection was nearly perfect in spite of minor irregularities of the taper visible in the photographs. By continuously reducing the incident angle, first, the guided wave continued to be totally reflected as shown in Fig. 2(b) at $\theta_i = 53°$, and then, between $\theta_i = 53°$ and 47°, the reflected beam became blurred and intermittently fanned out and faded away. Finally, at $\theta_i = 47°$ [Fig. 2(c)], the reflected beam disappeared completely and remained so at smaller incident angles [Fig. 2(d)]. If the disappeared reflected beam had been radiated into the air space above the film, we would have observed diffused glow near the film edge. This was simply not so as seen in the photographs. The light beam actually entered into the glass substrate, where it was negligibly scattered and thus became invisible.

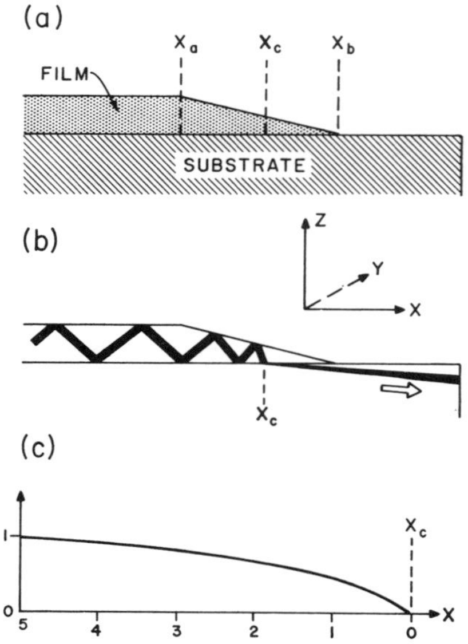

FIG. 1. (a) A thin film deposited on a substrate is tapered to nothing between X_a and X_b. Beyond X_c the waveguide becomes cutoff. (b) Ray optics show the guided wave entering into the substrate after reaching the cutoff point, X_c. (c) The curve shows a guided wave gradually converted into radiation field in substrate. The ordinate is the fraction of the original light intensity remained in the waveguide mode and the abscissa is the distance from the cutoff point, X_c, in units of 1 vacuum optical wavelength. The tapered film edge has a slope of 0.002.

The refractive indices of the film and the substrate are $n_1 = 2.342$ and $n_0 = 1.512$ at the 6328-Å laser wavelength. Let the guided wave propagate as $\exp(-i\omega t + i\beta s)$ along the light path s in the plane of the film. Here, ω is the laser angular

Reprinted with permission from *Appl. Phys. Lett.*, vol. 18, pp. 398-401, May 1, 1971.

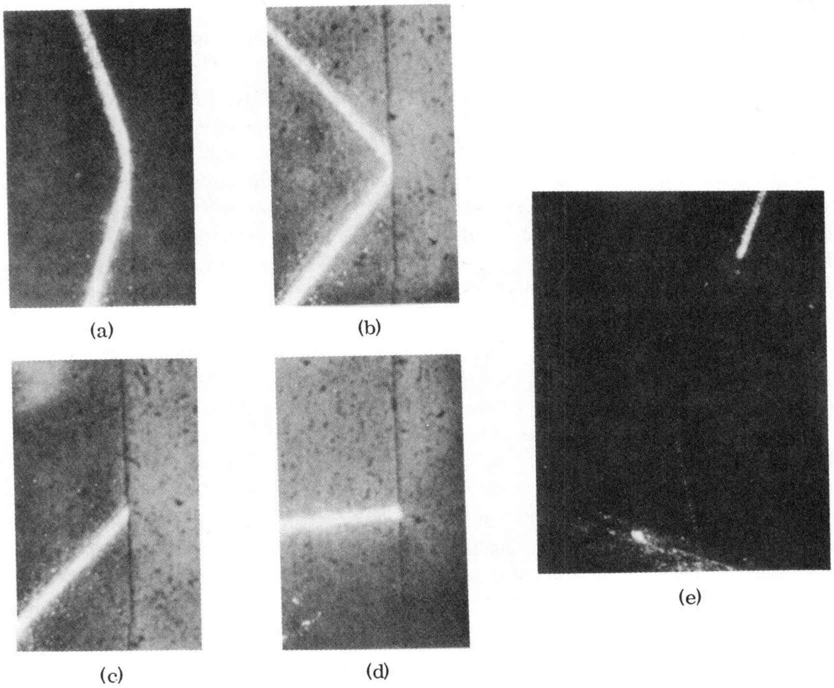

FIG. 2. In photographs (a) and (b), the guided light wave propagating in the plane of the film is totally reflected at a tapered film edge. In (c) and (d), it disappears after reaching the edge. The incident angles are (i) 75°, (ii) 53°, (iii) 47°, and (iv) 3°. In photograph (e), a laser beam which enters into one of the side surfaces of the substrate excites a streak of guided wave in the film. The bright spot in the photograph is the beam spot on the glass surface.

frequency and β is the phase constant. For $k=\omega/c$ (c = velocity of the light in vacuum), a ratio $\beta/k=1.984$ was measured by a prism-film coupler for the uniform part of the film and it indicates a film thickness of 1420 ± 50 Å. The film has only one waveguide mode, $m=0$, and it becomes cutoff[5] at the film thickness of 402 Å which occurs at X_c in Fig. 1(a). Beyond X_c, the guided mode cannot propagate. Near the cutoff, β/k approaches 1.512 which is[5] the refractive index of the substrate, n_0.

For the guided wave, the ratio β/k is the effective refractive index.[5] It is often convenient to think about light waves in a waveguide as those in an infinite medium by using β/k as the refractive index. As β/k varies with X in the tapered region of the film, any guided wave propagating in a direction other than $\theta_i=0$ will continuously change its direction. Letting β_1/k and β_2/k be the effective refractive indices at X_1 and X_2, respectively, and θ_1 and θ_2 be the corresponding angles measured in the plane of the film between the directions of propagation and the normal of the film edge (X_1, $X_2<X_c$), we find

$$\sin\theta_1/\sin\theta_2=(\beta_2/k)/(\beta_1/k), \tag{1}$$

which is, of course, the Snell law written in terms of the effective indices. For a guided wave to be totally reflected, the guided wave must turn around before reaching X_c, beyond which the waveguide becomes cut off. This requires

$$\sin\theta_i>\frac{(\beta/k)(\text{cutoff})}{(\beta/k)(\text{incident})}. \tag{2}$$

In our case, for a $\beta/k(\text{incident})=1.984$ and $\beta/k(\text{cutoff})=1.512$, we compute from (2), $\theta_i>49.5°$ which agrees with θ_i between 47° and 53° measured in Figs. 2(c) and 2(b). For a θ_i smaller than that specified in (2), the light wave enters into the cutoff region where it is converted into radiation modes of the substrate. The tapered film then serves as an output light-wave coupler.

We can understand the problem better by considering ray optics in Fig. 1(b). A guided wave can be considered[5] as a plane wave which bounces back and forth between the top and the bottom surfaces of the film forming a zigzag path.

As the light wave enters into the tapered region, the angle between the light ray and the Z axis becomes smaller and smaller, and eventually it is smaller than the critical angle of the film-substrate interface. The light ray then no longer bounces back into the film, and instead, it enters into the substrate as a refracted beam. Note that the critical angle of the film-air interface is smaller than that of the film-substrate interface; radiation into the air space above the film does not occur.

Several excellent papers have been published by Marcuse[4] concerning radiation modes. We have followed his theory and derived expressions for radiation modes and their coupling coefficients to the guided wave in an asymmetric waveguide which is considered here. To study the effect of the taper, we divide the taper into a large number of steps and carry out a series of numerical calculations from one step to another until the cutoff region is reached. The calculation shows that the guided wave is not converted into radiation modes until it reaches a distance of several optical wavelengths in front of the cutoff region. The curve in Fig. 1(c) shows the fraction of the original light intensity which remains in the waveguide mode as the guided wave approaches the cutoff point, X_c. The distance in X is expressed in units of 1 vacuum optical wavelength. The calculation was made for a tapered film edge of a slope 0.002. The theory also shows that the far-field pattern of the radiation in a plane parallel to Z should have a concentration of more than 80% of the light energy within an angle of 15° below the film-substrate interface. The field pattern changes only slightly by varying the slope of the taper from 0.01 to 0.001.

To test the tapered film as a light-wave coupler, we deposited ZnS film on a polished glass block, a part of which is masked during deposition to form a tapered film edge. The spacing between the mask and the top surface of the glass block determined the slope of the taper, and a slope as small as 0.001 was obtained in this way. In one experiment, a guided wave was launched into the film in the direction normal to the film edge and the excited radiation field was projected to a screen through one of the side surfaces of the glass block. It was very easy to couple all the light energy out of the film and the projected output radiation can be focused into a fine line by a spherical lens. The light intensity of the radiation field was concentrated within an angle of 15° in the X-Z plane, and that agrees with the calculation.

In another experiment a laser beam of nearly Gaussian intensity distribution was focused, through the side surface of the glass block, to the edge of the film. Figure 2(e) shows both the spot of the laser beam on the glass surface and a streak of the guided wave excited in the film. In this experiment, about 25% of the laser beam was fed into the film and the rest of the beam was reflected at the glass-film interface. The reflected beam was projected on a screen and we observed a dark line across the beam spot which indicated the coupling. We have also built couplers using an organic film on glass substrate, and an input coupling efficiency of about 40% has been obtained. For higher coupling efficiency, it may be necessary to control the intensity distribution of the incoming laser beam to match that of the radiation modes. The details of the theory and experiment will be published elsewhere.

In conclusion, using the concept of effective refractive indices we have explained the total reflection phenomenon observed near a tapered edge of a thin film. The use of the tapered film edge as a light-wave coupler was demonstrated experimentally.

[1]P.K. Tien, R. Ulrich, and R.J. Martin, Appl. Phys. Letters 14, 291 (1969).
[2]M.L. Dakss, L. Kuhn, P.F. Heidrich, and B.A. Scott, Appl. Phys. Letters 16, 523 (1970).
[3]H. Kogelnik and T.P. Sosnowski, Bell System Tech. J. 49, 1602 (1970).
[4]D. Marcuse, Bell System Tech. J. 48, 3187 (1969), 49, 273 (1970).
[5]P.K. Tien and R. Ulrich, J. Opt. Soc. Am. 60, 1325 (1970).

Part 5
Radiation Losses Caused by Bends and Surface Roughness

Bends in Optical Dielectric Guides

By E. A. J. MARCATILI

(Manuscript received March 3, 1969)

Light transmission through a curved dielectric rod of rectangular cross section embedded in different dielectrics is analyzed in closed, though approximate form. We distinguish three ranges:

(i) Small cross section guides such as a thin glass ribbon surrounded by air—Making its width 1 percent of the wavelength, most of the power travels outside of the glass; the attenuation coefficient of the guide is two orders of magnitude smaller than that of glass, and the radius of curvature that doubles the straight guide loss is around 10,000λ.

(ii) Medium cross section guide for integration optics—It is only a few microns on the side and capable of guiding a single mode either in low loss bends with short radii of curvature or in a high Q closed loop useful for filters. Q's of the order of 10^8 are theoretically achievable in loops with radii ranging from 0.04 to 1 mm, if the percentage refractive index difference between guide and surrounding dielectric lies between 0.1 and 0.01.

(iii) Large cross section guides—They are multimode and are used in fiber optics. Conversion to higher order modes are found more significant than radiation loss resulting from curvature.

I. INTRODUCTION

A dielectric rod, embedded in one or more dielectrics of lower refractive index, is the basic ingredient of three types of optical waveguide which differ only in their relative dimensions and consequently in their guiding properties.

The first is a small cross section guide which supports only the fundamental mode; most of the power travels in a lower loss external medium. Thus, the attenuation of the mode is smaller than if all the power flowed through the higher loss internal medium. Tiny rods, thin ribbons, or films made of glass or other substances embedded in either air or low loss liquids are typical examples.[1–3]

The second is a medium size guide capable of supporting only a few

Reprinted with permission from *Bell Syst. Tech. J.*, vol. 48, pp. 2103–2132, Sept. 1969.

modes; most of the power travels in the internal medium. Such a guide, (Fig. 1 of Ref. 10) has been proposed as the building block of passive and active components for integrated optical circuitry.[4–6] Lasers, modulators, directional couplers, and filters are some of the many devices which could be built in a single substrate utilizing the high precision techniques available from integrated circuitry; consequently they would be compact, mechanically stable, and reproducible.

The third, a large size guide (clad fiber) which can support many modes, is used typically in fiber optics.[7]

These basic guides, having round or rectangular cross section and straight axis, have been studied both analytically and through computer calculations.[8–13] Also the directional coupler (Fig. 2 of Ref. 10) obtained by running two guides of rectangular or circular cross sections parallel to each other, has been analyzed.[10,12,14]

To my knowledge, though, little is known quantitatively about the ability of any of the three types of guides to negotiate bends, or about the radiation losses in loops, such as the one depicted in Fig. 1 as part of a channel dropping filter. This paper should supply such information.

In Section II the boundary value problem is discussed, and the fundamental modes of each polarization are described. Section III contains a discussion of the results and numerical examples. Conclusions are drawn in Section IV and all the mathematical derivations are exiled to the appendix.

II. FORMULATION OF THE BOUNDARY VALUE PROBLEM

Figure 2 depicts, in perspective, the basic geometry of the curved guide with radius of curvature R. The cross section is a rectangle whose sides are a and b. The refractive index of the guide is n_1, and the refractive indices around the guide are n_2, n_3, n_4, and n_5, all of which are smaller than n_1. Furthermore, for reasons which become apparent later, we do not specify the refractive indices in the four shaded areas.

This boundary value problem is solved in closed, though approximate form in the appendix, by introducing the same simplification used in solving the problem of transmission in the straight guide.[10] That simplification arises from solving Maxwell's equations only for guide dimensions such that a small percentage of the total power flows through the shaded areas and consequently a negligible error is expected if one does not match properly the fields along their edges.

Two types of hybrid modes propagate through this curved guide;

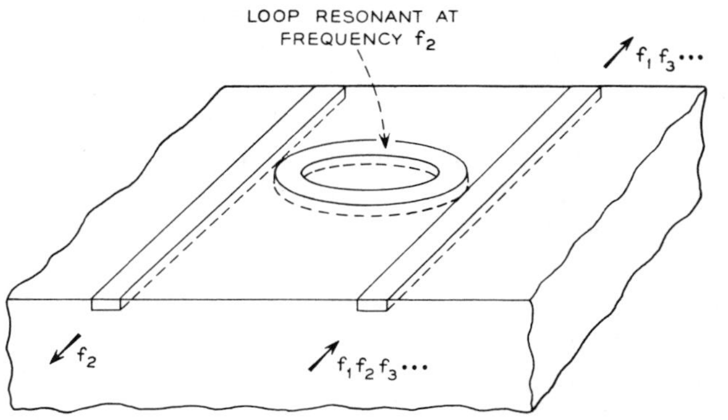

Fig. 1 — Channel dropping filter (ring type).

each one has six field components. But since some of the refractive indices n_2, n_3, n_4, and n_5 are chosen close to n_1, guidance occurs through total internal reflection only when the plane wavelets that make a mode impinge on the interfaces at grazing angles. Consequently, the only large field components are perpendicular to the curved z axis (Fig. 2). The modes are then of the TEM kind and we group them in two families, E^x_{pq} and E^y_{pq}. The main field components of the members of the first family are E_x and H_y, while those of the second are E_y and H_x.

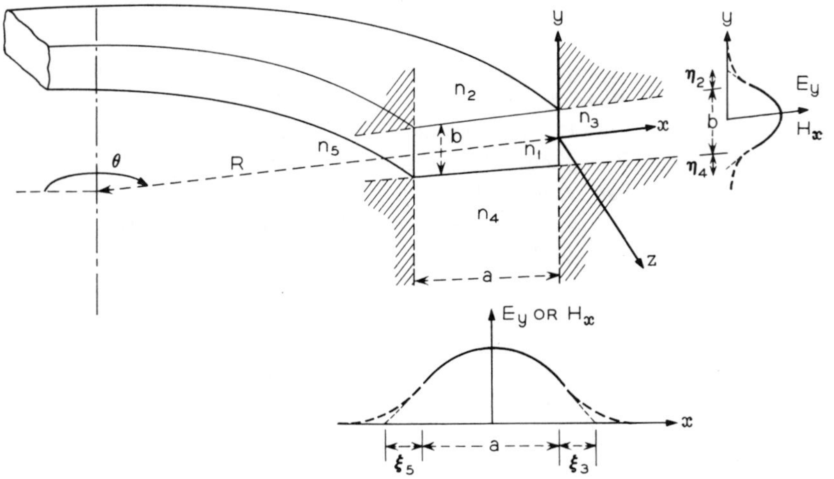

Fig. 2 — Curved dielectric guide.

Virtually every one of these components varies sinusoidally along x and y within the guiding medium 1 and decays exponentially in the surrounding media 2, 3, 4, and 5 (Fig. 2). The subindices p and q represent the number of extrema of each field component in the x and y directions, respectively. The field configurations of some members of the two families in straight guides are depicted in Fig. 5 of Ref. 10; section 2.1 describes the influence of a finite radius of curvature on those field configurations.

General expressions for the different phase and propagation constants in each medium of the curved guide are calculated in the appendix, for arbitrary modes and for $n_2 \neq n_3 \neq n_4 \neq n_5$. In the text, we consider only the fundamental modes of each family E^x_{11} and E^y_{11}; furthermore, we choose

$$n_3 = n_5 \tag{1}$$

and leave n_2 and n_4 arbitrary. This choice of refractive indexes encompasses the most interesting cases.

2.1 E^x_{11} *Mode*

We first study the E^x_{11} mode. As we said before, the main components are E_x along the x direction and H_y along y. Both components have a single maximum located within medium 1 and drop sinusoidally toward the edge of it. Outside of the medium, the decay is exponential.

The axial propagation constant is according to equation (47)

$$k_z = (k_1^2 - k_x^2 - k_y^2)^{\frac{1}{2}}, \tag{2}$$

where $k_1 = kn_1 = (2\pi/\lambda)n_1$ and λ is the free space wavelength, k_x is the propagation constant along x in media 1, 2, and 4, and k_y is the propagation constant along y in media 1, 3, and 5. This means that the electrical width of media 1, 2, and 4 is the same and equal to $k_x a$, and the electrical height of 1, 3, and 5 is also the same and equal $k_y b$.

The transverse propagation constant k_y is independent of the radius of curvature R and can be found from the transcendental equation (37)

$$k_y b = \pi - \tan^{-1}\left[\left(\frac{\pi}{k_y A_2}\right)^2 - 1\right]^{-\frac{1}{2}} - \tan^{-1}\left[\left(\frac{\pi}{k_y A_4}\right)^2 - 1\right]^{-\frac{1}{2}} \tag{3}$$

in which

$$A_{\substack{2\\4}} = \frac{\lambda}{2\left(n_1^2 - n_{\substack{2\\4}}^2\right)^{\frac{1}{2}}}. \tag{4}$$

If the height of the guide b is selected so large that

$$\frac{A_2 + A_4}{\pi b} \ll 1, \tag{5}$$

only a small percentage of the power carried by the mode travels in media 2 and 4; and equation (3) can be solved approximately, yielding

$$k_y = \frac{\pi}{b}\left(1 + \frac{A_2 + A_4}{\pi b}\right)^{-1}.$$

According to equation (49), the other transverse propagation constant

$$k_x = k_{x0}\left[1 + \frac{2c}{ak_{x0}} - i\,\frac{k_{z0}\alpha_c}{k_{x0}^2}\right] \tag{6}$$

is valid if

$$\begin{aligned} \frac{c}{ak_{x0}} &\ll 1 \\ \alpha_c R &\ll 1. \end{aligned} \tag{7}$$

The first term in equation (6), k_{x0}, is the propagation constant in the x direction of the guide without curvature; the second and third terms, which according to equation (7) must be small, are perturbations related to the change of field profile and to radiation loss, both of which are introduced by the curvature. More precisely, α_c is the attenuation coefficient of the curved guide, $\alpha_c R$ is the attenuation per radian, that is the attenuation in a length of guide equal to R, and c is a conversion loss coefficient such that, at a junction between a straight and a curved section of the same guide, c^2 measures the power that the fundamental mode in the straight section would couple to modes higher than the fundamental in the curved section. The fact that equation (6) is valid if $c \ll 1$ requires the radius of curvature R to be so large that the field profiles of the fundamental modes in the straight and curved guides are quite similar. Later in this section we consider formulas applicable when $c \cong 1$.

The axial propagation constant, k_{z0}, of the straight guide is related to k_{x0} and k_y by the expression

$$k_{z0} = (k_1^2 - k_{x0}^2 - k_y^2)^{\frac{1}{2}}; \tag{8}$$

and k_{x0} is the solution of the transcendental equation (55)

$$k_{x0}a = \pi - 2 \tan^{-1} \frac{n_3^2}{n_1^2} \left[\left(\frac{\pi}{k_{x0}A} \right)^2 - 1 \right]^{-\frac{1}{2}}. \tag{9}$$

The length

$$A = \frac{\lambda}{2(n_1^2 - n_3^2)^{\frac{1}{2}}} \tag{10}$$

is used as a normalizing dimension. What does it measure? If one assumes $b = \infty$, the guide becomes a slab of width a. If $a \leqq A$, only the fundamental mode is guided; if $a > A$, the slab is multimode.

Figure 3 is a graph of the electrical width, $k_{x0}a$, of the straight guide as a function of a/A. The solid curve is the solution of equation (9) assuming $n_1/n_3 = 1.5$, while the dotted one is the solution for $n_1/n_3 = 1$. For thin guides, $a/A \ll 1$, the electrical width is proportional to a; for thick guides, $a/A \gg 1$, the electrical width goes asymptotically to π.

The attenuation per radian $\alpha_c R$ and the conversion coefficient c, obtained from equations (50) and (51) with $n_3 = n_5$ are

$$\alpha_c R \doteq \frac{1}{2} \left(1 - \frac{n_3^2}{n_1^2} \right)^{-\frac{1}{2}} \left(\frac{n_3 k_{x0} a}{n_1} \right)^2 \left(\frac{A}{\pi a} \right)^3 \left[1 - \left(\frac{k_{x0}A}{\pi} \right)^2 \right]^{\frac{1}{2}}$$
$$\cdot \frac{\mathcal{R} \exp \left\{ -\frac{\mathcal{R}}{3} \left[1 - \left(\frac{k_{x0}A}{\pi} \right)^2 \left(1 + \frac{2c}{ak_{x0}} \right)^2 \right]^{\frac{3}{2}} \right\}}{1 - \left(1 - \frac{n_3^4}{n_1^4} \right) \left(\frac{k_{x0}A}{\pi} \right)^2 + 2 \frac{n_3^2 A}{n_1^2 a} \left[1 - \left(\frac{k_{x0}A}{\pi} \right)^2 \right]^{-\frac{1}{2}}} \tag{11}$$

and

$$c = \frac{1}{2k_{x0}a} \left(\frac{\pi a}{A} \right)^3 \frac{1}{\mathcal{R}}, \tag{12}$$

where

$$\mathcal{R} = \frac{2\pi^3 R}{k_{x0}^2 A^3} = 2 \frac{k_1^3}{k_{x0}^2} \left(1 - \frac{n_3^2}{n_1^2} \right)^{\frac{3}{2}} R. \tag{13}$$

The solid curves in Figs. 4 and 5 are graphs of the function

$$\alpha_c R \left(1 - \frac{n_3^2}{n_1^2} \right)^{\frac{1}{2}}$$

(which is proportional to the attenuation per radian) as a function of a/A using $\mathcal{R}$ as a parameter. In Fig. 4, we further assume that

$$\frac{n_1}{n_3} = 1 + \Delta$$

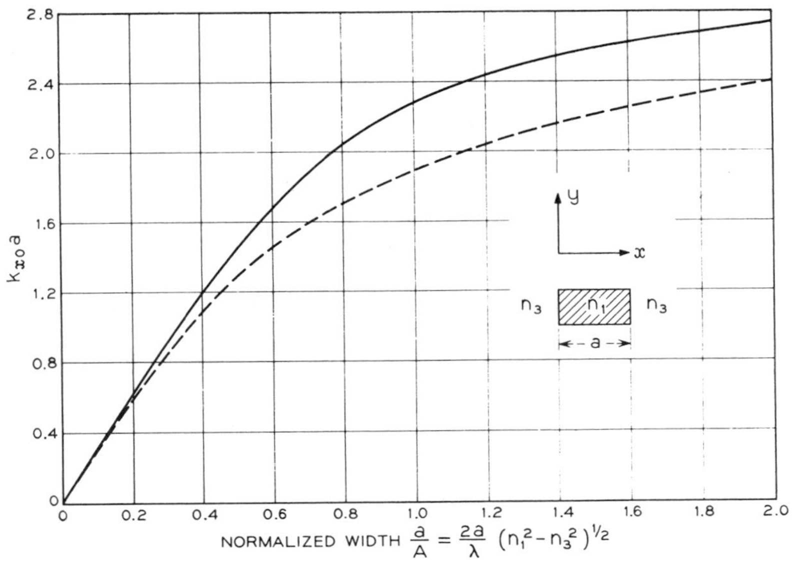

Fig. 3—Guide's electrical width. Solid line is for E_{11}^{z} mode with $n_1/n_3 = 1.5$; dashed line is for E_{11}^{x} mode with n_1/n_3 arbitrary, and for E_{11}^{z} mode with $n_1 \cong n_3$.

and

$$\Delta \ll 1;$$

in Fig. 5,

$$\frac{n_1}{n_3} = 1.5.$$

In the same figures each dashed line is a curve of constant conversion loss c. Since the calculations are valid for $c \ll 1$, we believe the solid curves are reliable to the left of the dotted curve $c = 0.3$ and grow progressively in error to the right of it.

To extend the use of this graph to arbitrarily large values of a/A, we calculate the loss per radian, equation (63), when $a/A \gg 1$ and $c \cong 1$. It is

$$\alpha_c R = \frac{n_3^2}{n_1^2}\left[1 - \left(\frac{n_3}{n_1}\right)^2\right]^{-\frac{1}{2}} \exp\left\{-\frac{\mathcal{R}}{3}\left[1 - \left(\frac{9\pi}{2\mathcal{R}}\right)^{\frac{2}{3}} + \frac{4n_3^2}{n_1^2\mathcal{R}}\right]^{\frac{3}{2}}\right\}; \tag{14}$$

the dotted lines in Figs. 4 and 5 represent this loss. The reader can smoothly extend the solid curves to the right of the dashed line, $c = 0.3$, so that they become asymptotic to the dotted lines. Thus, the whole range of guide width a from 0 to R has been covered.

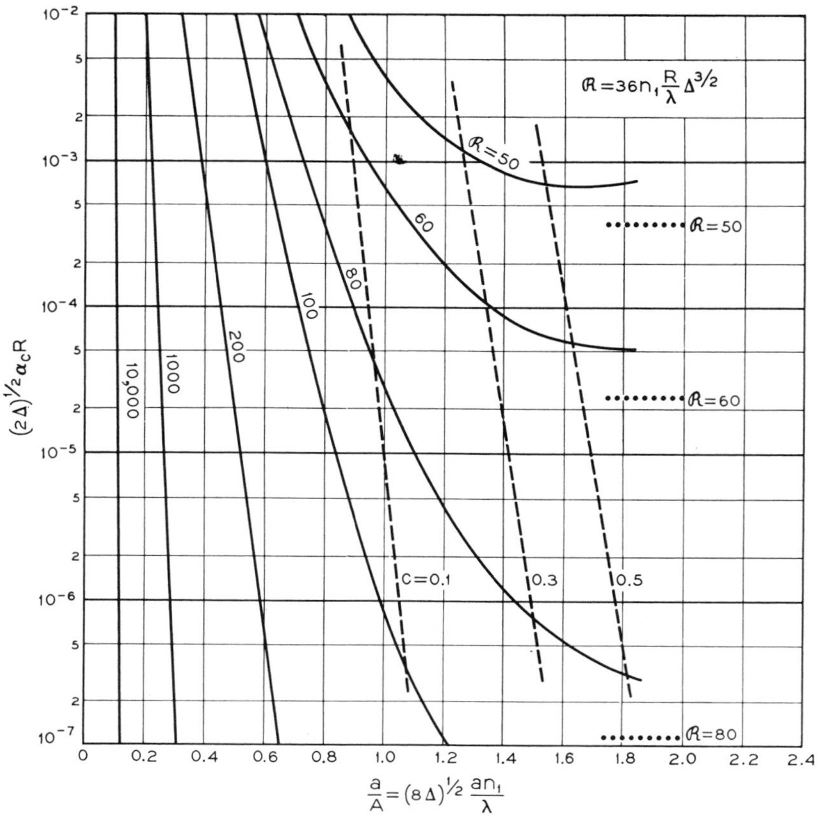

Fig. 4 — Attenuation per radian for E_{11}^x and E_{11}^y modes if $n_1/n_3 = 1 + \Delta$ and $\Delta \ll 1$.

To understand why these curves of constant R become asymptotic for $a/A \gg 1$, we have drawn in Fig. 6a a curved guide with a certain R; its width a is very large compared with A. Also the amplitudes of the field components E_x and H_y are plotted as functions of x and y.

Along x the field inside the guide behaves virtually as the Bessel function $J_\nu[k_1(R + x)]$ where ν is a very large number and outside of the guide decays exponentially. This guide has some radiation loss per radian.

Now, suppose that we start shrinking a without changing R. Since the field at $x = -a$ is very small, the radiation loss remains constant until a is made so short that the field at $x = 0$ and $x = -a$ are comparable (Fig. 6b). The field inside the guide varies almost sinusoidally, while outside decays exponentially and the attenuation per radian increases. If a is reduced even further (Fig. 6c) most of the power travels outside of the guide, and the loss increases even more. The field configuration along y is practically the same in the three cases (Fig. 6).

For resonant loops, such as the filter in Fig. 1, the intrinsic Q resulting from curvature radiation is more interesting than the attenuation α_c. They are related by the expression

$$Q_c = \frac{k_{z0}}{2\alpha_c}. \tag{15}$$

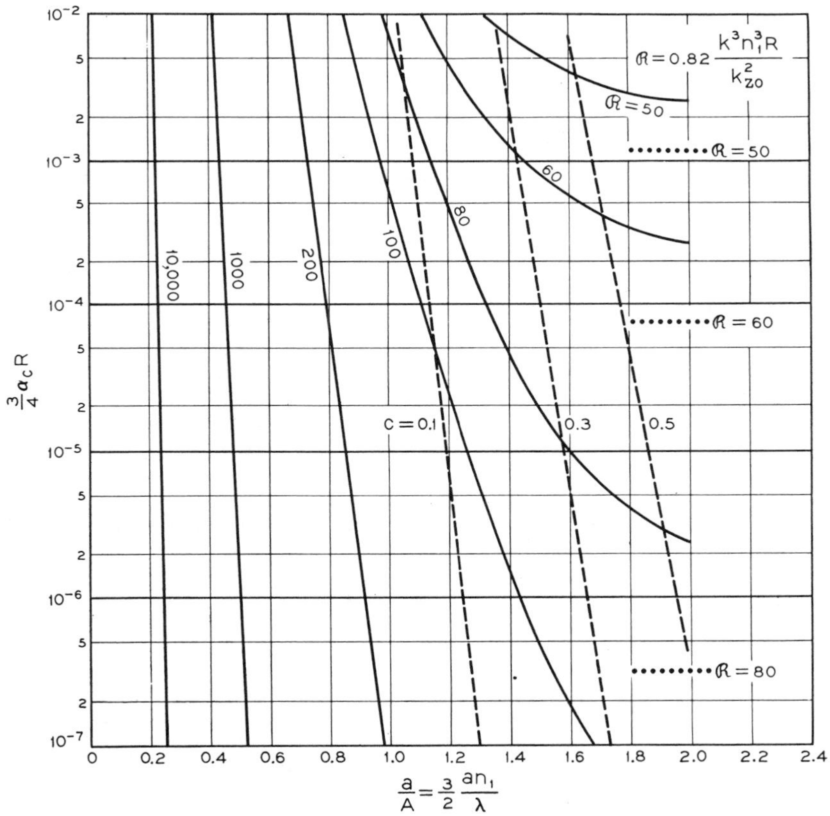

Fig. 5 — Attenuation per radian for E_{11}^x mode if $n_1/n_3 = 1.5$.

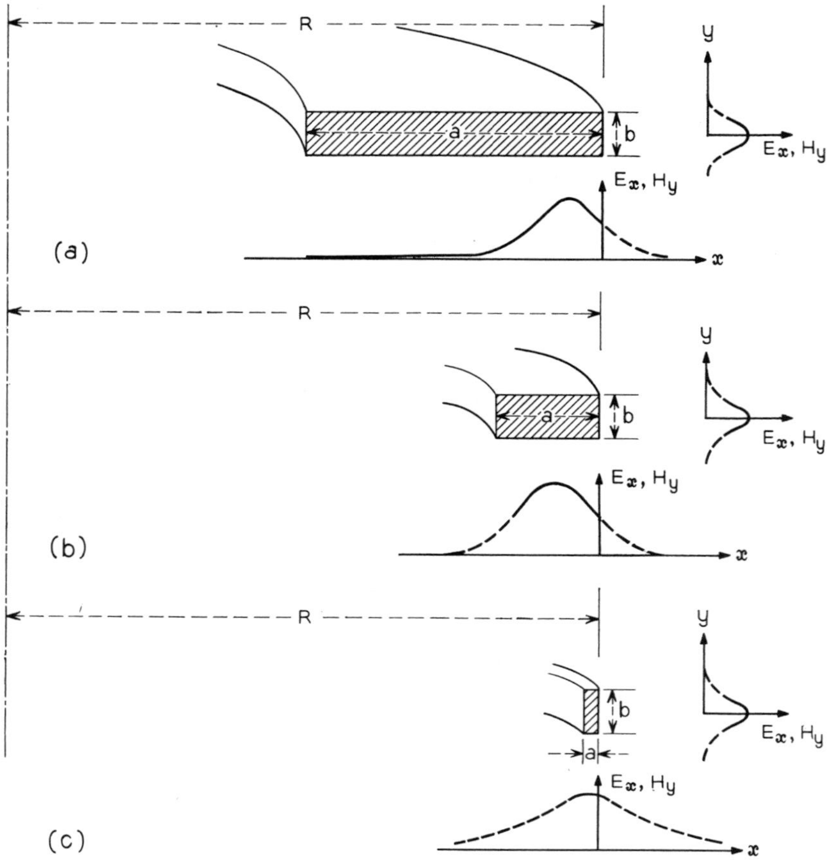

Fig. 6 — Field distribution as a function of guide width a with (a) $a/A \gg 1$, (b) $a/A \cong 1$, and (c) $a/A \ll 1$.

This function is plotted in Fig. 7, assuming

$$\frac{n_1}{n_3} = 1 + \Delta$$

and

$$\Delta \ll 1$$

and in Fig 8, assuming

$$\frac{n_1}{n_3} = 1.5,$$

using as before the normalized guide width a/A as variable and $\mathcal{R}$ as parameter. As in Figs. 4 and 5, the reader can easily match the solid and dotted curves. Further discussion of these curves is reserved for Section III.

The field components in media 2, 3, 4, and 5 decay almost exponentially away from the guiding rod, and the distances η_2, η_4, ξ_3, and ξ_5 over which the fields decrease by $1/e$ are

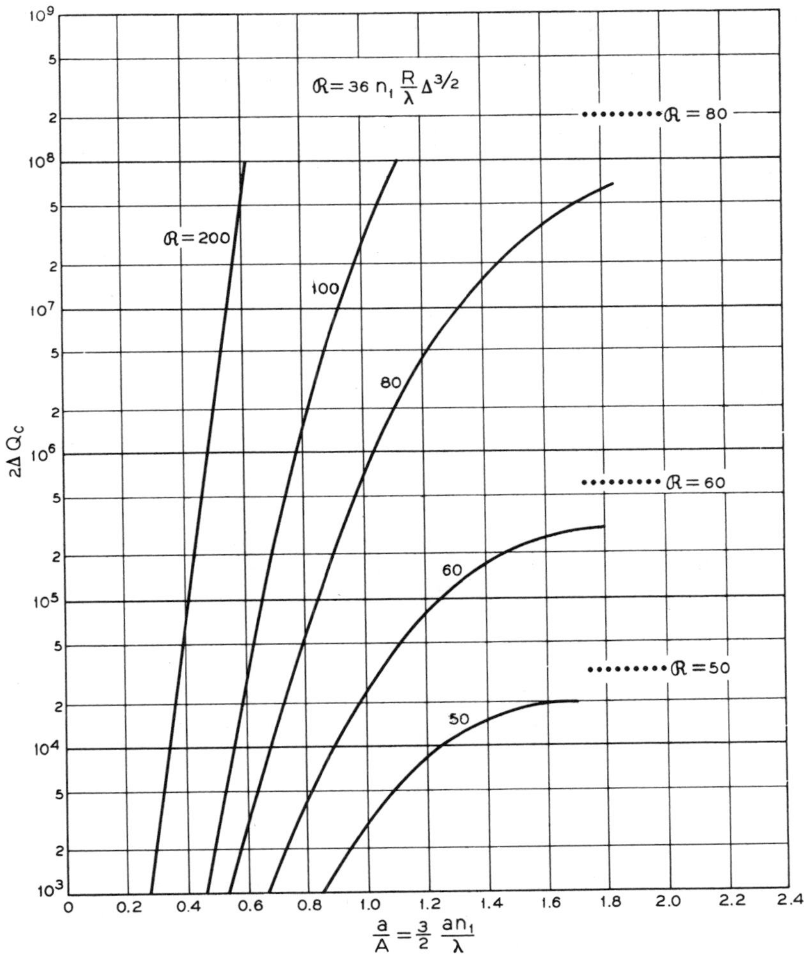

Fig. 7 — Intrinsic Q for $E_{11}{}^x$ and $E_{11}{}^y$ modes if $n_1/n_3 = 1 + \Delta$ and $\Delta \ll 1$.

$$\eta_{\substack{2\\4}} = \frac{1}{|k_{y\substack{2\\4}}|} = \frac{1}{\left(k_1^2 - k_{\substack{2\\4}}^2 - k_y^2\right)^{\frac{1}{2}}}, \tag{16}$$

$$\xi_3 = \xi_5 = \frac{1}{|k_{x3}|} = \frac{1}{(k_1^2 - k_3^2 - |k_x^2|)^{\frac{1}{2}}}. \tag{17}$$

2.2 E_{11}^y *Mode*

We now consider the E_{11}^y mode. The main components are E_y and H_x; they are qualitatively quite similar to components of the E_{11}^x mode, rotated 90°.

The propagation constant k_z is still given by equation (2); but now k_y

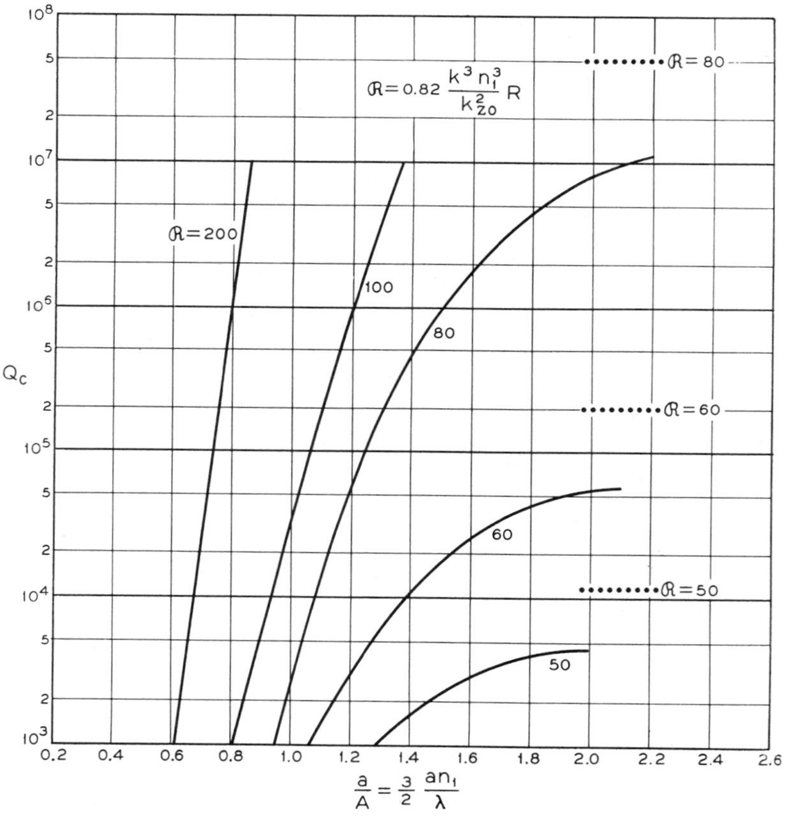

Fig. 8 — Intrinsic Q for $E_{11}{}^x$ mode if $n_1/n_3 = 1.5$.

is the solution of

$$k_y b = \pi - \tan^{-1} \frac{n_2^2}{n_1^2} \left[\left(\frac{\pi}{k_y A_2} \right)^2 - 1 \right]^{-\frac{1}{2}} - \tan^{-1} \frac{n_4^2}{n_1^2} \left[\left(\frac{\pi}{k_y A_4} \right)^2 - 1 \right]^{-\frac{1}{2}}. \tag{18}$$

The equivalent formula of any of those between equation (7) and (17) can be derived from that formula by substituting the ratio of refractive indexes by unity, but leaving them unchanged wherever they are subtracted from unity. For example, equation (11) becomes

$$\alpha_c R = \frac{1}{2} \left[1 - \left(\frac{n_3}{n_1} \right)^2 \right]^{-\frac{1}{2}} (k_{x0} a)^2 \left(\frac{A}{\pi a} \right)^3 \left[1 - \left(\frac{k_{x0} A}{\pi} \right)^2 \right]^{\frac{1}{2}}$$

$$\cdot \frac{\mathfrak{R} \exp \left\{ -\frac{\mathfrak{R}}{3} \left[1 - \left(\frac{k_{x0} A}{\pi} \right)^2 \left(1 + \frac{2c}{a k_{x0}} \right)^2 \right]^{\frac{3}{2}} \right\}}{1 - 2\left(1 - \frac{n_3^2}{n_1^2} \right) \left(\frac{k_{x0} A}{\pi} \right)^2 + 2 \frac{A}{a} \left[1 - \left(\frac{k_{x0} A}{\pi} \right)^2 \right]^{-\frac{1}{2}}}, \tag{19}$$

while c and $\mathfrak{R}$ given by equations (12) and (13) remain unchanged.

Figure 9 is a graph of the function $\alpha_c R[1 - (n_3/n_1)^2]^{\frac{1}{2}}$, valid for any ratio n_1/n_3. In particular, for $n_1/n_3 = 1 + \Delta$ and $\Delta \ll 1$, equations (19) and (11) become the same, and consequently these curves coincide with those in Fig. 4. This means that for $n_1 \cong n_3$, the E_{11}^x and E_{11}^y modes have the same loss.

Figure 10 is a graph of the intrinsic Q of a loop operating in the E_{11}^y mode which can be derived from equations (15) and (19). As before, in a resonant loop with $n_1/n_3 = 1 + \Delta$ and $\Delta \ll 1$, the E_{11}^x or E_{11}^y modes have the same Q's.

III. DISCUSSION AND EXAMPLES

The attenuation per radian of any dielectric guide of rectangular cross section and the Q_c resulting from curvature are strongly dependent on the radius of curvature. With the help of equation (17), the attenuation per radian equation (11) can be written

$$\alpha_c R = MR \exp \left(-\frac{1}{6\pi^2} \frac{\lambda_z^2 R}{|\xi_3|^3} \right), \tag{20}$$

where M is independent of R, λ_z is the guided wavelength along z, and ξ_3 is the length over which the field in medium 3 decays by $1/e$. According to Fig. 11, the function

$$R \exp \left(-\frac{1}{6\pi^2} \frac{\lambda_z^2 R}{|\xi_3|^3} \right)$$

becomes negligibly small, and consequently the attenuation per radian becomes negligibly small when

$$R > \frac{24\pi^2 \mid \xi_3 \mid^3}{\lambda_z^2}. \tag{21}$$

This simple criterion is developed further in Ref. 15.

We are interested, though, in a more detailed description of transmission through a bent dielectric guide. Given a guide with a certain radius of curvature (that is, given R and a/A), in general the loss per radian of the E_{11}^x mode is much larger than that of the E_{11}^y mode (compare, for example, Figs. 5 and 9 for $n_1/n_3 = 1.5$). That difference becomes negligible if $n_1/n_3 - 1 \ll 1$.

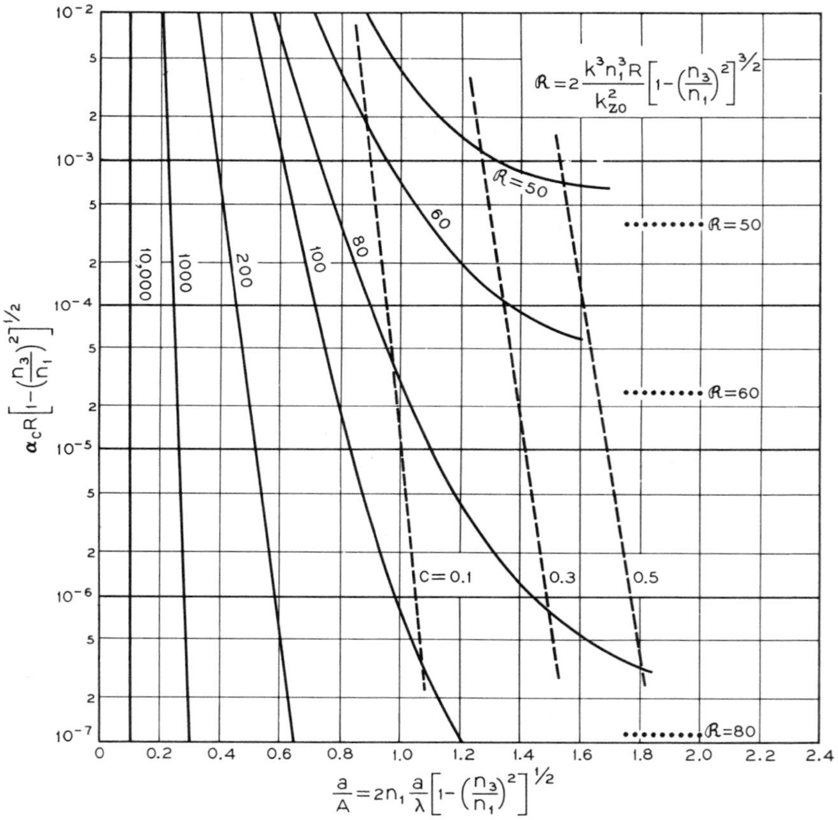

Fig. 9 — Attenuation per radian for $E_{11}{}^y$ mode and $n_1/n_3 > 1$.

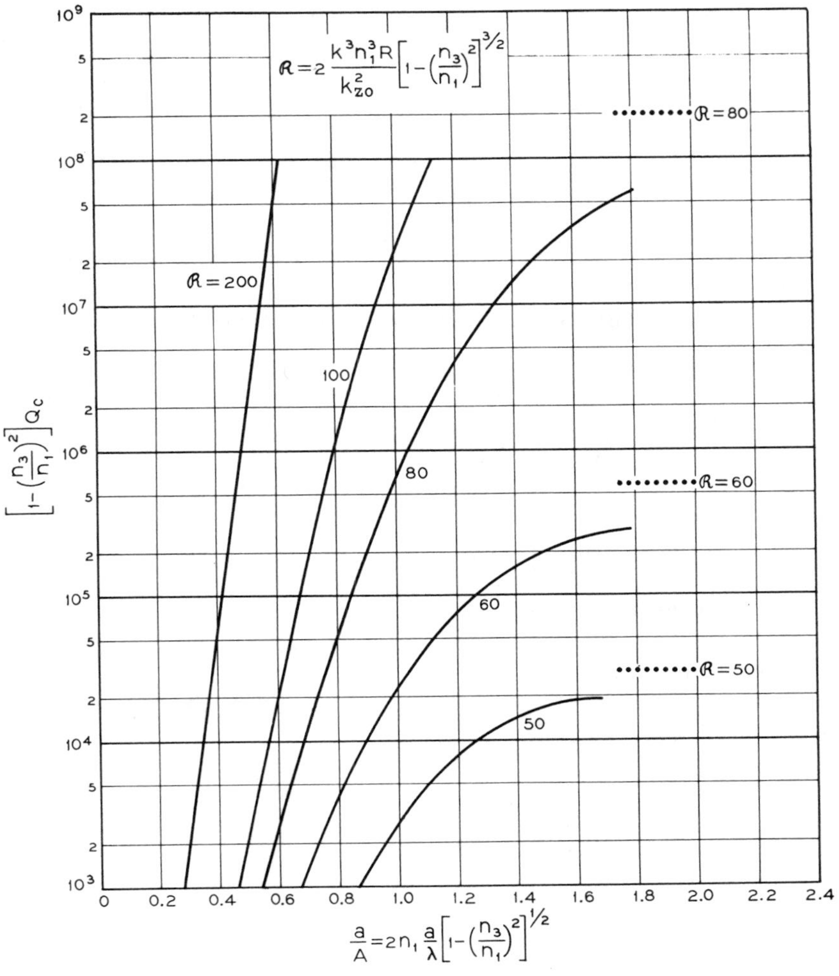

Fig. 10 — Intrinsic Q for E_{11}^y mode and $n_1/n_3 > 1$.

Let us consider separately the three types of guide: thin, medium and large.

3.1 *Thin or Low Loss Guides**

In thin guides the width a is so small that

* Low loss for straight guide.

$$\frac{a}{A} = \frac{2a(n_1^2 - n_3^2)^{\frac{1}{2}}}{\lambda} \ll 1. \tag{22}$$

The height b of the guide must be large so that only a little part of the power travels in the shaded areas of Fig. 2. Assuming that the guiding rod dielectric is lossy, its refractive index is

$$n_1 = n\left(1 + \frac{i\alpha}{kn}\right), \tag{23}$$

where n is real and α is the attenuation constant of a plane wave in that medium.

Substituting equations (22) and (23) in equations (2), (11), and (12), we obtain

$$k_z = k_{z0} + i\alpha_s + i\alpha_c . \tag{24}$$

The first term

$$k_{z0} = (k_3^2 - k_y^2)^{\frac{1}{2}} \begin{cases} 1 + \dfrac{1}{8}\left[k_3 a\left(1 - \dfrac{n_3^2}{n^2}\right)\right]^2 & \text{for } E_{11}^x \text{ mode} \\ 1 + \dfrac{1}{8}\left[k_3 a\left(\dfrac{n^2}{n_3^2} - 1\right)\right]^2 & \text{for } E_{11}^y \text{ mode} \end{cases} \tag{25}$$

is the phase constant. Since most of the power travels in the external medium, its value for either mode is close to kn_3. The conversion loss term c is negligible.

The imaginary part of equation (24) is the attenuation constant, and is made of two terms. The first term

$$\alpha_s = \frac{\alpha}{2} n n_3 k^2 a^2 \left(\frac{n^2}{n_3^2} - 1\right) \begin{cases} \left(\dfrac{n_3}{n}\right)^6 & \text{for } E_{11}^x \text{ mode} \\ 1 & \text{for } E_{11}^y \text{ mode} \end{cases} \tag{26}$$

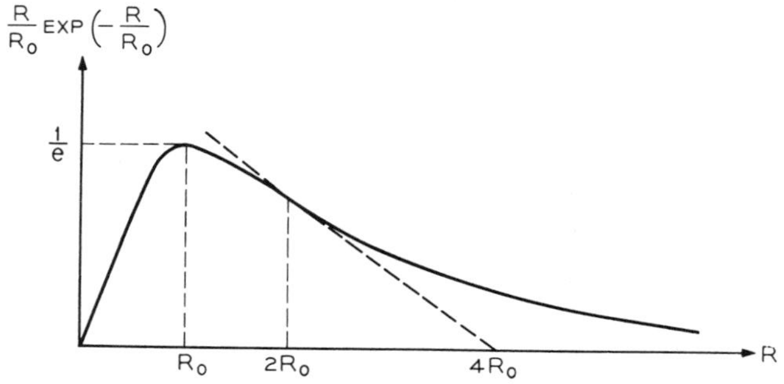

Fig. 11 — Plot of $R/R_0 \exp(-R/R_0)$ and tangent at inflection point.

is the attenuation that each mode would have if the guide were straight.[16] The second term

$$\alpha_c = \frac{k_3^3 a^2}{8}\left(\frac{n^2}{n_3^2} - 1\right)^2 \cdot \begin{cases} \left(\frac{n_3}{n}\right)^4 \exp\left\{-\frac{k_3^4 a^3 R}{12}\left(\frac{n_3}{n}\right)^6\left(\frac{n^2}{n_3^2} - 1\right)^3\left[1 - \frac{1}{2}\left(\frac{k_y}{k_3}\right)^2\right]\right\} & \text{(for } E_{11}^x \text{ mode)} \\ \exp\left\{-\frac{k_3^4 a^3 R}{12}\left(\frac{n^2}{n_3^2} - 1\right)^3\left[1 - \frac{1}{2}\left(\frac{k_y}{k_3}\right)^2\right]\right\} & \text{(for } E_{11}^y \text{ mode)} \end{cases} \tag{27}$$

is the attenuation resulting from the radiation introduced by the curvature. The E_{11}^y mode is more tightly bound to the guiding rod and consequently has more straight loss and less curvature loss than the E_{11}^x mode.

From equations (26) and (27), the radius of curvature R_d that doubles the straight guide loss is

$$R_d = \frac{12}{k_3}\left[\frac{\alpha n}{2\alpha_s n_3\left(\frac{n^2}{n_3^2} - 1\right)}\right]^{\frac{3}{2}}\left[1 - \frac{1}{2}\left(\frac{k_y}{k_3}\right)^2\right]^{-1} \cdot \begin{cases} \left(\frac{n_3}{n}\right)^3 \log\left[\frac{kn}{4\alpha}\left(\frac{n^2}{n_3^2} - 1\right)\right] & \text{(for } E_{11}^x \text{ mode).} \\ \log\left[\frac{kn_3^2}{4\alpha n}\left(\frac{n^2}{n_3^2} - 1\right)\right] & \text{(for } E_{11}^y \text{ mode).} \end{cases} \tag{28}$$

Example 1: Consider a thin ribbon guide made of glass surrounded by air and assume that $n = 1.5$, $n_3 = 1$, $\alpha = 0.1$ nepers per m, and $b = \infty$. From equations (26) and (28) we calculate the values in Table I.

It is doubly advantageous to use the E_{11}^x mode rather than the E_{11}^y because (*i*) the thickness required for equal radiation loss and straight guide loss is roughly $(n/n_3)^3$ times larger, and (*ii*) R_d is about $(n/n_3)^3$ times smaller.

If the height b of the ribbon is finite, k_y/kn_3 is no longer zero and the radii are, according to equation (28), $[1 - \frac{1}{2}(k_y/k_3)^2]^{-1}$ times longer than those in Table I.

3.2 *Medium Size Guide for Integrated Optical Circuitry*

It is likely that guides for integrated optical circuitry will be possible to fabricate only with $n_1 \cong n_3$. The radiation loss per radian and the Q_c of

TABLE I—VALUES CALCULATED FROM EQUATIONS (26) AND (28)

	E_{11}^y Mode		E_{11}^x Mode	
α_s (nepers/m)	$\frac{a}{\lambda}$	$\frac{R_d}{\lambda}$	$\frac{a}{\lambda}$	$\frac{R_d}{\lambda}$
0.01	0.05	1.9×10^3	0.17	6.3×10^2
0.001	0.016	6.2×10^4	0.055	2×10^4
0.0001	0.005	2×10^6	0.017	6.5×10^5

loops made with these guides can be obtained from Figs. 4 and 7, considering abscissas around $a/A = 1$. For both modes, E_{11}^y and E_{11}^x, most of the power travels within the guiding rod.*

In general, the losses are very sensitive to the radius of curvature. They are also sensitive to the guide's width to the left of the dashed curve $c = 0.5$, but fairly insensitive to the right of it.

Example 2: Let us design a guide:

(*i*) The attenuation per radian resulting from radiation loss is

$$\alpha_c R = 0.01 \text{ nepers} = 0.087 \text{ dB}.$$

(*ii*) Its width a is the maximum compatible with single mode guidance in the infinitely high slab, that is

$$\frac{a}{A} = \frac{2a}{\lambda}(n_1^2 - n_3^2)^{\frac{1}{2}} = 1.$$

(*iii*) We assume $b = \infty$ and $n_3 = n_1(1 - \Delta)$, where $\Delta \ll 1$ and $n_1 = 1.5$.

From Fig. 4 we derive the guide dimensions for different values of Δ:

Δ	$\frac{a}{\lambda}$	$\frac{R}{\lambda}$
0.1	0.745	30
0.01	2.36	1,060
0.001	7.45	37,000

Unless Δ is 0.01 or larger, the radius of curvature R becomes uncomfortably large for integrated optical circuitry. Furthermore, if b is finite, k_y is no longer zero, and the radii become $[1 - (k_y/k_3)^2]^{-1}$ times larger than those in the table above.

* This is not true if $b/B_{\frac{2}{4}} \ll 1$. Then k_{z0} must be calculated from equation (8).

Example 3: We design a resonant loop (Fig. 1) such that its Q_c resulting from radiation is equal to the Q resulting from transmission loss in typical glass ($n_1 = 1.5$, $\alpha = 0.1$ neper/m at $\lambda = 1\mu$); that is,

$$Q = Q_c = 5 \times 10^7.$$

Furthermore, let us assume as in Example 2 that $a/A = 1$, $n_3 = n_1(1 - \Delta)$, and $b = \infty$. With the help of Fig. 7 we derive

Δ	$\frac{a}{\lambda}$	$\frac{R}{\lambda}$
0.1	0.745	57
0.01	2.36	1,550
0.001	7.45	42,000

Again, unless Δ is larger than 0.01, the radius of curvature becomes unwieldily large for integrated optical circuitry.

Instead of using a loop as the resonant circuit of Fig. 1, it is possible to make $a = R$, and the loop becomes a pillbox (Fig. 12). This structure may be simpler to fabricate. For this case, also from Fig. 4, using the refractive indices of the previous example, we obtain

Δ	$\frac{R}{\lambda}$
θ.1	42
0.01	1,170
0.001	32,000

The pillbox resonator requires a 30 percent shorter radius than the ring resonator. As before, if b is finite, the radii are $[1 - (k_y/k_3)^2]^{-1}$ times longer than those in the last two tables.

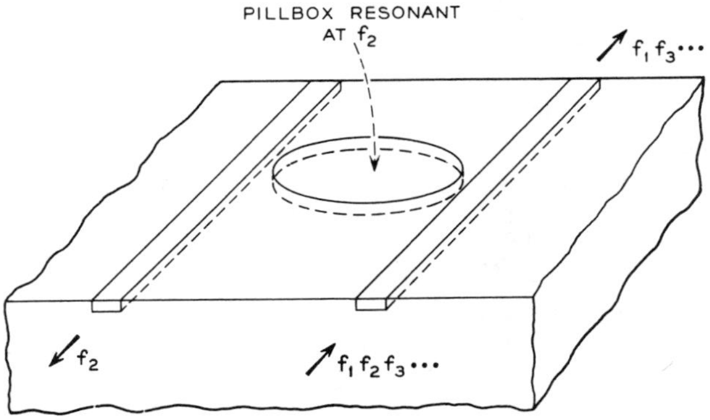

Fig. 12 — Channel dropping filter (pillbox type).

3.3 *Large Guides for Fiber Optics*

The large guide is multimode, $a/A \gg 1$, and the radius for small mode conversion is derived from equations (11) and (12), making $k_{x0}a = \pi$ and $k_{z0} = 2\pi n_1/\lambda$. Then

$$c = \pi n_1^2 \frac{a^3}{\lambda^2 R}.$$

For a power conversion $c^2 = 0.01$, and $n_1 = 1.5$, we have

$\frac{a}{\lambda}$	$\frac{R}{\lambda}$
5	8,900
10	71,000

The conversion loss is many orders of magnitude larger than the loss radiated by the fundamental mode because of the curvature. Radiation loss of higher order modes can be found in equations (51) and (63).

In general, clad fibers are of circular cross section; consequently our calculations do not strictly apply. Nevertheless, a guide of circular cross section and another of equal area but square cross section must have quite comparable attenuation per radian unless mode degeneracy occurs, but this is quite unlikely.

Though we have been talking throughout of light guides, it is obvious that all the calculations are equally applicable to microwave guides.

IV. CONCLUSIONS

Relations between radiation losses resulting from curvature, geometry, and electric characteristics of the bent dielectric guide are summarized in Figs. 4, 5, and 7 through 10 and they are discussed and exemplified in Section III.

The main qualitative results are that for a given radius of curvature R, the radiation loss can be reduced

(*i*) by increasing the difference between the refractive index n_1 of the guide and those of the media toward the outside, n_3, and inside, n_5, of the curved guide axis (Fig. 2);

(*ii*) by increasing the guide width a. Nevertheless, once a is bigger than

$$\left(\frac{R\lambda^2}{\pi n_1^2}\right)^{\frac{1}{3}},$$

(where λ is the free space wavelength), there is little reduction of the loss;

(*iii*) by choosing the height of the guide large enough to confine the fields as much as possible within the guide in the direction normal to the plane of curvature.

In general, the radiation losses are small if

$$R > \frac{24\pi^2 \mid \xi_3 \mid^3}{\lambda^2},$$

where ξ_3 is the length over which the field decays by $1/e$ in medium 3 (Fig. 2).

Thin ribbons of glass, surrounded by air and oriented as in Fig. 6c, operate better with the electric field perpendicular to the ribbon's plane. Choosing the thickness $a = 0.055\lambda$, the attenuation of the straight guide is 1 percent of the attenuation in glass, and the radius of curvature which doubles that low attenuation is $20{,}000\lambda$.

The dielectric guide for integrated optical circuitry seems suitable to negotiate bends and to make resonant loops of small radii of curvature and small radiation losses. For example, for

$$n_1 = 1.5$$

$$a = \frac{\lambda}{2n_1\left(1 - \frac{n_3^2}{n_1^2}\right)^{\frac{1}{2}}} \quad \text{(single mode guide)}$$

a 1 percent attenuation (0.087 dB) resulting from radiation in a length of guide equal to R is achieved with the following values

$1 - \frac{n_3}{n_1}$	$\frac{a}{\lambda}$	$\frac{R}{\lambda}$
0.1	0.745	30
0.01	2.36	1060
0.001	7.45	37000

The smaller $n_1 - n_3$, the larger the radius of curvature. For $\lambda = 0.63\mu$, if one wants to keep R below 1 mm, the difference between the internal and external refractive indices must be larger than 0.01.

Large cross section dielectric guides capable of supporting many modes are far more sensitive to mode conversions than to radiation losses. For the fundamental mode, the power conversion loss at the junction between a straight and a curved section of a multimode guide is

$$c^2 = \left(\pi n_1^2 \frac{a^3}{\lambda^2 R}\right)^2.$$

For $n_1 = 1.5$, $a = 6.3\mu$, and $\lambda = 0.63\mu$, the radius of curvature R that produces a power conversion c^2 of 0.01 is 45 mm. The radiation loss in a length of guide equal to R is many orders of magnitude below 0.01.

APPENDIX

Field Analysis of the Curved Guide

Figure 2 shows the geometry and dielectric distribution of the curved guide. In this appendix two families of modes are found, E^x_{pq} and E^y_{pq}; each is studied separately.

A.1 E^x_{pq} *Modes: Polarization Along x*

The field components in each region should be written as integral expressions, but, as discussed in Section II, the power propagating through the shaded areas is neglected, and the field matching is performed only along the sides of region 1. Consequently, those field components do not need to be so general. As a matter of fact, the simplest field components in the mth of the five areas are[16]

$$H_{xm} = \frac{1}{k_m^2 - k_{ym}^2} \frac{\partial^2 H_{ym}}{\partial x\, \partial y},$$

$$H_{ym} = e^{-i\nu\theta + i\omega t}$$

$$\cdot \begin{cases} M_1 J_\nu[(k_1^2 - k_{y1}^2)^{\frac{1}{2}}(R + x) + \psi_1] \cos(k_{y1} y + \Omega_1) & \text{for } m = 1 \\ M_{\substack{2\\4}} J_\nu\left[\left(k_{\substack{2\\4}}^2 - k_{y\substack{2\\4}}^2\right)^{\frac{1}{2}}(R + x) + \psi_{\substack{2\\4}}\right] \exp\left(\mp i k_{y\substack{2\\4}} y\right) & \text{for } m = \substack{2\\4} \\ M_3 H_\nu^{(2)}[(k_3^2 - k_{y3}^2)^{\frac{1}{2}}(R + x)] \cos(k_{y3} y + \Omega_3) & \text{for } m = 3 \\ M_5 J_\nu[(k_5^2 - k_{y5}^2)^{\frac{1}{2}}(R + x)] \cos(k_{y1} y + \Omega_5) & \text{for } m = 5 \end{cases},$$

$$H_{zm} = \frac{i}{k_m^2 - k_{ym}^2} \frac{\nu}{R + x} \frac{\partial H_{ym}}{\partial y},$$

$$E_{xm} = -\frac{\omega\mu}{k_m^2 - k_{ym}^2} \frac{\nu}{(R + x)} H_{ym},$$

$$E_{ym} = 0,$$

$$E_{zm} = \frac{-i\omega\mu}{k_m^2 - k_{ym}^2} \frac{\partial H_{ym}}{\partial x}, \tag{29}$$

in which M_m is the amplitude of the field in the mth medium; ψ_m and Ω_m are constants that locate the field maxima in region m; ω is the angular frequency; ϵn_m^2 and μ, the permittivity and permeability of each medium, are related by $k_m^2 = k^2 n_m^2 = \omega^2 \epsilon\mu n_m^2$; k_{ym} is the propagation constant along y in medium m; and J_ν and $H_\nu^{(2)}$ are Bessel and Hankel functions, respectively.

Strictly speaking, the H_y component in media 1, 2, and 4 should be written as a sum of Bessel functions of the first and second kind, but later on they are approximated by circular functions; therefore, we do not make any mistake using only the Bessel function of the first kind with an arbitrary phase constant in the argument.

We consider only guide geometries for which the guide wavelengths measured in the x and y directions in medium 1 are large compared with the wavelength measured in the z direction. This means that (i)

$$\frac{\partial H_{ym}}{\partial x} \ll \frac{\nu}{R}\,, \tag{30}$$

and, as a consequence, the field component H_{x1} is very small compared with H_z and is neglected; (ii) the propagating modes are basically of the TEM type.

In order to match the remaining components along the boundaries of medium 1, the field components in media 1, 2, and 4 must have the same dependence along x, while the field components in media 1, 3, and 5 must have the same dependence along y. Therefore

$$k_{y1} = k_{y3} = k_{y5} = k_y\,, \tag{31}$$

$$k_1^2 - k_y^2 = k_2^2 - k_{y2}^2 = k_4^2 - k_{y4}^2\,, \tag{32}$$

$$\psi_1 = \psi_2 = \psi_4 = \psi, \quad \text{and} \quad \Omega_1 = \Omega_3 = \Omega_5 = \Omega. \tag{33}$$

Furthermore, the field matching yields the following four equations from which two characteristic equations will be derived

$$\tan\left(k_y \frac{b}{2} + \Omega\right) = i\,\frac{k_{y2}}{k_y}\,, \qquad \tan\left(k_y \frac{b}{2} - \Omega\right) = i\,\frac{k_{y4}}{k_y} \tag{34}$$

$$\frac{J_\nu(\rho_{13})}{J_\nu'(\rho_{13})} = \frac{\rho_3}{\rho_{13}}\,\frac{H_\nu^{(2)}(\rho_3)}{H_\nu^{(2)\prime}(\rho_3)}\,, \quad \text{and} \quad \frac{J_\nu(\rho_{15})}{J_\nu'(\rho_{15})} = \frac{\rho_5}{\rho_{15}}\,\frac{J_\nu(\rho_5)}{J_\nu'(\rho_5)} \tag{35}$$

where

$$\left.\begin{aligned} \rho_{13} &= R(k_1^2 - k_y^2)^{\frac{1}{2}} + \psi, & \rho_{15} &= (R-a)(k_1^2 - k_y^2)^{\frac{1}{2}} + \psi \\ \rho_3 &= R(k_3^2 - k_y^2)^{\frac{1}{2}}, \quad \text{and} & \rho_5 &= (R-a)(k_5^2 - k_y^2)^{\frac{1}{2}}. \end{aligned}\right\} \tag{36}$$

Similar to what happens with the straight guide, equations (34) and (35) are the boundary conditions of two independent problems far simpler than the one depicted in Fig. 2. Thus, for a dielectric slab infinite in the x and z directions and with dimensions and refractive indices as depicted in Fig. 13a, the boundary conditions for modes with no E_y component coincide with equation (34). Similarly, for a bent slab infinite in the y direction as shown in Fig. 13b, the boundary conditions for modes with a negligible H_x component coincide with equation (35).

The elimination of Ω between the two expressions of equation (34) yields the characteristic equation for the plane slab[10]

$$k_y b = q\pi - \tan^{-1} \frac{1}{\left[\left(\frac{\pi}{A_2 k_y}\right)^2 - 1\right]^{\frac{1}{2}}} - \tan^{-1} \frac{1}{\left[\left(\frac{\pi}{A_4 k_y}\right)^2 - 1\right]^{\frac{1}{2}}}, \tag{37}$$

in which

$$A_{\substack{2\\4}} = \frac{\lambda}{2\left(n_1^2 - n_{\substack{2\\4}}^2\right)^{\frac{1}{2}}}; \tag{38}$$

the $\tan^{-1}$ functions are to be taken in the first quadrant, and the arbitrary integer q is the order of the mode, that is, the number of extrema of each field component within the guiding rod in the y direction.

The transcendental equation (37) has an approximate closed form solution already found in Ref. 10

$$k_y \cong \frac{q\pi}{b}\left(1 + \frac{A_2 + A_4}{\pi b} + \cdots\right)^{-1}, \tag{39}$$

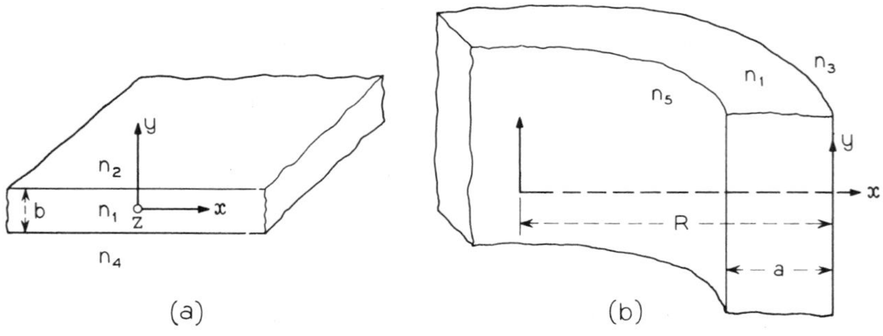

Fig. 13 — Guiding dielectric slabs.

which is valid only when b is so large that

$$\frac{A_2 + A_4}{\pi b} \ll 1 \tag{40}$$

and consequently the parenthesis is close to unity.

The field components in media 2 and 4 decay exponentially by $1/e$ in lengths η_2 and η_4, which are deduced from equation (32) to be

$$\eta_{\substack{2\\4}} = \frac{1}{\left| k_{\nu \substack{2\\4}} \right|} = \frac{1}{\left(k_1^2 - k_{\substack{2\\4}}^2 - k_\nu^2\right)^{\frac{1}{2}}}. \tag{41}$$

Let us consider the solution of the characteristic equation of the bent slab (Fig. 13b). For guided modes, both the arguments and the order of the Bessel and Hankel functions involved in equation (35) are large compared with unity, and consequently they can be replaced by their Watson's first term approximations,[17]

$$\left.\begin{aligned} J_\nu(\rho) &= \left[\frac{2}{\pi(\rho^2 - \nu^2)^{\frac{1}{2}}}\right]^{\frac{1}{2}} \begin{cases} \dfrac{1}{2} \exp\left[-\dfrac{(\nu^2 - \rho^2)^{\frac{3}{2}}}{3\nu^2}\right] & \text{for} \quad \nu > \rho \\ \sin\left[\dfrac{(\rho^2 - \nu^2)^{\frac{3}{2}}}{3\nu^2} + \dfrac{\pi}{4}\right] & \text{for} \quad \rho > \nu \end{cases} \\ Y_\nu(\rho) &= -\left[\frac{2}{\pi(\rho^2 - \nu^2)^{\frac{1}{2}}}\right]^{\frac{1}{2}} \begin{cases} \exp\left[\dfrac{(\nu^2 - \rho^2)^{\frac{3}{2}}}{3\nu^2}\right] & \text{for} \quad \nu > \rho \\ \cos\left[\dfrac{(\rho^2 - \nu^2)^{\frac{3}{2}}}{3\nu^2} + \dfrac{\pi}{4}\right] & \text{for} \quad \rho > \nu. \end{cases} \end{aligned}\right\} \tag{42}$$

These expressions are valid if

$$\frac{\nu^2}{(\rho^2 - \nu^2)^{\frac{3}{2}}} \ll 1. \tag{43}$$

Introducing these approximations for the Bessel functions in both equations (35) and eliminating ψ between them, we obtain the characteristic equation for the bent slab

$$\begin{aligned} \frac{1}{3\nu^2} &[(\rho_{13}^2 - \nu^2)^{\frac{3}{2}} - (\rho_{15}^2 - \nu^2)^{\frac{3}{2}}] \\ &= p\pi - \tan^{-1}\left(\frac{n_3^2}{n_1^2}\left[\frac{\rho_{13}^2 - \nu^2}{\nu^2 - \rho_3^2}\right]^{\frac{1}{2}}\left\{1 + i \exp\left[-\frac{2}{3}\frac{(\nu^2 - \rho_3^2)^{\frac{3}{2}}}{\nu^2}\right]\right\}\right) \\ &\qquad - \tan^{-1}\frac{n_5^2}{n_1^2}\left(\frac{\rho_{15}^2 - \nu^2}{\nu^2 - \rho_5^2}\right)^{\frac{1}{2}}, \end{aligned} \tag{44}$$

in which p is an arbitrary integer bigger than zero which determines the order of the mode in the x direction, and the $\tan^{-1}$ functions are to be taken in the first quadrant.

Let us rewrite this equation substituting ρ_3, ρ_5, ρ_{13}, and ρ_{15} by the values given in equation (36); furthermore, let

$$A_{\substack{3\\5}} = \frac{\lambda}{2\left(n_1^2 - n_{\substack{3\\5}}^2\right)^{\frac{1}{2}}}, \tag{45}$$

$$\nu = k_z R \tag{46}$$

and

$$k_z = (k_1^2 - k_y^2 - k_x^2)^{\frac{1}{2}}. \tag{47}$$

Because of these two last definitions, k_z, k_x, and k_y are the axial and the transverse propagation constants at $x = 0$. The characteristic equation (44) then becomes

$$\frac{Rk_x^3}{3k_z^2}\left[1 - \left(1 - \frac{2ak_z^2}{k_x^2 R}\right)^{\frac{3}{2}}\right]$$

$$= p\pi - \tan^{-1}\frac{n_3^2}{n_1^2}\frac{1 + i\exp\left\{-\frac{2}{3}\frac{\pi^3 R}{k_z^2 A_3^3}\left[1 - \left(\frac{k_x A_3}{\pi}\right)^2\right]^{\frac{3}{2}}\right\}}{\left[\left(\frac{\pi}{k_x A_3}\right)^2 - 1\right]^{\frac{1}{2}}}$$

$$- \tan^{-1}\frac{n_5^2}{n_1^2}\left[\frac{\left[1 - \frac{a}{R}\right]^2 (k_1^2 - k_y^2) - k_z^2}{k_z^2 - \left(1 - \frac{a}{R}\right)^2 (k_5^2 - k_y^2)}\right]^{\frac{1}{2}}. \tag{48}$$

To solve this equation for k_x we expand the left side and the second $\tan^{-1}$ in powers of $1/R$ and the first $\tan^{-1}$ in powers of the exponential. Assuming R is large and keeping the first term of each perturbation calculation, the solution of equation (48) is

$$k_x = k_{x0}\left(1 + \frac{2c}{ak_{x0}} - i\frac{k_{z0}\alpha_c}{k_{x0}^2}\right), \tag{49}$$

where

$$c = \frac{1}{2k_{x0}a}\left(\frac{\pi a}{A_3}\right)^3 \frac{1}{\mathcal{R}} \frac{1 + 2F_5}{1 + F_3 + F_5} \tag{50}$$

and

$$\alpha_c = \frac{k_{x0}^2}{k_{z0}}\left[1 - \left(\frac{k_{x0}A_3}{\pi}\right)^2\right]F_3 \frac{\exp\left\{-\frac{\mathcal{R}}{3}\left[1 - \left(\frac{k_{x0}A_3}{\pi}\right)^2\left(1 + \frac{2c}{ak_{x0}}\right)^2\right]^{\frac{3}{2}}\right\}}{1 + F_3 + F_5}, \tag{51}$$

in which

$$F_{\substack{3\\5}} = \left(\frac{n_{\substack{3\\5}}}{n_1}\right)^2 \frac{A_{\substack{3\\5}}}{\pi a\left[1 - \left(\frac{k_{x0}A_{\substack{3\\5}}}{\pi}\right)^2\right]^{\frac{1}{2}}} \frac{1}{1 - \left[1 - \left(\frac{n_{\substack{3\\5}}^2}{n_1^2}\right)\right]\left(\frac{k_{x0}A_{\substack{3\\5}}}{\pi}\right)^2}, \tag{52}$$

$$\mathcal{R} = \frac{2\pi^3 R}{k_{x0}^2 A_3^3} = 2(n_1^2 - n_3^2)^{\frac{3}{2}} \frac{k^3 R}{k_{x0}^2}, \tag{53}$$

$$k_{z0} = (k_1^2 - k_y^2 - k_{x0}^2)^{\frac{1}{2}}, \tag{54}$$

and k_{x0} is the solution of the equation

$$k_{x0}a = p\pi - \tan^{-1}\frac{n_3^2}{n_1^2}\frac{1}{\left[\left(\frac{\pi}{k_{x0}A_3}\right)^2 - 1\right]^{\frac{1}{2}}} - \tan^{-1}\frac{n_5^2}{n_1^2}\frac{1}{\left[\left(\frac{\pi}{k_{x0}A_5}\right)^2 - 1\right]^{\frac{1}{2}}}. \tag{55}$$

This is the physical interpretation of equation (49): the transverse propagation constant k_x measured at $x = 0$ is made of three terms. The first term, k_{x0}, is the transverse propagation constant of the guide without curvature; the second and third terms are perturbations related to the change of field profile and radiation introduced by the curvature. It is easy to find that c^2 is the mode conversion loss that would exist at a junction between a straight guide and a curved one, and α_c is the attenuation coefficient of the curved guide.

The field components in media 3 and 5 decay almost exponentially away from the guide. The length ξ_3, over which the intensity in medium 3 decays by $1/e$, is derived as in equation (41) to be

$$\xi_3 = \frac{1}{|k_{x3}|} = \frac{1}{(k_1^2 - k_3^2 - |k_x^2|)^{\frac{1}{2}}} \tag{56}$$

and only approximately

$$\xi_5 = \frac{1}{|k_{x5}|} = \frac{1}{(k_1^2 - k_5^2 - |k_x^2|)^{\frac{1}{2}}}. \tag{57}$$

All these equations have been derived under the assumption that inequality (43) is satisfied; this means that the field configuration of the curved guide is very close to that of the straight guide. In other words, $c \ll 1$. For a given R, if one chooses the width a of the guide large enough, these inequalities are not satisfied, the previous results are no longer applicable, and a new solution is needed. We proceed to find it.

Let us assume as a limiting case that in Fig. 2

$$a = R. \tag{58}$$

The characteristic equation derived from the first equation of (35), making $\psi = 0$, is

$$\frac{(\rho_{13}^2 - \nu^2)^{\frac{3}{2}}}{3\nu^2} = (p - \tfrac{1}{4})\pi - \tan^{-1}\frac{n_3^2}{n_1^2}\left(\frac{\rho_{13}^2 - \nu^2}{\nu^2 - \rho_3^2}\right)^{\frac{1}{2}} \cdot\left\{1 + i\exp\left[-\frac{2}{3}\frac{(\nu^2 - \rho_3^2)^{\frac{3}{2}}}{\nu^2}\right]\right\}. \tag{59}$$

Following similar steps to those taken to solve equation (44), we substitute ρ_{13}, ρ_3, and ν by the values given in equations (36) and (46); we obtain

$$\frac{R(k_x')^3}{3(k_z')^2} = (p - \tfrac{1}{4})\pi - \tan^{-1}\frac{n_3^2}{n_1^2}\cdot\frac{1 + i\exp\left\{-\dfrac{2}{3}\dfrac{\pi^3 R}{(k_z')^2 A_3^3}\left[1 - \left(\dfrac{k_x' A_3}{\pi}\right)^2\right]^{\frac{3}{2}}\right\}}{\left[\left(\dfrac{\pi}{k_x' A_3}\right)^2 - 1\right]^{\frac{1}{2}}} \tag{60}$$

The primes distinguish the symbols from those used previously.

To solve this equation we notice that for small losses it must be that

$$\frac{k_x' A_3}{\pi} \ll 1. \tag{61}$$

Therefore, the $\tan^{-1}$ can be replaced by its argument and the approximate solution of equation (60) is

$$k_x' = k_{x0}'\left[1 - i\frac{k_{z0}'\alpha_c}{(k_{x0}')^2}\right], \tag{62}$$

where

$$\alpha_c = \frac{n_3^2}{n_1^2}\frac{k_{z0}'}{kR(n_1^2 - n_3^2)^{\frac{1}{2}}}$$

$$\cdot \exp\left(-\frac{\mathcal{R}'}{3}\left\{1 - \left[\frac{6\pi(p-\frac{1}{4})}{\mathcal{R}'}\right]^{\frac{2}{3}}\left[1 - \frac{2}{3}\frac{n_3^2}{n_1^2}\left(\frac{6}{\pi^2(p-\frac{1}{4})^2\mathcal{R}'}\right)^{\frac{1}{3}}\right]\right\}^{\frac{3}{2}}\right), \tag{63}$$

$$k'_{z0} = [k_1^2 - k_y^2 - (k'_{x0})^2]^{\frac{1}{2}}, \tag{64}$$

$$k'_{x0} = \frac{\pi}{A_3}\left[\frac{6\pi(p-\frac{1}{4})}{\mathcal{R}'}\right]^{\frac{1}{3}}\left\{1 - \frac{1}{3}\frac{n_3^2}{n_1^2}\left[\frac{6}{\pi^2(p-\frac{1}{4})^2\mathcal{R}'}\right]^{\frac{1}{3}}\right\}, \tag{65}$$

and

$$\mathcal{R}' = \frac{2\pi^3 R}{(k'_{z0})^2 A_3^3} = 2(n_1^2 - n_3^2)^{\frac{3}{2}}\frac{k^3 R}{(k'_{z0})^2}. \tag{66}$$

The field components outside the guide decay to $1/e$ in a length

$$\xi'_3 = \frac{1}{|k'_{x3}|} = \frac{1}{[k_1^2 - k_3^2 - (k'_{x0})^2]^{\frac{1}{2}}}. \tag{67}$$

A.2 E^y_{pq} *Modes: Polarization Along* y

The field components and propagation constants can be derived from those in Section A.1 by changing E into H, μ into $-\epsilon$, and vice versa. Except for their polarizations, the E^x_{pq} and E^y_{pq} modes are very similar.

The formulas equivalent to equations (37) and (41) are

$$k''_y b = q\pi - \tan^{-1}\frac{n_2^2}{n_1^2}\frac{1}{\left[\left(\frac{\pi}{A_2 k''_y}\right)^2 - 1\right]^{\frac{1}{2}}} - \tan^{-1}\frac{n_4^2}{n_1^2}\frac{1}{\left[\left(\frac{\pi}{A_4 k''_y}\right)^2 - 1\right]^{\frac{1}{2}}} \tag{68}$$

$$\eta''_{\substack{2\\4}} = \frac{1}{|k''_{y\substack{2\\4}}|} \quad \frac{1}{\left(k_1^2 - k_{\substack{2\\4}}^2 - (k''_y)^2\right)^{\frac{1}{2}}}. \tag{69}$$

The double prime distinguish these symbols from those used before.

The equivalent formula to any of those between equation (45) and (67) can be derived from that formula by substituting the ratio of refractive indexes by unity, but leaving the differences between squares of indexes unchanged. For example, the formula equivalent to equation (52) for E^y_{pq} modes is

$$F''_{\substack{3\\5}} = \frac{A_{\substack{3\\5}}}{\pi a\left[1 - \left(\frac{k''_{x0}A_{\substack{3\\5}}}{\pi}\right)^2\right]^{\frac{1}{2}}}\;\frac{1}{1 - \left(1 - \frac{n_{\substack{3\\5}}^2}{n_1^2}\right)\left(\frac{k_{x0}A_{\substack{3\\5}}}{\pi}\right)^2}. \tag{70}$$

REFERENCES

1. Miller, S. E., U. S. Patent 3434774, applied for February 2, 1965 granted March 25, 1969.
2. Karbowiak, A. E., "New Type of Waveguide for Light and Infrared Waves," Elec. Letters, *1,* No. 2 (April 1965), p. 47.
3. Wolff, P. A., unpublished work.
4. Miller, S. E., "Integrated Optics: An Introduction," B.S.T.J., this issue, pp. 2059–2069.
5. Kaplan, R. A., "Optical Waveguide of Macroscopic Dimension in Single-Mode Operation," Proc. IEEE, *51,* No. 8 (August 1963), p. 1144.
6. Schineller, E. R., "Summary of the Development of Optical Waveguides and Components," Wheeler Laboratories, Report #1471, April 1967.
7. Kapany, N. S., *Fiber Optics,* New York: Academic Press, 1967.
8. Stratton, J. A., *Electromagnetic Theory,* New York: McGraw-Hill, 1941, pp. 524–527.
9. Snitzer, E., "Cylindrical Dielectric Waveguide Modes," J. Opt. Soc. Amer., *51,* No. 5 (May 1961), pp. 491–498.
10. Marcatili, E. A. J., "Dielectric Rectangular Waveguide and Directional Coupler for Integrated Optics," B.S.T.J., this issue, pp. 2071–2102.
11. Schlosser, W., and Unger, H. G., "Partially Filled Waveguides and Surface Waveguides of Rectangular Cross-Section," *Advances in Microwaves,* New York: Academic Press, 1966, pp. 319–387.
12. Bracey, M. F., Cullen, A. L., Gillespie, E. F. F., and Staniforth, J. A., "Surface-Wave Research in Sheffield," IRE Trans. Antennas and Propagation, *AP7* (December 1959), pp. S219–S225.
13. Goell, J. E., "A Circular-Harmonic Computer Analysis of Rectangular Dielectric Waveguides," B.S.T.J., this issue, pp. 2133–2160.
14. Jones, A. L., "Coupling of Optical Fibers and Scattering in Fibers," J. Opt. Soc. Amer., *55,* No. 3 (March 1965), pp. 261–271.
15. Marcatili, E. A. J., and Miller, S. E., "Improved Relations Describing Directional Control in Electromagnetic Wave Guidance," B.S.T.J., this issue, pp. 2161–2188.
16. Stratton, J. A., *Electromagnetic Theory,* New York: McGraw-Hill, 1941, pp. 361.
17. Magnus, W., Oberhettinger, F., and Soni, R. P., *Formulas and Theorems for the Special Functions of Mathematical Physics,* New York: Springer-Verlag, 1966, p. 144.

Optical Waveguide Scattering and Griffith Microcracks

Abstract—Experiments with low bulk loss liquid guides indicate that scattering losses can be attributed to the glass substrates. Order-of-magnitude computations point to trapped air as the source of scatter.

A series of experiments not originally designed to observe optical surface wave scattering has nevertheless demonstrated some interesting features in that regard. Waveguides were formed by injecting nitrobenzene into a trough as shown in Fig. 1. The reagent grade nitrobenzene ($n = 1.55$) entered by capillarity into a channel that was fabricated by shadow-masking part of a glass substrate ($n = 1.52$) during RF sputtering of barium crown glass films ($n = 1.61$). Tight clamping of the substrate over the channel maintained the fluid thickness to that of the glass film (2.5 μ). Under appropriate index matching conditions, the channel generally provides a good transition from glass to fluid guides as had been previously observed with CS_2 ($n = 1.62$). Clean room procedures were observed during preparation of the samples.

Fig. 2 is a photograph of a 1-cm portion of a guided wave centered about the glass–fluid guide transition. This region is evidenced by a bright patch and a number of illuminated features. The incident laser beam is coupled into the glass film with a prism coupler [1], [2] and enters the junction from the left. Substantial scattering occurs at the junction because of a significant difference in index between the fluid and the glass film. The attenuation due to scattering of a TE_0 mode in the glass film was observed by photographic means to be of comparable value to that of the liquid guide. Direct transmission measurements on a wedge-shaped cell of nitrobenzene indicated, however, that bulk scattering losses were less than 0.01 dB/cm. These results imply that the substrates were responsible for the scattering.

Glasses are known to exhibit an interesting property that offers possible explanation for at least a portion of the scattering observed in our experiments as well as for the scattering observed with sputtered films. The property involves the appearance of cracks known as Griffith microcracks at the surface of glass. Ernsberger has studied the properties of these cracks because of their relation to glass strength, and in a series of informative papers [3]–[5] he outlines some of their properties. Unfortunately, information about them must be obtained by indirect means. In his experiments, Ernsberger expands the cracks over small regions for optical viewing by first increasing the surface stress with an ion exchange involving the replacement of sodium by lithium and then etching the surface.

Some of the properties of the Griffith microcracks that Ernsberger relates include the following. The cracks extend perpendicularly inward from the surface a distance on the order of a micron and their length is probably on the same order. The width of the cracks, on the other hand, is a small fraction of this dimension, although this dimensionality is apparently not known for certain. Electron microscopic studies are limited by the relatively low density of the cracks and there may be other factors that limit surface replication. In one set of measurements, he counted a crack density of $5.6 \times 10^8/\mathrm{m}^2$. Both the density and the orientations are quite random. The cracks themselves are mechanically induced as has been verified by repeating the ion exchange technique after surface etching and noting an absence of crack formation. The forces required to induce the cracks are apparently quite small. Surface contact between two pieces of glass, for example, will induce a high density of cracks. Of importance for optical thin-film considerations is the fact that the cracks are not induced spontaneously and they may be eliminated by surface treatment.

Manuscript received December 28, 1970.

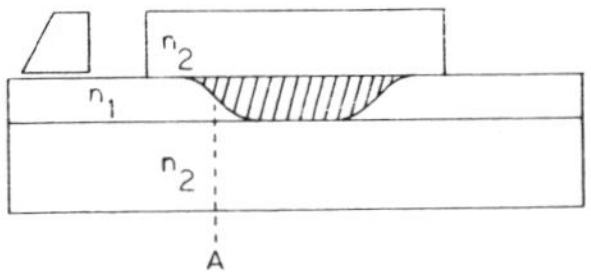

Fig. 1. Structure used to excite and guide optical waves into nitrobenzene (shaded) of index $n_3 = 1.55$; glass film index $n_1 = 1.61$; $n_2 = 1.52$.

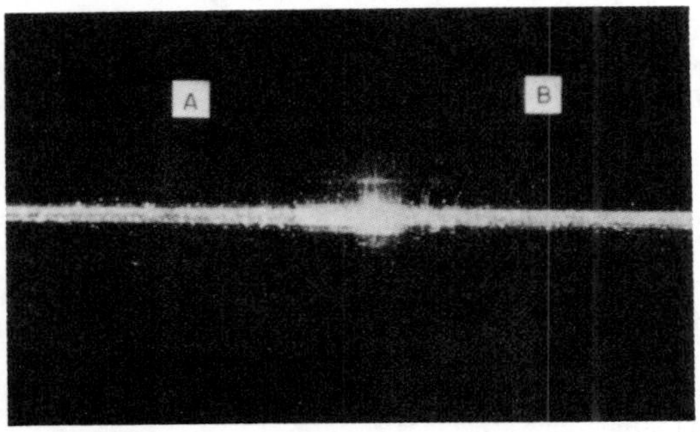

Fig. 2. Photograph of a 1-cm portion of the TE_0 guided wave as viewed through the top substrate. The bright spot corresponds to point A of Fig. 1. $\lambda_0 = 0.6328\ \mu$.

If we consider the effect of cracks of this kind on optical propagation, then it is clear first from their low density (compared with $1/\lambda^2$) and second from their random orientation with respect to the direction of propagation that the scattering from crack to crack may be considered to be incoherent. In terms of observed values of the power attenuation constant $\alpha = dP/dz$, we may therefore define an effective scattering cross section of a microcrack σ as

$$\sigma = P_c/p_i = \alpha t_e D \qquad (1)$$

where P_c is the average power radiated by a crack, p_i is the average incident power density, t_e is the effective width of the waveguide (power per unit width/p_i), and D is the surface density of cracks. The numerical value for (1) obtained using a value for α of 1 dB/cm (which we have observed with our best glass films and others have reported [6]), an effective thickness of 2.88 μ, and a two-substrate density corresponding to that observed by Ernsberger is $\sigma = 5.9 \times 10^{-14}\ \mathrm{m}^2$. Our own estimate of the crack density based on observation of a low-exposure photograph is approximately an order of magnitude greater than that of Ernsberger and yields a cross section $\sigma = 5.9 \times 10^{-15}$.

To relate σ to the crack geometry, a detailed analysis involving the three-dimensional continuous spectrum is required [2]. An analysis of this kind is planned, but for the present we arrive at an order-of-magnitude figure through the following considerations.

1) An equivalent current density J established by the Maxwell relation

$$\nabla \times \boldsymbol{H} = i\omega\varepsilon_0(n^2 + 2n\Delta n)\boldsymbol{E}_i = i\omega\varepsilon_0 n^2 \boldsymbol{E}_i + \boldsymbol{J} \qquad (2)$$

is assumed to be the source of scatter [7], [8].

2) The average power per crack coupled out of the surface wave beam is taken to be that of a short dipole of average current density J radiating in a homogeneous environment, viz. [9],

$$P_c = \sqrt{\mu/\varepsilon}\,\pi J^2(\Delta V)^2/3\lambda^2 \qquad (3)$$

Reprinted from *Proc. IEEE* (Lett.), vol. 59, July 1971, pp. 1123–1124.

where ΔV is the volume of the crack. Although (3) is an overestimate, it is probably fairly good considering the closeness of refractive index values for film and substrate, orientation, and other factors.

3) The average incident field is used to compute $\boldsymbol{J}$ in (2). Employing well-known expressions for the surface wave fields, a power computation results in

$$E_i^{2'} = 2(\mu/\varepsilon_0)^{1/2} p_i/n_s. \tag{4}$$

Combining (1) through (4), the approximate scattering cross section may be expressed

$$\sigma = 3.3 \times 10^2 [n\Delta n \Delta V/\lambda_0^2]^2. \tag{5}$$

We now assume ΔV to have dimensions corresponding to the product of the average penetration depth of the evanescent field ($1.4\ \lambda_0$), a length 10^{-6}, and a width c. The two expressions for σ based on observed attenuation and density measurements then provide the following expressions for the crack width:

$$c = 37.1/\Delta n \text{ Å} \qquad \text{or} \qquad c = 11.6/\Delta n \text{ Å}. \tag{6}$$

If the crack is assumed to be filled with nitrobenzene, then $\Delta n \simeq 3 \times 10^{-2}$ and the crack width turns out by this computation to have a width of from 400 to 1200 Å, which seems somewhat excessive. On the other hand, if the crack remains unfilled, then $\Delta n \simeq 0.5$ and crack widths of 23 to 75 Å are obtained from (6). This figure seems quite reasonable and would tend to suggest that covered-over Griffith microcracks have been a source of scattering in thin film waveguides.

ACKNOWLEDGMENT

The authors wish to thank W. D. Scott for bringing the microcracks to their attention.

J. H. HARRIS
D. P. GIARUSSO
R. SHUBERT
Dep. Elec. Eng.
Univ. Washington
Seattle, Wash. 98105

REFERENCES

[1] J. H. Harris and R. Shubert, "Optimum power transfer from a beam to surface wave," in *Conf. Abstracts, URSI Spring Meeting* (Washington, D. C., Apr. 1969), p. 71.
[2] J. H. Harris, R. Shubert, and J. N. Polky, "Beam coupling to films," *J. Opt. Soc. Amer.*, vol. 60, Aug. 1970, pp. 1007–1016.
[3] F. M. Ernsberger, "A study of the origin and frequency of occurrence of Griffith microcracks on glass surfaces," in *Research into Glass.* Pittsburgh, Pa.: PPG Industries–Science Press, 1967, pp. 72–81.
[4] ——, "Strength controlling structure in glass," *ibid.*, pp. 104–112.
[5] ——, "Strength and strengthening of glass," in *Research into Glass.* Pittsburgh, Pa.: PPG Industries–Science Press, 1970, pp. 21–37.
[6] J. E. Goell and R. D. Standley, "Integrated optical circuits," *Proc. IEEE*, vol. 58, Oct. 1970, pp. 1504–1512.
[7] D. Marcuse, "Mode conversion caused by surface imperfections of a dielectric slab waveguide," *Bell Syst. Tech. J.*, vol. 48, Dec. 1969, pp. 3187–3215.
[8] A. W. Snyder, "Excitation and scattering of modes on a dielectric or optical fiber," *IEEE Trans. Microwave Theory Tech.*, vol. MTT-17, Dec. 1969, pp. 1138–1144.
[9] S. Ramo, J. R. Whinnery, and T. Van Duzer, *Fields and Waves in Communication Electronics.* New York: Wiley, 1965, p. 645.

Properties of Irregular Boundary of RF Sputtered Glass Film for Light Guide

Abstract—**The correlation lengths and the average deviation of the irregular boundary of glass film prepared by the RF sputtering were estimated by comparing the measured directivity of the scattered light with the calculated one.**

Both of the longitudinal and the transverse correlation lengths were 0.2–1 μm and the average amplitude was about 70 Å.

The scattering loss [1] caused by the irregular boundaries of the dielectric slab light guide depends upon the amplitude of boundary deviation a, the longitudinal and transverse correlation lengths B_z and B_y, respectively, the slab thickness $2b$, and the relative refractive index differences between the film and the outer medium Δ_{12} and Δ_{13}, respectively, [2], [3].

To estimate a, B_z, and B_y, we calculated [3] first the directivity of the scattered light from the irregular boundaries with an assumption of the correlation function on the irregular boundary as follows [2]:

$$a^2 \exp\left(-\left|z - z'\right|/B_z\right) \exp\left(-2b/B_y\right). \quad (1)$$

An example of the calculated directivity is shown in Fig. 1. Thus the given directivities are strongly affected by the B_z, B_y, and $2b$. They also depend upon the propagation modes which are not shown here for the sake of simplicity. If we can measure the directivity of the scattered light, we may determine the B_z and B_y, compared with the calculated data. We also made a theoretical calculation on the scattering loss [2], [3]. Further, if we can measure the scattering loss, we may finally determine the a compared with the calculated value, making use of the values B_y and B_z determined above.

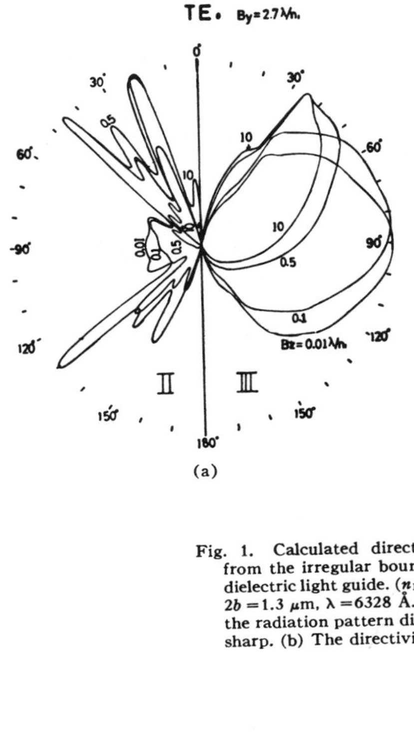

(a)

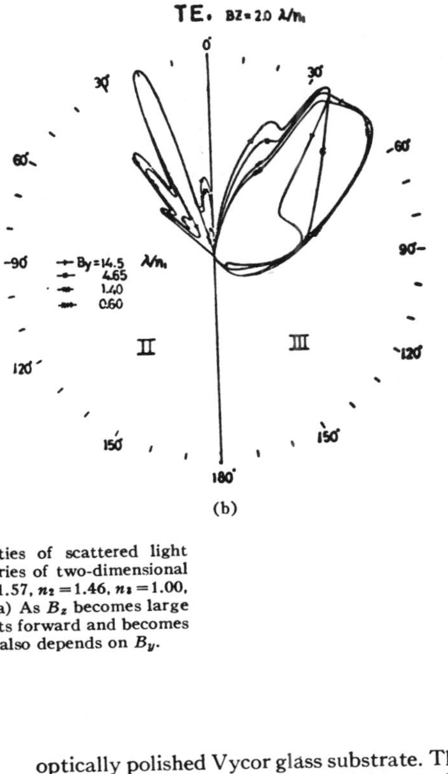

(b)

Fig. 1. Calculated directivities of scattered light from the irregular boundaries of two-dimensional dielectric light guide. ($n_1 = 1.57$, $n_2 = 1.46$, $n_3 = 1.00$, $2b = 1.3$ μm, $\lambda = 6328$ Å.) (a) As B_z becomes large the radiation pattern directs forward and becomes sharp. (b) The directivity also depends on B_y.

The measured directivities for four modes of propagation are shown in Fig. 2. An He–Ne laser was used for the source. The sample was prepared by the RF sputtering of Corning 7059 glass on the surface of optically polished Vycor glass substrate. The film thickness was 1.3 μm, which was determined by the angle measurements of incident light to the prism coupler [4]. The calculated curves shown in Fig. 2 are chosen from many curves with parameters of B_z and B_y. These curves fit for the measured curves of Fig. 2 over four modes with one combination of B_z and B_y. From this comparison we determined that both

Manuscript received February 8, 1972.

Reprinted from *Proc. IEEE* (Lett.), vol. 60, June 1972, pp. 744–745.

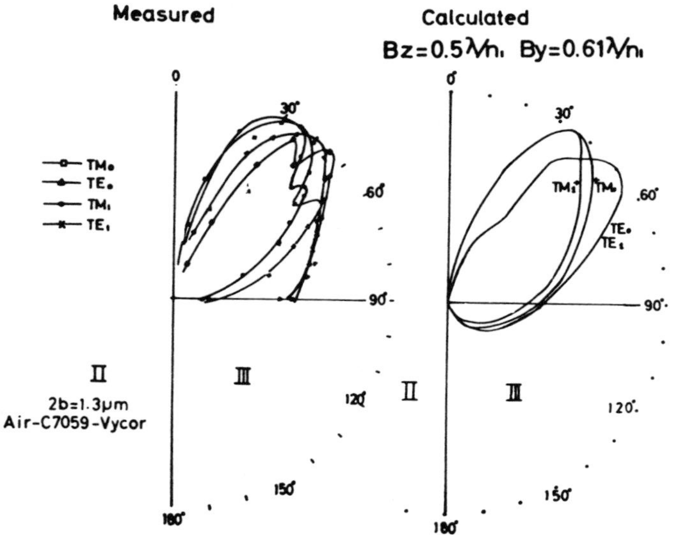

Fig. 2. Measured directivities and corresponding calculated directivities.

values of B_z and B_y were 0.5 λ/n_1 to 3 λ/n_1, where λ and n_1 =1.57 are the wavelength and the refractive index of the film, respectively.

The scattering loss of the guide was measured as 0.4 dB/cm [5]. From these data, we estimated that the rms amplitude a of the boundary deviation of the RF sputtered film was about 70 Å.

In this experiment, the scattered light is estimated to come mainly from the film–air surface; however, the effect of the Griffith crack [6] may be taken into account.

ACKNOWLEDGMENT

The authors are indebted to Prof. M. Kawakami, Prof. Y. Sakai, and Prof. T. Sekiguchi of the Tokyo Institute of Technology for their encouragement of and discussions concerning this research. They also wish to thank Y. Takao and Dr. T. Tokue of T. D. K. Co., Ltd., for helpful support.

Y. SUEMATSU
K. FURUYA
M. HAKUTA
K. CHIBA
Dep. Electronics
Tokyo Inst. of Technol.
O-okayama, Meguro-ku,
Tokyo, Japan 152

REFERENCES

[1] D. Marcuse, "Mode conversion caused by surface imperfections of a dielectric slab waveguide," *Bell Syst. Tech. J.*, vol. 48, p. 3187, Dec. 1969.
[2] Y. Suematsu and K. Furuya, "Propagation mode and scattering loss of two-dimensional dielectric waveguide with gradual distribution of refractive-index," to be published.
[3] ——, "Far-field radiation pattern caused by random wall distortion of dielectric waveguides and determination of correlation length," presented at the monthly meeting of QE IECE of Japan, Jan. 1972. Paper QE71-55.
[4] Y. Suematsu, Y. Sasaki, H. Noda, E. Asai, and M. Hakuta, "Measurements of refractive index and film thickness of glass films by use of propagation constants," *J. Inst. Electron. Commun. Eng. Jap.*, to be published.
[5] Y. Suematsu *et al.*, presented at the Nat. Conv. of the Inst. Electron. Commun. Japan. Paper 733, 1971.
[6] J. H. Harris, D. P. GiaRusso, and R. Shubert, "Optical waveguide scattering and Griffith microcracks," *Proc. IEEE* (Lett.), vol. 59, pp. 1123–1124, July 1971.

Mode Conversion Caused by Surface Imperfections of a Dielectric Slab Waveguide

By DIETRICH MARCUSE

(Manuscript received May 8, 1969)

This paper contains a perturbation theory which is applicable to the scattering losses suffered by guided modes of a dielectric slab waveguide as a consequence of imperfections of the waveguide wall. The development of the theory occupies the bulk of the paper. Numerical results appear in Sections VI and VIII to which a reader less interested in the theory is referred.

The theory allows us to conclude that random deviations of the waveguide wall in the order of 1 percent, for guides designed to guide an optical wave of $\lambda_0 = 1\mu$ wavelength, can cause scattering losses of 10 percent per centimeter or 0.46 dB per centimeter. A systematic sinusoidal deviation of the waveguide wall can cause total exchange of energy from the lowest order to the first order guided mode in a distance of approximately 1 cm if the amplitude of the sinusoidal deviation from perfect straightness is only 0.5 percent of the thickness of the guide. An rms deviation of one of the waveguide walls of 9Å causes a radiation loss of 10 dB per kilometer (index difference 1 percent, guide width 2.5μ).

I. INTRODUCTION

The problem of how to transmit laser light over large distances or carry it short distances inside the laboratory has renewed the interest in dielectric waveguides.[1-5] Such waveguides usually used in the form of clad fibers or as strips of a medium of larger dielectric constant embedded in another dielectric medium are capable, in principle, of guiding electromagnetic radiation. By proper dimensioning, a dielectric waveguide can be made to transmit only one guided mode. In this respect mode guidance by dielectric waveguides resembles mode guidance by hollow metallic waveguides. Hollow metallic tubes can be constructed to allow only one mode to propagate so that mode conver-

Reprinted with permission from *Bell Syst. Tech. J.*, vol. 48, pp. 3187–3215, Dec. 1969.

sion (except for conversion to the reflected dominant mode) becomes impossible. Such truly single mode operation is impossible for dielectric waveguides since these guides can always lose electromagnetic energy to the continuous spectrum of unguided modes.

The possible solutions of Maxwell's equations for a dielectric waveguide consist of a discrete spectrum of a finite number of guided modes plus a continuum of waveguide modes.[6] The guided modes have field configurations which concentrate the electromagnetic energy inside and in the immediate vicinity of the structure. The continuum of unguided modes extends to infinite distances from the waveguide and consists of a superposition of incident and reflected waves. A convenient way of visualizing the physical significance of the continuum of unguided modes is as follows. If a plane wave is incident on the dielectric waveguide at an arbitrary angle, part of it penetrates the dielectric structure while some portion is reflected. The resulting superposition field of incident and reflected waves satisfies Maxwell's equations and the boundary conditions at the dielectric waveguide and as such can be viewed as a mode of the structure, but the energy of this mode is not concentrated near the waveguide and there are no specific restrictions on the projection of the propagation vector in the direction of the guide axis.

A perfect dielectric waveguide can transmit any of its guided modes without converting energy to any of the other possible guided modes or to the continuous spectrum. But any imperfection of the guide, such as a local change of its index of refraction or a deviation from perfect straightness or an imperfection of the interface between two regions with different index of refraction, couples the particular guided mode to all other guided modes as well as to all the modes of the unguided continuum. Imperfections of this type are unavoidable. They transfer energy from the desired guided mode to unwanted guided modes and the radiation field of the continuum of unguided modes, thus increasing the loss of the desired guided mode.

This paper gives a simple, approximate theory of the losses of dielectric waveguides, caused by imperfections of the boundary between the inner region of higher dielectric index and the surrounding outer region of the dielectric waveguide. Even though the method of analysis used here can be used to describe any arbitrary dielectric waveguide, we limit the discussion to a simple case. We describe the effects of mode conversion for a dielectric slab surrounded by vacuum, assuming for simplicity, that there is no variation of the dimensions or properties of the rod as well as the field distribution in one co-ordinate direction. The restriction of demanding $\partial/\partial_y = 0$ for one of the co-ordinates y is no limitation on the method of analysis but is imposed strictly for convenience. It simplifies the analysis considerably without drastically changing the conclusions. The tolerance requirements based on our analysis are rather stringent. They show the order of magnitude of the losses which can be expected from deviations from perfect geometry. Additional variations in the direction considered perfect in this paper is unlikely to improve any of the loss predictions.

II. TE MODES OF A DIELECTRIC SLAB

Let us consider the transverse electric modes of the dielectric slab of Fig. 1. True to the simplifying assumption discussed in Section I, we assume

$$\frac{\partial}{\partial_y} = 0 \tag{1}$$

with y being the co-ordinate perpendicular to the x and z directions, but parallel to the slab. The only nonvanishing field components are E_y, H_x, and H_z.

Leaving the z and time dependence

$$e^{i(\omega t - \beta z)} \tag{2}$$

understood, we obtain the following modes of the ideal structure as a solution of Maxwell's equations satisfying the boundary conditions.

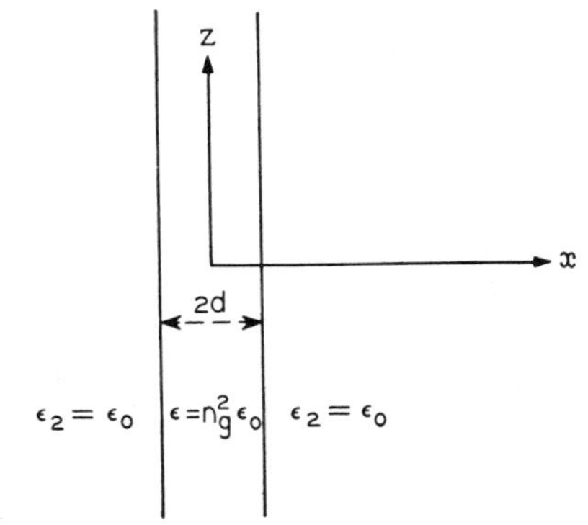

Fig. 1 — Geometry of a dielectric slab waveguide.

2.1 *Even Guided Modes*

For even guided modes

$$\mathcal{E}_y = A^{(e)} \cos \kappa x \quad \text{for} \quad |x| \leqq d, \tag{3a}$$

$$\mathcal{E}_y = A^{(e)} \cos \kappa d e^{-\gamma(x-d)} \quad \text{for} \quad x \geqq d, \tag{3b}$$

$$\mathcal{H}_x = -\frac{i}{\omega\mu}\frac{\partial \mathcal{E}_y}{\partial z}, \tag{4}$$

$$\mathcal{H}_z = \frac{i}{\omega\mu}\frac{\partial \mathcal{E}_y}{\partial x}, \tag{5}$$

The field component $\mathcal{E}_y$ satisfies the wave equation

$$\frac{\partial^2 \mathcal{E}_y}{\partial x^2} + \frac{\partial^2 \mathcal{E}_y}{\partial z^2} + n_0^2 k^2 \mathcal{E}_y = 0. \tag{6}$$

The value of the index of refraction n_0 is different inside and outside of the dielectric slab. For simplicity, we assume

$$n_0 = 1 \quad \text{for} \quad |x| > d. \tag{7}$$

The other constants are related as follows

$$k^2 = \omega^2 \epsilon_0 \mu_0, \tag{8}$$

$$\kappa = (n^2 k^2 - \beta^2)^{\frac{1}{2}}, \tag{9}$$

$$\gamma = (\beta^2 - k^2)^{\frac{1}{2}}. \tag{10}$$

The propagation constant β is obtained as a solution of the eigenvalue equation

$$\tan \kappa d = \frac{\gamma}{\kappa}. \tag{11}$$

The mode amplitude A can be expressed in terms of the power P carried by the mode.

$$P = \frac{1}{2} \operatorname{Re} \int_{-\infty}^{\infty} (-\mathcal{E}_y \mathcal{H}_x^*)\, dx = \frac{\beta}{\omega\mu} \int_0^{\infty} |\mathcal{E}_y|^2\, dx. \tag{12}$$

P is the power per unit length (unit length in y-direction) flowing along the z-axis. We obtain for the amplitude coefficient

$$A^{(e)2} = \frac{2\omega\mu}{\beta d + \dfrac{\beta}{\gamma}} P. \tag{13}$$

2.2 *Even Modes of the Continuum*

The continuum of unguided modes of even symmetry is given by the equations:

$$\mathcal{E}_y = B^{(e)} \cos \sigma x \quad \text{for} \quad | x | \leqq d, \tag{14a}$$

$$\mathcal{E}_y = C^{(e)} e^{i\rho x} + D^{(e)} e^{-i\rho x} \quad \text{for} \quad x \geqq d. \tag{14b}$$

The other field components follow again from equations (4) and (5) and $\mathcal{E}_y$ is a solution of equation (6). The constants are related to each other by the equations

$$\sigma = (n^2 k^2 - \beta^2)^{\frac{1}{2}}, \tag{15}$$

$$\rho = (k^2 - \beta^2)^{\frac{1}{2}}. \tag{16}$$

The radial propagation constant ρ can assume all values from 0 to ∞. The continuous mode spectrum starts at $\beta = k$ and continuous to $\beta = 0$ at which point we have $\rho = k$. Larger values of ρ are obtained for imaginary values of β corresponding to modes of the continuum exhibiting a cutoff behavior.

The boundary conditions do not lead to an eigenvalue equation for β but they determine $C^{(e)}$ and $D^{(e)}$ in relation to $B^{(e)}$.

$$C^{(e)} = \tfrac{1}{2} B^{(e)} e^{-i\rho d} \left(\cos \sigma d + i \frac{\sigma}{\rho} \sin \sigma d \right), \tag{17}$$

$$D^{(e)} = C^{(e)*}, \tag{18}$$

(the asterisk indicates the complex conjugate quantity).

The normalization of the modes of the continuum involves a δ-function. Instead of equation (12) we use

$$P\, \delta(\rho - \rho') = \frac{\beta}{\omega\mu} \int_0^\infty \mathcal{E}_y(\rho)\, \mathcal{E}_y^*(\rho')\, dx. \tag{19}$$

With this normalization we get

$$B^{(e)2} = \frac{2\omega\mu P}{\pi\beta \left(\cos^2 \sigma d + \frac{\sigma^2}{\rho^2} \sin^2 \sigma d \right)}. \tag{20}$$

2.3 *Odd Guided Modes*

In a manner similar to that for obtaining the preceding equations we obtain the equations for the odd guided modes

$$\mathcal{E}_y = A^{(0)} \sin \kappa x \quad \text{for} \quad x \leqq d, \tag{21a}$$

$$\mathcal{E}_y = A^{(0)} \sin \kappa d e^{-\gamma(x-d)} \quad \text{for} \quad x \geqq d. \tag{21b}$$

Equations (4) through (10) apply to the odd modes unaltered. The eigenvalue equation is given by

$$\tan \kappa d = -\frac{\kappa}{\gamma}, \tag{22}$$

and the mode normalization is

$$A^{(0)^2} = \frac{2\omega\mu}{\beta d + \dfrac{\beta}{\gamma}} P. \tag{23}$$

2.4 *Odd Modes of the Continuum*

As in Section 2.3 we obtain the equations for the odd modes of the continuum

$$\mathcal{E}_y = B^{(0)} \sin \sigma x \quad \text{for} \quad |\, x \,| \leqq d, \tag{24a}$$

$$\mathcal{E}_y = C^{(0)} e^{i\rho x} + D^{(0)} e^{-i\rho x} \quad \text{for} \quad x \geqq d, \tag{24b}$$

$$C^{(0)} = \tfrac{1}{2} B^{(0)} e^{-i\rho d} \left(\sin \sigma d - i \frac{\sigma}{\rho} \cos \sigma d \right), \tag{25}$$

$$D^{(0)} = C^{(0)*}, \tag{26}$$

$$B^{(0)^2} = \frac{2\omega\mu P}{\pi\beta \left(\sin^2 \sigma d + \dfrac{\sigma^2}{\rho^2} \cos^2 \sigma d \right)}. \tag{27}$$

All these modes are orthogonal to one another. The even modes are orthogonal to all the odd modes, the guided modes are orthogonal to all the modes of the continuum, and all guided modes as well as all modes of the continuum are orthogonal among each other. The orthogonality of the modes of the continuum among each other was already expressed by equation (19). Labeling the discrete modes by indices and dropping the vector component label y we can express the orthogonality of the discrete modes by the equation

$$P\, \delta_{nm} = \frac{\beta_m}{2\omega\mu} \int_{-\infty}^{\infty} \mathcal{E}_n \mathcal{E}_m^* \, dx. \tag{28}$$

III. MODE COUPLING CAUSED BY IMPERFECTIONS

We want to study the losses which the lowest order guided mode suffers because of imperfections of the waveguide wall. A dielectric waveguide with wall imperfections is shown in Fig. 2.

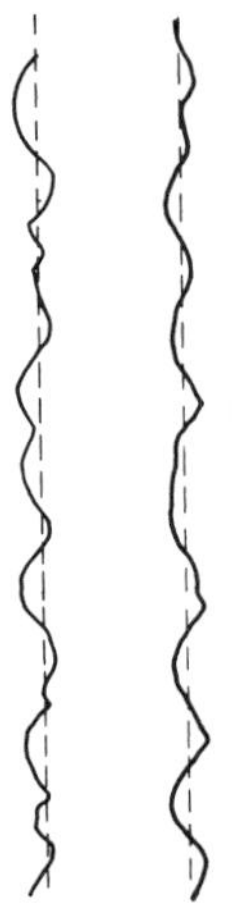

Fig. 2 — Dielectric slab waveguide with wall distortions.

The waveguide with wall imperfections is mathematically described by a refractive index distribution

$$n^2(x, z) = n_0^2(x, z) + \Delta n^2(x, z). \tag{29}$$

The index distribution

$$n_0^2(x, z) = \begin{cases} n_g^2 & | x | < d \\ 1 & | x | > d \end{cases} \tag{30}$$

describes the ideal dielectric waveguide whose TE modes were given in the Section II. The additional term Δn^2 describes how the guide deviates from its perfect shape. Consider a deviation shown in Fig. 3. The corresponding distribution Δn^2 is (n_g = index of refraction of the dielectric material of the guide)

$$\Delta n^2 = \begin{cases} 0 \begin{cases} x < d & \text{if} \quad d < f(z) \\ x < f(z) & \text{if} \quad d > f(z) \end{cases} \\ n_g^2 - 1 \qquad d < x < f(z) \quad \text{if} \quad d < f(z) \\ -(n_g^2 - 1) \qquad f(z) < x < d \quad \text{if} \quad d > f(z) \\ 0 \begin{cases} x > f(z) & \text{if} \quad d < f(z) \\ x > d & \text{if} \quad d > f(z) \end{cases} \end{cases} . \tag{31}$$

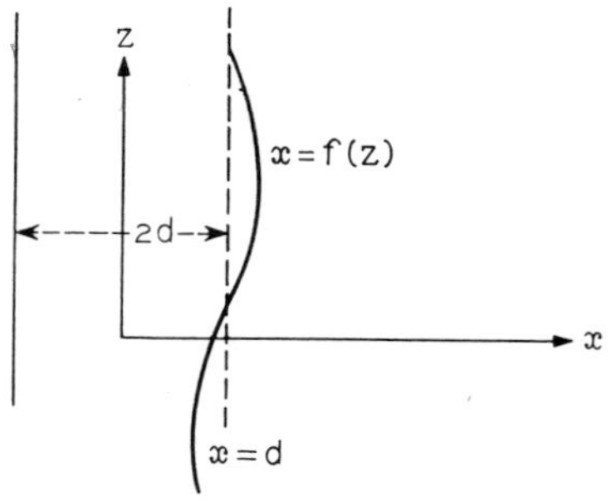

Fig. 3 — Illustration of the wall distortion function $f(z)$.

The field distribution E_y of this waveguide is a solution of

$$\frac{\partial^2 E_y}{\partial x^2} + \frac{\partial^2 E_y}{\partial z^2} + (n_0^2 + \Delta n^2)k^2 E_y = 0 \tag{32}$$

with H_x and H_z given by equations (4) and (5). The modes of the perfect waveguide form a complete orthogonal set for all TE modes with no variation in the y-direction. It is, therefore, possible to express any field distribution on the waveguide with imperfect walls by the expansion

$$E_y = \sum_n C_n(z)\, \mathcal{E}_n + \sum \int_0^\infty g(\rho, z)\, \mathcal{E}(\rho)\, d\rho. \tag{33}$$

The first summation extends over all even and odd modes of the discrete spectrum of guided modes. The integral extends over all modes of the continuum, and the summation sign in front of the integral indicates summation over even and odd modes. The expansion coefficients C_n and $g(\rho)$ are unknown functions of z.

To obtain a coupled system of differential equations for the expansion coefficients we substitute equation (33) into equation (32). Multiplying the resulting equation by

$$\frac{\beta_m}{2\omega\mu}\, \mathcal{E}_m^*,$$

integrating over x from $-\infty$ to $+\infty$, and using the orthogonality relations and the fact that $\mathcal{E}_n$ and $\mathcal{E}(\rho)$ are the (discrete and continuous) modes of the perfect guide leads to

$$\frac{\partial^2 C_m}{\partial z^2} - 2i\beta_m \frac{\partial C_m}{\partial z} = F_m(z) \tag{34}$$

with

$$F_m(z) = -\frac{\beta_m k^2}{2\omega\mu P}\left[\sum_n C_n(z)\int_{-\infty}^{\infty} \mathcal{E}_m^* \, \Delta n^2 \, \mathcal{E}_n \, dx + \sum \int_0^{\infty} d\rho g(\rho, z) \int_{-\infty}^{\infty} \mathcal{E}_m^* \, \Delta n^2 \, \mathcal{E}(\rho) \, dx\right]. \tag{35}$$

Similarly multiplying by

$$\frac{\beta'}{2\omega\mu} \mathcal{E}^*(\rho')$$

leads to

$$\frac{\partial^2 g(\rho')}{\partial z^2} - 2i\beta' \frac{\partial g(\rho')}{\partial z} = G(\rho' z) \tag{36}$$

with

$$G(\rho', z) = -\frac{\beta' k^2}{2\omega\mu P}\left[\sum_n C_n(z)\int_{-\infty}^{\infty} \mathcal{E}^*(\rho') \, \Delta n^2 \, \mathcal{E}_n \, dx + \sum \int_0^{\infty} d\rho g(\rho, z) \int_{-\infty}^{\infty} \mathcal{E}^*(\rho') \, \Delta n^2 \, \mathcal{E}(\rho) \, dx\right]. \tag{37}$$

No n-label on the power term P is necessary since we assume that all the normal modes are normalized to the same amount of power. The actual power carried by each mode relative to the power of the other modes is given by the C_n coefficients. Solutions of equations (34) and (35) with appropriate initial conditions provide us with exact solutions of the imperfect waveguide. It is interesting to note that this method of solution does not require the consideration of boundary conditions.

The normal modes $\mathcal{E}_n$ and $\mathcal{E}(\rho)$ were assumed to have the time and z-dependence of equation (2); this means they represent waves traveling in the positive z-direction. However, the solutions of equations (34) and (36) introduce waves traveling in positive as well as negative z-direction. To see this, let us assume that $\Delta n^2 = 0$ so that $F_n(z) = 0$. The equation

$$\frac{\partial^2 C_n}{\partial z^2} - 2i\beta_n \frac{\partial C_n}{\partial z} = 0 \tag{38}$$

has the solution

$$C_n(z) = A + Be^{2i\beta_n z} \tag{39}$$

with constant A and B. The product of A with $\mathcal{E}_n$ results in a wave traveling in the positive z-direction but the product of B exp $(2i\beta_n z)$

with $\mathcal{E}_n$ results in a wave traveling in the negative z-direction. So, even though we started out with waves traveling in the positive z-direction the expansion (33) contains partial waves traveling in positive as well as negative z-direction.

For the purpose of obtaining perturbation solutions of equations (34) and (36), an integral form of these equations is more useful. Treating equations (34) and (36) as inhomogeneous differential equations, we can immediately write the following integral equations

$$C_m = A_m + B_m e^{2i\beta_m z} + \frac{1}{2i\beta_m} \int_0^z [e^{2i\beta_m(z-\zeta)} - 1] F_m(\zeta)\, d\zeta, \tag{40}$$

$$g(\rho', z) = C(\rho') + D(\rho') e^{2i\beta' z} + \frac{1}{2i\beta'} \int_0^z [e^{2i\beta'(z-\zeta)} - 1] G(\rho', \zeta)\, d\zeta. \tag{41}$$

It is important to know which part of equations (40) and (41) is associated with waves traveling in the positive or negative z-direction. Therefore, we introduce the notation.

$$C_m = C_m^{(+)} + C_m^{(-)} \tag{42}$$

with

$$C_m^{(+)}(z) = A_m - \frac{1}{2i\beta_m} \int_0^z F_m(\zeta)\, d\zeta, \tag{43}$$

$$C_m^{(-)}(z) = \left\{ B_m + \frac{1}{2i\beta_m} \int_0^z e^{-2i\beta_m \zeta} F_m(\zeta)\, d\zeta \right\} e^{2i\beta_m z}. \tag{44}$$

The superscript $(+)$ indicates the coefficient which after substitution into equation (33) produces waves traveling in positive z-direction, while $(-)$ indicates the part which produces waves traveling in negative z-direction. A similar notation and resulting equations is used for $g(\rho', z)$; however, the corresponding equations are obvious and are therefore omitted.

The constants A_m, B_m, and so on, occurrng in equations (43), (44), and the corresponding equations for $g(\rho', z)$ must be determined from initial conditions. We always assume that the lowest order guided mode is incident on the imperfect waveguide at $z = 0$. Using the subscript 0 for this incident mode we get immediately from equation (43)

$$C_m^{(+)} = 0 \quad \text{for} \quad m \neq 0 \qquad \text{at} \quad z = 0$$

or

$$A_m = 0 \quad \text{for} \quad m \neq 0, \tag{45}$$

but

$$A_0 = 1. \tag{46}$$

We imagine that at $z = L$ the waveguide is connected to a perfect guide so that at that point there are no waves traveling in negative z-direction. This leads to the condition

$$B_m = -\frac{1}{2i\beta_m}\int_0^L e^{-2i\beta_m\zeta}F_m(\zeta)\,d\zeta \tag{47}$$

for all values of m. The power loss ΔP of the incident mode due to mode conversion is given by

$$\frac{\Delta P}{P} = \sum_{n=1}^{\infty}[|C_n^{(+)}(L)|^2 + |C_n^{(-)}(0)|^2] + \sum\int_0^{\infty}[|g^{(+)}(\rho, L)|^2 + |g^{(-)}(\rho, 0)|^2]\,d\rho. \tag{48}$$

Equation (48) states that the total power lost by mode conversion from the incident mode escapes at $z = L$ in spurious modes traveling in position z-direction and at $z = 0$ in spurious modes traveling in negative z-direction. The factor P is the normalized power factor of equations (12) and (19); it is the power incident in mode 0. Notice that because of equations (45) and (47) only the integral terms of equations (43) and (44) (taken from $z = 0$ to $z = L$) enter into equation (48).

The integral equations (43) and (44) can only be solved approximately. We perform first order perturbation theory by using $C_m(0)$ instead of $C_m(z)$ and $g(\rho, 0)$ instead of $g(\rho, z)$ in equations (35) and (37). Furthermore, we realize that $C_m^{(-)}(0)$ for all m is a quantity of first order and will therefore be neglected in equations (35) and (37). The same is true for $C_m^{(+)}(0)$ with $m \neq 0$. In the spirit of first order perturbation theory we use therefore

$$C_m = \delta_{0m} \tag{49}$$

and

$$g(\rho) = 0 \tag{50}$$

in equations (35) and (37).

The perturbation theory is feasible not only when $n_g^2 - 1 \ll 1$ but also when $n_g^2 - 1$ is arbitrarily large but the geometrical deviation of the guide walls from perfect straightness is slight. In either case we obtain from equations (35) and (37) the simple approximations

$$F_m(z) = -\frac{\beta_m k^2}{2\omega\mu P}(n_g^2 - 1)\{[f(z) - d]\,\mathcal{E}_0(d, z)\,\mathcal{E}_m^*(d, z)$$

$$- [h(z) + d]\,\mathcal{E}_0(-d, z)\,\mathcal{E}_m^*(-d, z)\}, \tag{51}$$

$$G(\rho, z) = -\frac{\beta k^2}{2\omega\mu P}(n_g^2 - 1)\{[f(z) - d]\,\mathcal{E}^*(\rho, d, z)\,\mathcal{E}_0(d, z)$$

$$- [h(z) + d]\,\mathcal{E}^*(\rho, -d, z)\,\mathcal{E}_0(-d, z)\}. \tag{52}$$

The function $f(z)$ describes the dielectric-air interface in the vicinity of $x = d$, while $h(z)$ describes it near $z = -d$. We assumed that $f(z)$ and $h(z)$ depart so little from $x = d$ and $x = -d$ that the functions $\mathcal{E}(x, z)$ could be replaced by $\mathcal{E}(\pm d, z)$.

IV. EVALUATION OF THE SPURIOUS MODE AMPLITUDES

We begin the discussion of the consequences of our scattering theory by calculating the coefficients $C_m^{(+)}$ and $g^{(+)}$. We obtain [from equations (43) and (51) with the help of equations (3a) and (13) for the even modes] the following

$$C_{me}^{(+)}(L) = \frac{Lk^2}{2i}(n_g^2 - 1)\frac{\cos\kappa_0 d\,\cos\kappa_m d}{\left[\left(\beta_0 d + \frac{\beta_0}{\gamma_0}\right)\left(\beta_m d + \frac{\beta_m}{\gamma_m}\right)\right]^{\frac{1}{2}}}(\varphi_m - \psi_m). \tag{53}$$

The coefficients φ_m and ψ_m are defined by

$$\varphi_m = \frac{1}{L}\int_0^L [f(z) - d]e^{-i(\beta_0 - \beta_m)z}\,dz \tag{54}$$

and

$$\psi_m = \frac{1}{L}\int_0^L [h(z) + d]e^{-i(\beta_0 - \beta_m)z}\,dz. \tag{55}$$

These are the Fourier coefficients of the functions $f(z) - d$ and $h(z) + d$ which are expanded in a domain

$$0 \leqq z \leqq L.$$

The amplitude of the mth even mode depends on the Fourier components of the wall function whose "spatial frequency" Γ is

$$\Gamma_m = \frac{2\pi}{\Lambda_m} = \beta_0 - \beta_m\,. \tag{56}$$

The corresponding expression for the even modes of the continuous spectrum is:

$$g_e^{(+)}(\rho, L) = \frac{Lk^2}{2i(\pi)^{\frac{1}{2}}}(n_g^2 - 1)\frac{\cos \kappa_0 d \cos \sigma d[\varphi(\beta) - \psi(\beta)]}{\left[\left(\beta_0 d + \frac{\beta_0}{\gamma_0}\right)\beta\left(\cos^2 \sigma d + \frac{\sigma^2}{\rho^2}\sin^2 \sigma d\right)\right]^{\frac{1}{2}}} \tag{57}$$

with [$\beta = \beta(\rho)$ see equation (16)]

$$\varphi(\beta) = \frac{1}{L}\int_0^L [f(z) - d]e^{-i(\beta_0 - \beta)z}\, dz, \tag{58}$$

$$\psi(\beta) = \frac{1}{L}\int_0^L [h(z) + d]e^{-i(\beta_0 - \beta)z}\, dz. \tag{59}$$

The corresponding expressions for the odd modes are

$$C_{n0}^{(+)}(L) = \frac{Lk^2}{2i}(n_g^2 - 1)\frac{\cos \kappa_0 d \sin \kappa_n d}{\left[\left(\beta_0 d + \frac{\beta_0}{\gamma_0}\right)\left(\beta_n d + \frac{\beta_n}{\gamma_n}\right)\right]^{\frac{1}{2}}}(\varphi_n + \psi_n), \tag{60}$$

$$g_0^{(+)}(\rho, L) = \frac{Lk^2}{2i(\pi)^{\frac{1}{2}}}(n_g^2 - 1)\frac{\cos \kappa_0 d \sin \sigma d[\varphi(\beta) + \psi(\beta)]}{\left[\left(\beta_0 d + \frac{\beta_0}{\gamma_0}\right)\beta\left(\sin^2 \sigma d + \frac{\sigma^2}{\rho^2}\cos^2 \sigma d\right)\right]^{\frac{1}{2}}}. \tag{61}$$

The Fourier coefficients φ and ψ are given by equations (54), (55), (58), and (59) except that β_n and β are now the propagation constants of the odd modes.

The corresponding expressions for $C^{(-)}$ and $g^{(-)}$ are obtained by replacing β_m with $-\beta_m$ and β with $-\beta$ in equations (54), (55), (56), (58), and (59).

V. SINUSOIDAL WALL DEFLECTIONS

As a specific example, let us assume that the wall imperfections have sinusoidal shape. Then

$$f(z) - d = a \sin \theta z \tag{62}$$

and

$$h(z) + d = -a \sin (\theta z + \alpha). \tag{63}$$

The phase factor α allows us to consider either a waveguide whose width varies sinusoidally

$$\alpha = 0, \tag{64}$$

or one whose direction changes sinusoidally

$$\alpha = \pi. \tag{65}$$

We obtain from equation (54) with

$$\theta = \beta_0 - \beta_m \tag{66}$$

the Fourier component

$$\varphi_m = \frac{a}{2i} \tag{67}$$

and from equation (55)

$$\psi_m = -\frac{a}{2i} e^{i\alpha}. \tag{68}$$

A term of the order $a/L \ll 1$ was omitted in equations (67) and (68). It is apparent that only one spurious mode is excited by the sinusoidal wall deflection since condition (66) can be satisfied for only one value of β_m. If condition (66) is not satisfied, φ_m and ψ_m are of the order of $a/L \ll 1$. The fractional power scattered into one spurious guided mode due to a sinusoidal wall irregularity is [from equations (48), (53), (67) and (68)]

$$\left(\frac{\Delta P}{P}\right)_{eg} \quad \frac{L^2 a^2 k^4}{4} (n_g^2 - 1)^2 \frac{\cos^2 \kappa_0 d \cos^2 \kappa_m d}{\left(\beta_0 d + \frac{\beta_0}{\gamma_0}\right)\left(\beta_m d + \frac{\beta_m}{\gamma_m}\right)} \cos^2 \frac{\alpha}{2} \tag{69}$$

for even modes or [from equations (48) and (60)]

$$\left(\frac{\Delta P}{P}\right)_{0g} \quad \frac{L^2 a^2 k^2}{4} (n_g^2 - 1)^2 \frac{\cos^2 \kappa_0 d \sin^2 \kappa_m d}{\left(\beta_0 d + \frac{\beta_0}{\gamma_0}\right)\left(\beta_m d + \frac{\beta_m}{\gamma_m}\right)} \sin^2 \frac{\alpha}{2} \tag{70}$$

for odd modes. However only one even or one odd mode can be excited by one particular sinusoidal wall deviation since it is impossible to satisfy the "resonance" condition (66) for more than one mode simultaneously.

If $\alpha = 0$, that is if the width of the guide changes sinusoidally, only even modes can be excited while sinusoidal deviations from straightness ($\alpha = \pi$) couple the even fundamental mode only to odd spurious modes. It must also be noticed that for a long period length

$$\Lambda = \frac{2\pi}{\theta} \tag{71}$$

equation (66) can be satisfied only for forward scattering modes. To couple to backward scattering modes, the period length D must be approximately equal to half the wavelength of the guided modes. The fact that only one spurious mode is coupled to the incident mode by sinusoidal wall imperfections (it can be shown that the coupling to the continuous mode spectrum is also weak if one guided mode can couple strongly) allows us to give a much better description of the coupling process.

Since the mode amplitudes C_m can change only slowly in the distance of one wavelength we can neglect the second derivative of C_m in equation (34). Labeling the incident mode 0 and the one coupled spurious mode 1 we can write the equation system (34) in the following form

$$\frac{\partial C_0}{\partial z} = -\kappa_{01} C_1 , \tag{72}$$

$$\frac{\partial C_1}{\partial z} = \kappa_{01}^* C_0 , \tag{73}$$

with

$$\kappa_{01} = \frac{k^2 a}{2} (n_g^2 - 1) \frac{\cos \kappa_0 d \, \cos \kappa_1 d}{\left[\left(\beta_0 d + \frac{\beta_0}{\gamma_0}\right)\left(\beta_1 d + \frac{\beta_1}{\gamma_1}\right)\right]^{\frac{1}{2}}} \exp\left(i \frac{\alpha}{2}\right) \cos \frac{\alpha}{2}. \tag{74}$$

The coupling coefficient κ_{01} of equations (74) holds for coupling from an even mode 0 to an even mode 1. The case of coupling from an even mode 0 to an odd mode 1 can be treated similarly. In fact, except for an unimportant phase factor, we get it from $(1/L)[(\Delta P/P)_{eo}]^{\frac{1}{2}}$ of equarion (70). In equation (72) we omitted a term with C_0 on the right-hand side, and similarly a term with C_1 was omitted in equation (73). These terms would be multiplied by sinusoidally varying functions and would describe the local change of phase velocity as the guide dimensions vary. These terms give no contribution if we use an average over C_0 and C_1 over the mechanical period length of equation (71).

Assuming $C_0 = 1$, $C_1 = 0$ at $z = 0$ the equation system (72) and (73) has the solution

$$C_0 = \cos |\kappa_{01}| z, \tag{75}$$

$$C_1 = \left(\frac{\kappa_{01}^*}{\kappa_{01}}\right)^{\frac{1}{2}} \sin |\kappa_{01}| z. \tag{76}$$

Total exchange of energy is possible between the two coupled modes.

The distance D over which all the energy is exchanged is given by

$$D = \frac{\pi}{2\,|\,\kappa_{01}\,|}. \tag{77}$$

Finally, we need the power loss to the modes of the continuous spectrum. From equations (48), (57), (61), (62), and (63) we obtain

$$\left(\frac{\Delta P}{P}\right)_c = \frac{a^2k^4}{\pi}(n_g^2 - 1)^2 \frac{\cos^2 \kappa_0 d}{\beta_0 d + \dfrac{\beta_0}{\gamma_0}}$$

$$\cdot \int_0^\infty \left[\frac{\cos^2 \sigma d \cos^2 \dfrac{\alpha}{2}}{\beta\left(\cos^2 \sigma d + \dfrac{\sigma^2}{\rho^2}\sin^2 \sigma d\right)} + \frac{\sin^2 \sigma d \sin^2 \dfrac{\alpha}{2}}{\beta\left(\sin^2 \sigma d + \dfrac{\sigma^2}{\rho^2}\cos^2 \sigma d\right)}\right]$$

$$\cdot \frac{\sin^2 [\theta - (\beta_0 - \beta)] \dfrac{L}{2}}{[\theta - (\beta_0 - \beta)]^2}\, d\rho. \tag{78}$$

The integration can be performed easily if one realizes that for large values of L only a very narrow region in the β range near $\beta - \beta_0 - \theta$ contributes to the integral. We consider all functions in the integrand as constant in this very narrow range and take them out of the integral with the exception of

$$\left\{\frac{\sin [\theta - (\beta_0 - \beta)] \dfrac{L}{2}}{\theta - (\beta_0 - \beta)}\right\}^2.$$

This remaining integral can easily be performed if we use equation (16) to obtain

$$d\rho = -\frac{\beta}{\rho}\, d\beta.$$

Following this procedure yields

$$\left(\frac{\Delta P}{P}\right)_c = \frac{La^2k^4}{2}(n_g^2 - 1)^2 \frac{\cos^2 \kappa_0 d}{\beta_0 d + \dfrac{\beta_0}{\gamma_0}}$$

$$\cdot \left\{\frac{\rho \cos^2 \sigma d \cos^2 \dfrac{\alpha}{2}}{\rho^2 \cos^2 \sigma d + \sigma^2 \sin^2 \sigma d} + \frac{\rho \sin^2 \sigma d \sin^2 \dfrac{\alpha}{2}}{\rho^2 \sin^2 \sigma d + \sigma^2 \cos^2 \sigma d}\right\}. \tag{79}$$

The parameters σ and ρ follow from equations (15) and (16) with

$$\beta = \beta_0 - \theta. \tag{80}$$

Equation (79) holds only for $\beta < k$; we get $\Delta P/P = 0$ for $\beta > k$. The most interesting aspect of equation (79) is its linear dependence on L. The scattering loss due to the modes of the continuous spectrum acts like a true loss process. By contrast, the corresponding equation (69) for the loss to guided modes is proportional to L^2 because coupling to a guided mode does not result in loss of energy but results in energy exchange between the two coupled modes. Energy loss to one of the guided modes is followed by energy gain when the energy exchange has reversed itself.

VI. NUMERICAL EXAMPLES FOR SINUSOIDAL IMPERFECTIONS

A few numerical examples resulting from equations (74) and (77) are listed in Table I. Two different values of the index of refraction n_g have been assumed, and for each value of the index three different values of $kd = 2\pi(d/\lambda_0)$ have been chosen so that one, two, or three guided modes can exist simultaneously. The mode with β_0 is the lowest

TABLE I—NUMERICAL EXAMPLES FOR SINUSOIDAL IMPERFECTIONS

n_g	kd	$\beta_0 d$	$\beta_1 d$	$\beta_2 d$	$\frac{aD}{d^2}$	Remarks
1.5	1.3	1.729	—	—	—	Single mode operation
	1.8	2.495	1.916	—	6.98	$0 - 1$ coupling $\alpha = \pi$
	3.0	4.336	3.831	3.051	6.17	$0 - 1$ coupling $\alpha = \pi$
					5.52	$0 - 2$ coupling $\alpha = 0$
1.01	8.0	8.041	—	—	—	Single mode operation
	15.0	15.113	15.022	—	42.54	$0 - 1$ coupling $\alpha = \pi$
	23.0	23.199	23.112	23.002	36.28	$0 - 1$ coupling $\alpha = \pi$
					43.69	$0 - 2$ coupling $\alpha = 0$

order even guided mode which is assumed to be incident on the waveguide with sinusoidal wall imperfections. This mode couples to the first odd mode with β_1 or the next even mode with β_2. The values for the normalized, dimensionless quantitity $(aD)/d^2$ [a = amplitude of the sinusoidal wall deviation according to equation (62) and (63), d = half width of the guide, and D = energy exchange length] have been obtained with the assumption that equation (66) is satisfied for the two modes which are coupled together. Coupling from mode 0 to mode 1 is considered only for the case of sinusoidal straightness deviations of the waveguide ($\alpha = \pi$) while coupling between even modes 0 to 2 is considered only for sinusoidal changes of the thickness of the waveguide ($\alpha = 0$). It is immediately apparent from Table I that the energy exchange length D is shorter for a guide with larger values of the refractive index.

To obtain a feeling for the numbers involved in this mode coupling phenomenon, let us assume that $n_g = 1.5$ and that the free space wavelength is $\lambda_0 = 1\mu$. The value of $kd = 1.8$ corresponds to $d = 0.286\mu$. To achieve total exchange of energy between modes 0 and 1 in $D = 1$ cm requires the extremely small amplitude $a = 5.72\ 10^{-5}\mu$ or $a = 0.572$ Å!† The length of the mechanical period in this example is $\Lambda = 3.1\mu$.

Next, let us assume that the index of refraction is $n_g = 1.01$. Using again, $\lambda_0 = 1\mu$, we obtain from $kd = 15.0$ the value $d = 2.39\mu$ for the half width of the waveguide. Requiring again, $D = 1$ cm, we find $a =$ 243 Å.

We can look at this problem in a different way. It is unlikely that any optical waveguide has a strictly sinusoidal deviation from perfect straightness. In fact, the numbers just presented show that it would be impossible to produce such a waveguide intentionally. However, we have seen [equation (53)] that the mode conversion between two guided modes is produced by a Fourier component of the actual deviation function. It is therefore not necessary to have a strictly sinusoidal straightness deviation. Any arbitrary deviation from straightness can be decomposed into a Fourier series and the Fourier component at the mechanical frequency which satisfies equation (66) is responsible for the coupling. In the more general case of arbitrary straightness deviations, there can be no complete exchange of energy between any two modes since power loss to other guided modes and the continuous

† A mechanical period of a fraction of an Angstrom is somewhat unphysical due to the granular nature of matter. However, this result can be restated to say that complete power conversion occurs in 0.1 mm if the amplitude is $a = 57.2$ Å.

spectrum of modes compete with each other since all of them are coupled simultaneously.

We can now ask the question: What amplitude of the mechanical straightness deviation is required to transfer 10 percent of power from mode 0 to mode 1 in a distance of $L = 1$ cm? Again, we use the previous examples. From equation (76) [or directly from equations (53) and (77)] we obtain

$$\frac{\Delta P}{P} = |\kappa_{01}|^2 L^2 = \frac{\pi^2}{4} \frac{L^2}{D^2}.$$

For the first example we obtain with $n_g = 1.5$, $\Delta P/P = 0.1$, $d = 0.286\mu$, and $aD/d^2 = 6.98$ the value $a = 0.115$ Å.† This result shows that if the Fourier component of the mechanical straightness deviation with a period length of 3.1μ is $a = 0.12$ Å (measured over a distance of 1 cm) the power loss caused by mode conversion to the first odd mode is 10 percent.

For the second example, we use again $n_g = 1.01$, $\Delta P/P = 0.1$, $d = 2.39\mu$, and $aD/d^2 = 42.54$ and obtain $a = 48.8$ Å. The important Fourier component in this case has a period of $\Lambda = 135\mu$. The power loss to the modes of the continuous spectrum caused by a sinusoidal change in thickness of the waveguide (which is very similar to its effect as a straightness deviation) can be calculated from equation (79) with $\alpha = 0$.

Let us consider only one case, $n_g = 1.01$, $kd = 15$, $\Lambda/d = 25$. For these values we obtain from equation (79)

$$\frac{d^3}{a^2 L} \frac{\Delta P}{P} = 4.6 \times 10^{-2}.$$

Assuming again $\Delta P/P = 0.1$ for a guide length $L = 1$ cm, we obtain with $d = 2.39\mu$

$$a = 5.46 \times 10^{-2}\mu = 546 \text{ Å}.$$

This number can be compared to the value $a = 48.8$ Å which gave 10 percent loss by conversion to one guided mode. However, for a meaningful comparison, we must remember that all the Fourier components of a Fourier expansion of the guide imperfections scatter power into the modes of the continuous spectrum. The total loss would have to be obtained by integrating the scattering loss over the spectral dis-

† Again it is more reasonable to restate this example to say that 10 percent loss occurs over a distance of $L = 0.1$ mm if $a = 12$ Å.

tribution of the Fourier components of the mechanical Fourier spectrum. Instead of doing this integration we use a different approach in Section VII.

VII. STATISTICAL TREATMENT OF WALL IMPERFECTIONS†

Equation (48) gives the relative loss of a guided mode caused by a definite (deterministic) distortion of the boundary of a dielectric waveguide. A quantity that may be even more interesting is the average of equation (48) taken over an ensemble of statistically identical systems.

For simplicity, let us assume that one wall of the waveguide is perfect while the other is randomly distorted. If both walls are randomly distorted, with no correlation between the distortions on opposite walls the loss value doubles compared to the case of only one wall being distorted. If the distortions on opposite sides of the waveguide are perfectly correlated the amount of loss is at most increased four times. So to simplify the discussion we assume

$$h(z) + d = 0. \tag{81}$$

In order to be able to calculate $\langle \Delta P/P \rangle_{av}$, we must evaluate

$$\langle | \varphi_m |^2 \rangle_{av} = \frac{1}{L^2} \int_0^L dz \int_0^L dz' R(z - z') e^{-i(\beta_0 - \beta_m)(z - z')} \tag{82}$$

We assumed that the correlation function

$$R(z - z') = \langle [f(z) - d][f(z') - d] \rangle_{av} \tag{83}$$

depends only on the difference between the coordinates z and z' but not on their individual values.

A change of integration variables allows us to write

$$\langle | \varphi_m |^2 \rangle_{av} = \frac{2}{L^2} \int_0^L (L - u) R(u) \cos (\beta_0 - \beta_m) u \, du. \tag{84}$$

To obtain equation (84) we made use of the fact that $R(u)$ is an even function.

The particular form of $R(u)$ depends on the statistics of the wall imperfections. However, all correlation functions have two features in common. They all have their maximum value at $u = 0$ and decrease to zero as $u \rightarrow \infty$. If $R(u)$ would not become 0 as $u \rightarrow \infty$ there would be a

† An excellent statistical treatment of random coupling effects in metallic waveguides can be found in Ref. 7.

systematic distortion of the waveguide boundary instead of the assumed random behavior. To get an idea of what one might expect, we assume the following form for the correlation function

$$R(u) = A^2 \exp\left(-\frac{|u|}{B}\right). \tag{85}$$

A is the rms deviation of the wall from perfect straightness and B is the correlation length. Using equation (85) we obtain from equation (84)

$$\langle | \varphi_m |^2 \rangle_{\text{av}} = \frac{2A^2}{L} \frac{1}{(\beta_0 - \beta_m)^2 + \frac{1}{B^2}} \left\{ \frac{1}{B} + \frac{(\beta_0 - \beta_m)^2 - \frac{1}{B^2}}{L\left((\beta_0 - \beta_m)^2 + \frac{1}{B^2}\right)} \right\} \tag{86}$$

where we neglected terms with exp $(-L/B)$ assuming that L/B is sufficiently large. In fact if

$$L \gg B, \tag{87}$$

equation (86) can be simplified further:

$$\langle | \varphi_m |^2 \rangle_{\text{av}} = \frac{2A^2}{BL} \frac{1}{(\beta_0 - \beta_m)^2 + \frac{1}{B^2}}. \tag{88}$$

Using equation (88) we obtain, from equation (53) for the ensemble average of the square magnitudes of the even guided modes,

$$\langle | C_{me} |^2 \rangle_{\text{av}} = \frac{A^2 k^4 L}{2B} (n_g^2 - 1)^2 \cdot \frac{\cos^2 \kappa_0 d \cos^2 \kappa_m d}{\left((\beta_0 - \beta_m)^2 + \frac{1}{B^2}\right)\left(\beta_0 d + \frac{\beta_0}{\gamma_0}\right)\left(\beta_m d + \frac{\beta_m}{\gamma_m}\right)}. \tag{89}$$

The corresponding expression for the odd modes is very similar except that $\cos^2 \kappa_m d$ is replaced by $\sin^2 \kappa_m d$ and β_m, κ_m, and γ_m are the parameters of the odd modes.

The total loss caused by coupling to all guided modes supported by the dielectric waveguide is the sum over all $\langle | C_m |^2 \rangle_{\text{av}}$ for even as well as odd modes traveling in positive ($\beta_m = +| \beta_m |$) as well as negative ($\beta_m = -| \beta_m |$) z-direction. It is noteworthy that equation (89) is proportional to L and not to L^2. The conversion to spurious guided modes by random imperfections appears as a true loss to the incident mode.

The losses due to the modes of the continuous spectrum are obtained from equations (48), (57), (61), (81) and (88) (with $\beta_m = \beta$):

$$\left\langle \frac{\Delta P}{P} \right\rangle_{\text{av}} = \frac{A^2 k^4 L}{2\pi B} (n_g^2 - 1)^2 \int_{-k}^{k} \left[\frac{\rho \cos^2 \kappa_0 d}{\left((\beta_0 - \beta)^2 + \frac{1}{B^2}\right)\left(\beta_0 d + \frac{\beta_0}{\gamma_0}\right)} \right.$$

$$\left. \cdot \left(\frac{\cos^2 \sigma d}{\rho^2 \cos^2 \sigma d + \sigma^2 \sin^2 d} + \frac{\sin^2 \sigma d}{\rho^2 \sin^2 \sigma d + \sigma^2 \cos^2 \sigma d} \right) \right] d\beta. \quad (90)$$

The relation between β, σ, and ρ is given by equations (15) and (16) while β_0, κ_0, and γ_0 are related by equations (9) and (10) and their value is obtained by solving equation (11). The integral in equation (90) is extended over β from $-k$ to k, the range of real values of the propagation constant (in z-direction) of the modes of the continuous spectrum. Equation (90) thus includes the losses due to forward as well as backward scattered radiation. The radiation modes with imaginary values of β can carry power away from the waveguide only strictly perpendicular to its axis. This power loss, if any, is not included in equation (90).

VIII. NUMERICAL RESULTS FOR THE STATISTICAL CASE

Figures 4 through 9 show numerical evaluations of equations (89) and (90). These figures can be grouped into two classes. Figures 4 through 6 are drawn for a dielectric waveguide whose index of refraction is $n_g = 1.01$. Figures 7 through 9 apply to a waveguide with $n_g = 1.5$. Within each of these two classes, the kd value was chosen to allow for three different cases. Figures 4 and 7 apply to waveguides which can support only the lowest order guided mode. In this case there is power lost only to the modes of the continuous spectrum. Figures 5 and 8 apply to waveguides supporting two guided modes and Figs. 6 and 9 apply to waveguides supporting three guided modes. Each figure shows the normalized loss caused by scattering into modes of the continuous spectrum as solid lines and the loss to the possible guided modes as dotted lines. Also shown are the ratios of backward to forward scattered power as solid lines for the modes of the continuum and as dotted lined for the guided modes. The total power lost to the lowest order guided modes is the sum of the losses to the continuum and the spurious guided modes.

Several remarkable features of these loss curves are worthy of a comment. The losses caused by the modes of the continuum as well as by the guided modes peak at certain values of the correlation length B. The location of these peaks are different, however, for the continuum and guided modes.

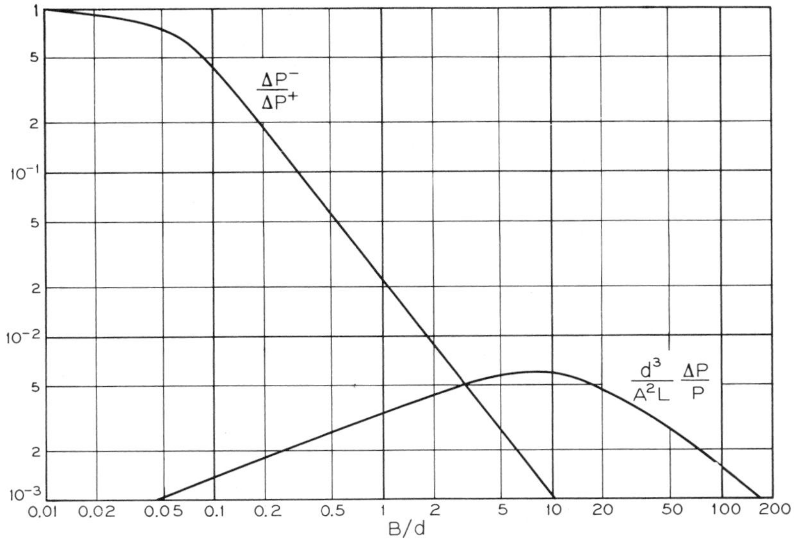

Fig. 4 — Normalized radiation loss (d^3/A^2L) $(\Delta P/P)$ and ratio of backward to forward scattered power $\Delta P^-/\Delta P^+$ as functions of the normalized correlation length B/d for $n_g = 1.01$ and $kd = 8.0$. Single guided mode operation (d = half width of waveguide, A = rms deviation of one waveguide wall, L = Length of waveguide section, n_g = index of refraction of waveguide, k = free space propagation constant).

The losses to the guided modes increase with increasing number of guided modes supported by the waveguide. However, the losses caused by the continuum of modes also increase as an increasing number of guided modes can be supported. This increase is less rapid, however, as one might expect because of the dependence of equation (90) on the fourth power of k. The fourth power dependence on frequency (or inverse wavelength) is typical for Rayleigh scattering by small particles, and it is not surprising that we encounter it here.

Finally, it is apparent from the curves showing the ratio of back-scattered to forward scattered power that forward scattering is predominant for large values of the correlation length. The ratio of $\Delta P^-/\Delta P^+$ levels off for large values of B. In some of the curves the leveling of the $\Delta P^-/\Delta P^+$ curves occurs out of the diagram but it is a common feature of all the curves. For small values of the correlation length there is as much scattering in the forward as in the backward direction.

For many practical applications, a waveguide supporting only one

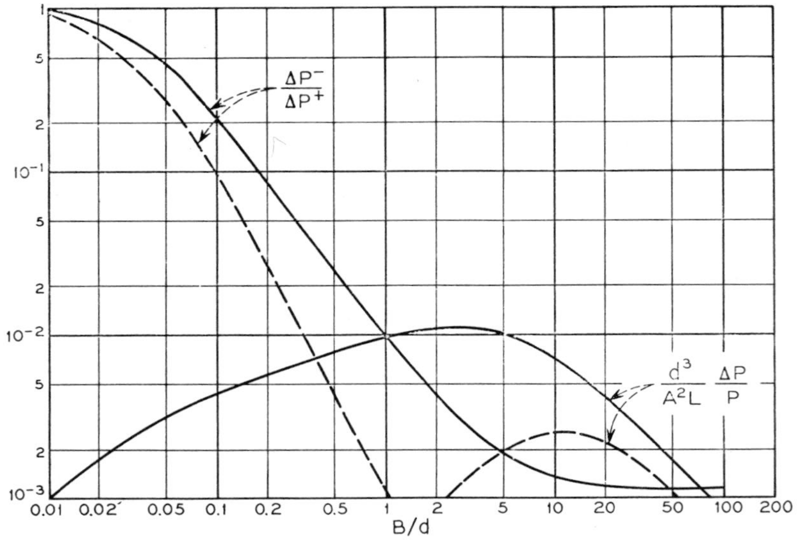

Fig. 5 — Normalized power loss and ratio of foreward to backward scattered power for radiation (solid curves) and spurious guided modes (dashed curves). Two guided modes ($n_g = 1.01$, $kd = 15$).

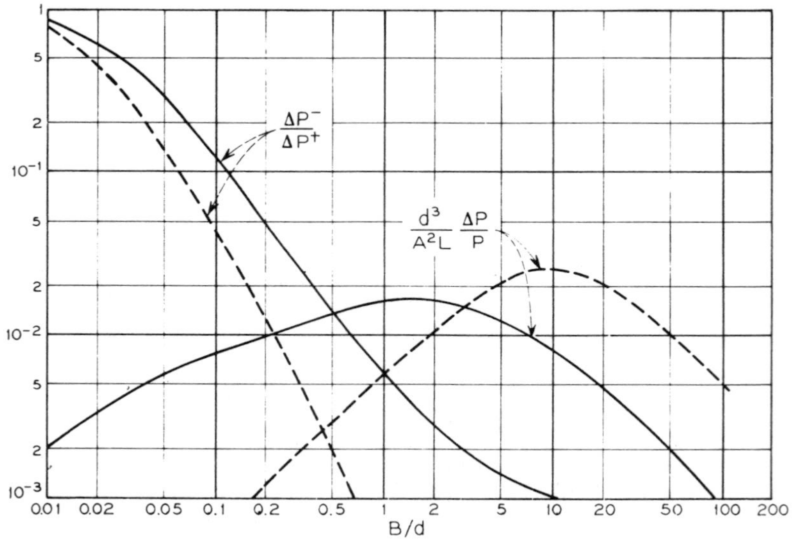

Fig. 6 — Similar to Fig. 5. Three guided modes ($n_g = 1.01$, $kd = 23$).
— — — — two guided mode loss; ——— continuum loss.

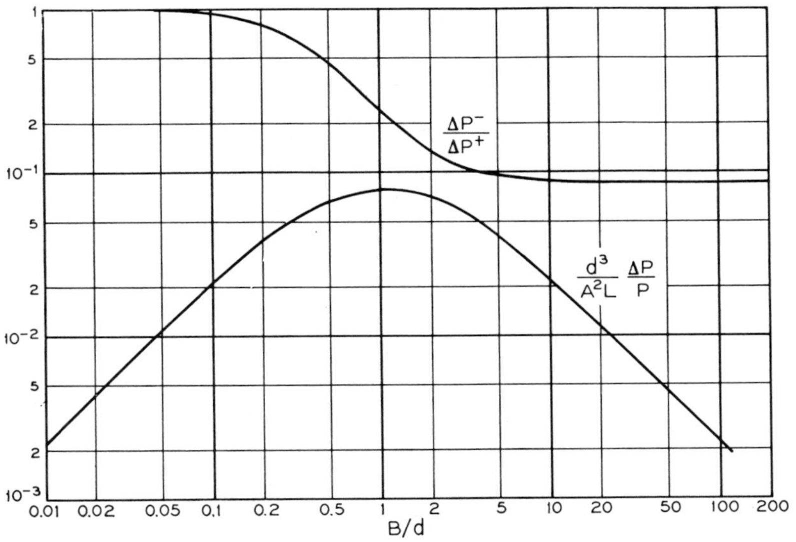

Fig. 7 — Similar to Fig. 4. One guided mode ($n_g = 1.5$, $kd = 1.3$).

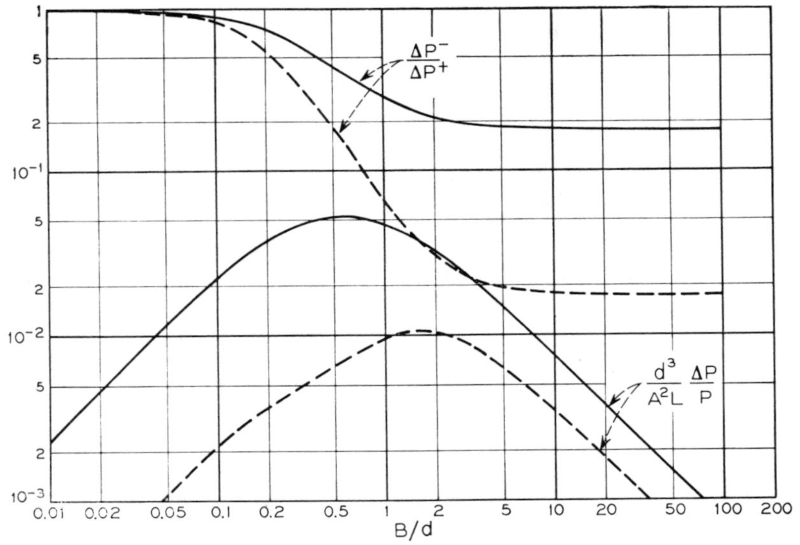

Fig. 8 — Similar to Fig. 5. Two guided modes ($n_g = 1.5$, $kd = 1.8$).
— — — — one guided mode loss; ——— continuum loss.

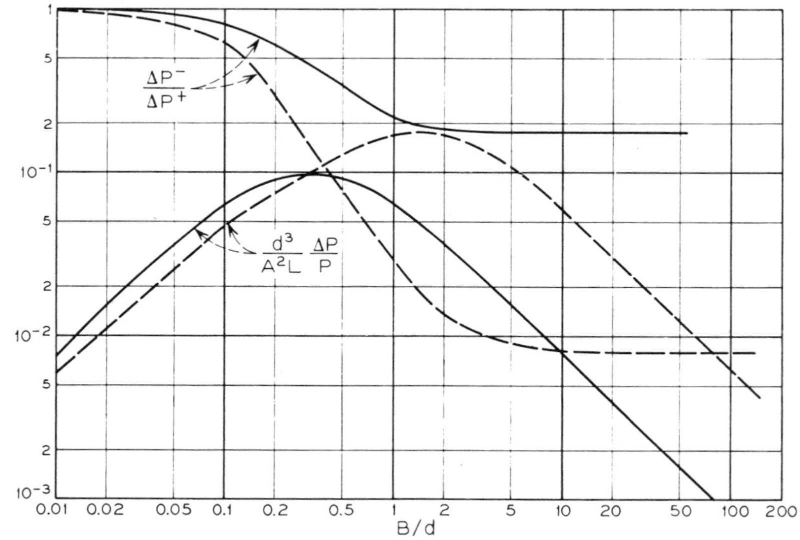

Fig. 9 — Similar to Fig. 5. Three guided modes ($n_g = 1.5$, $kd = 3$). — — — — two guided mode loss; continuum loss.

guided mode may be of most interest. Let us assume $\lambda_0 = 1\mu$. Figure 4, holding for $kd = 8.0$ and $n_g = 1.01$, applies to a waveguide whose half width is $d = 1.27\mu$. Taking the worst possible case of $B/d = 9$ or $B = 11.4\mu$, we find from Fig. 4

$$\frac{d^3}{A^2L}\frac{\Delta P}{P} = 6 \times 10^{-3}.$$

If we want to know how much rms deviation A of one wall of the guide would be required to cause a 10 percent loss ($\Delta P/P = 0.1$) in one centimeter of waveguide ($L = 1$ cm) we find $A = 5.85 \times 10^{-2}\mu = 585$ Å. The ratio of A over d gives an idea of the relative tolerance requirements:

$$\frac{A}{d} = 4.6 \times 10^{-2} = 4.6\%.$$

If the waveguide were to conform to the conditions of Fig. 6, we would have for $\lambda_0 = 1\mu$ a half width $d = 3.66\mu$. The losses caused by the two spurious modes are of the same order of magnitude as the radiation losses caused by the continuous spectrum. For $B/d = 10$ or $B = 36.6\mu$ we get a total loss of

$$\frac{d^3}{A^2L}\frac{\Delta P}{P} = 3.4 \times 10^{-2}.$$

To cause $\Delta P/P = 0.1$ for $L = 1$ cm requires that

$$A = 1.2 \times 10^{-1}\mu \quad \text{or} \quad \frac{A}{d} = 3.28\%.$$

The relative tolerance requirements are, therefore, approximately the same in both examples.

As a last example let us use Fig. 9 corresponding ($\lambda_0 = 1\mu$) to a waveguide with $n_g = 1.5$ and a half width $d = 0.477\mu$. For $B/d = 1.3$ or $B = 6.2\mu$ we find for the total loss

$$\frac{d^3}{A^2L}\frac{\Delta P}{P} = 2.3 \times 10^{-1}.$$

We get $\Delta P/P = 0.1$ with $L = 1$ cm for

$$A = 2.18 \times 10^{-3}\mu = 21.8 \text{ Å} \quad \text{or} \quad \frac{A}{d} = 0.457\%.$$

The perturbation theory, strictly speaking, holds only for small values of $\Delta P/P$. However, it is reasonable to expect that the power scattered into the radiation modes escapes sufficiently rapidly so that no appreciable amount of power reconversion from the radiation field to the guided mode occurs. The incremental power loss, $\Delta P/P = -\alpha L$, is therefore the same for any section of the guide so that we obtain the total scattering loss into the continuum of radiation modes $P = P_0e^{-\alpha L}$. We may now ask how much rms deviation is required to cause a radiation loss of 10 dB/km or $\alpha = 2.3 \text{ km}^{-1} = 2.3 \times 10^{-5} \text{ cm}^{-1}$. Using $B/d = 10$, corresponding to the top of the loss curve of Fig. 4, we obtain the equation

$$\frac{d^3}{A^2} \times 2.3 \times 10^{-5} = 6 \times 10^{-3}$$

so that ($\lambda = 1\mu$, $n_g = 1.01$, $kd = 8.0$, $d = 1.27 \times 10^{-4}$ cm)

$$\frac{A}{d} = 6.98 \times 10^{-4} \quad \text{or} \quad A = 8.86 \times 10^{-8} \text{ cm} = 8.86 \text{ Å}.$$

This figure dramatizes the stringent tolerance requirements of dielectric waveguides for long distance optical communications. In fact, such tolerances seem impossible to obtain. One can only hope that the correlation length can be kept far from the worst possible value of $B/d = 10$

(in this example) so that these extremely stringent tolerance requirements might be eased.

IX. CONCLUSION

We have analyzed the losses suffered by the lowest order symmetric mode propagating on a dielectric slab waveguide caused by imperfections of the waveguide boundaries. The analysis was simplified by assuming that there is no change in either the dielectric slab or the guided and unguided fields in one direction parallel to the slab. This assumption causes all our conclusions to be optimistic since variation of the slab in this direction can only cause additional losses. However, we expect that the results of this analysis give at least the correct order of magnitude of the actual scattering losses.

The statistical analysis was limited to a study of the effects which an exponential correlation function might have on the waveguide losses. The actual form of the correlation function may be quite different from this assumed exponential shape.† Conclusions regarding loss predictions are further hampered by a lack of knowledge of the expected correlation length.

However, our analysis does lead one to conclude that scattering losses suffered by optical fibers or other dielectric waveguide structures may be very serious. Deviations of the waveguide wall in the order of a few percent can cause a power loss of 10 percent or 0.46 dB/cm if the wall imperfection can be described by an exponential correlation function with a correlation length to guide half width ratio of approximately $B/d = 10$. An rms deviation of $A = 9$ Å causes a radiation loss of 10 dB/km if the free space wavelength is $\lambda_0 = 1\mu$ and the guide has an index of refraction of $n_g = 1.01$ (with vacuum on the outside). The width of the slab in this last example is $2d = 2.54\mu$.

The mode coupling and radiation loss theory has been experimentally confirmed at microwave frequencies. A report on these measurements is given in Ref. 8.

---

† Several other correlation functions have been tried and it was found that the results are insensitive to the particular choice of the function for values of B/d less than the value corresponding to the loss peak. In particular, the maximum loss value and the position of this loss peak were the same for different correlation functions. However, the loss values for B/d larger than the value corresponding to the maximum of the curve are very strongly dependent on the choice of the correlation function.

REFERENCES

1. Kapany, N. S., "Fiber Optics," New York: Academic Press, 1967.
2. Jones, A. L., "Coupling of Optical Fibers and Scattering in Fibers," J. Opt. Soc., *55*, No. 3 (March 1965), pp. 261–271.
3. Marcatili, E. A. J., unpublished work.
4. Miller, S. E., unpublished work.
5. McKenna, J., "The Excitation of Planar Dielectric Waveguides at p-n Junctions, I," B.S.T.J., *46*, No. 7 (September 1967), pp. 1491–1526.
6. Collin, R. E., "Field Theory of Guided Waves," New York: McGraw-Hill, 1960.
7. Rowe, H. E., and Warters, W. D., "Transmission in Multimode Waveguide with Random Imperfections," B.S.T.J., *41*, No. 3 (May 1962), pp. 1031–1170.
8. Marcuse, D., and Derosier, R. M., "Mode Conversion caused by Diameter Charges of a Round Dielectric Waveguide," B.S.T.J., this issue, pp. 3217–3232.

Radiation Losses of Dielectric Waveguides in Terms of the Power Spectrum of the Wall Distortion Function

By DIETRICH MARCUSE

(Manuscript received July 23, 1969)

In an earlier paper I described a perturbation theory of the radiation losses of a dielectric slab waveguide. The statistical treatment of the radiation losses was based on the correlation function of the wall distortion. This paper discusses the results of the radiation loss theory in terms of the power spectrum of the function describing the thickness of the slab. We found that only those mechanical frequencies θ *of the power spectrum contribute to the radiation loss that fall into the range* $\beta_0 - \mathrm{k} < \theta < \beta_0 + \mathrm{k}$. *(*$\beta_0$ *= propagation constant of guided mode,* k *= free space propagation constant.) The mechanical frequencies near both end points of this mechanical frequency range contribute more to the radiation loss than the region well inside of this range.*

We also discuss the far-field radiation pattern caused by a strictly sinusoidal wall distortion.

I. INTRODUCTION

In an earlier paper I developed a perturbation theory of the mode conversion effects between guided modes and of the radiation losses of a given guided mode caused by deviations from perfect straightness of the waveguide wall.[1] For simplicity, the discussion had been limited to a waveguide in the form of an infinitely extended dielectric slab.

The statistical discussion had been based on the description of the wall distortion by means of a correlation function. In Ref. 1 an exponential correlation function had been assumed. However, it has been established that the shape of the correlation function has little influence on the radiation losses.

It is possible to base the discussion of radiation losses not on correlation functions, but on the mechanical power spectrum of the wall distortion function. This study provides information as to how the various

Reprinted with permission from *Bell Syst. Tech. J.*, vol. 48, pp. 3233–3242, Dec. 1969.

mechanical frequencies of the wall distortion function contribute to the radiation losses.

The analysis of Ref. 1 was based on the use of radiation modes of the dielectric slab which represent standing waves in directions transverse to the propagation direction of the guided modes. The question naturally arises how a superposition of these standing waves can result in radiation flowing away from the rod. This question is answered by examining the far field radiation pattern caused by a sinusoidal distortion of one wall of the dielectric waveguide. This paper gives the relation between the length of the mechanical period, the wavelength of the guided mode, and the direction of the main lobe of the radiation.

II. RADIATION LOSS AND POWER SPECTRUM

The amplitudes of the modes of the continuous spectrum were derived in Ref. 1, equations (65) and (69). We have

$$g_e(\rho, L) = \frac{Lk^2}{2i(\pi)^{\frac{1}{2}}}(n_v^2 - 1)\frac{\rho(\cos\kappa_0 d\cos\sigma d)[\varphi(\theta) - \psi(\theta)]}{\left[\beta\left(\beta_0 d + \frac{\beta_0}{\gamma_0}\right)(\rho^2\cos^2\sigma d + \sigma^2\sin^2\sigma d)\right]^{\frac{1}{2}}} \tag{1}$$

for the even modes, and

$$g_0(\rho, L) = \frac{Lk^2}{2i(\pi)^{\frac{1}{2}}}(n_v^2 - 1)\frac{\rho(\cos\kappa_0 d\sin\sigma d)[\varphi(\theta) + \psi(\theta)]}{\left[\beta\left(\beta_0 d + \frac{\beta_0}{\gamma_0}\right)(\rho^2\sin^2\sigma d + \sigma^2\cos^2\sigma d)\right]^{\frac{1}{2}}} \tag{2}$$

for the odd modes. The functions

$$\varphi(\theta) = \frac{1}{L}\int_0^L [f(z) - d]e^{-i\theta z}\,dz, \tag{3}$$

$$\psi(\theta) = \frac{1}{L}\int_0^L [h(z) + d]e^{-i\theta z}\,dz, \tag{4}$$

with

$$\theta = \beta_0 - \beta \tag{5}$$

are the Fourier transforms of the wall distortion functions $f(z) - d$ and $h(z) + d$. [$x = f(z)$ is the boundary of the dielectric-air interface, $x = d$ describes the wall of the perfect guide, and $x = h(z)$ is the distorted boundary near $x = -d$.]

The meaning of the constants appearing in equations (1) to (5) is:

β_0 = propagation constant of guided mode (propagating in z-direction),
β = component of the propagation constant of the continuum mode in z-direction,
k = propagation constant in free space,
L = length of guide section with wall distortions,
n_g = dielectric constant of slab,

$$\rho = (k^2 - \beta^2)^{\frac{1}{2}} \tag{6}$$

$$\sigma = (n_g^2 k^2 - \beta^2)^{\frac{1}{2}}, \tag{7}$$

$$\kappa_o = (n_g^2 k^2 - \beta_0^2)^{\frac{1}{2}}, \tag{8}$$

$$\gamma_0^2 = (\beta_0^2 - k^2)^{\frac{1}{2}}. \tag{9}$$

The y-component of the electric radiation field caused by the wall distortions is given by

$$E_y = \int_0^\infty [g_e(\rho, L)\mathcal{E}_e(\rho, z) + g_0(\rho, L)\mathcal{E}_0(\rho, z)]\, d\rho. \tag{10}$$

The functions $\mathcal{E}_e$ and $\mathcal{E}_o$ are the even and odd radiation modes. The ratio of scattered power to incident guided mode power is obtained from

$$\frac{\Delta P}{P} = \int_{-k}^{k} (| g_e(\rho, L) |^2 + | g_0(\rho, L) |^2) \frac{\beta}{\rho}\, d\beta. \tag{11}$$

For simplicity we assume that one wall of the slab is perfect

$$h(z) = -d, \tag{12}$$

so that

$$\psi(\theta) = 0, \tag{13}$$

the relative scattering loss, follows from equations (1), (2), and (10)

$$\frac{\mathrm{d}}{\mathrm{L}} \frac{\Delta P}{P} = \int_{-k}^{k} \frac{1}{d^2} L \mid \varphi(\theta) \mid^2 I(\beta)\, d\beta \tag{14a}$$

with

$$I(\beta) = \frac{(kd)^4}{4\pi} (n_g^2 - 1)^2 \frac{\cos^2 \kappa_0 d}{\beta_0 d + \dfrac{\beta_0}{\gamma_0}} (\rho d) \left[\frac{\cos^2 \sigma d}{(\rho d)^2 \cos^2 \sigma d + (\sigma d)^2 \sin^2 \sigma d}\right.$$

$$\left. + \frac{\sin^2 \sigma d}{(\rho d)^2 \sin^2 \sigma d + (\sigma d)^2 \cos^2 \sigma d}\right]. \tag{14b}$$

Since $\varphi(\theta)$ is the Fourier component of the wall distortion function its absolute square value

$$|\varphi(\theta)|^2 \tag{15}$$

is the "power spectrum" of $f(z) - d$. It is apparent from equation (14) that $\Delta P/P$ depends on the power spectrum of the wall distortion function. Incidentally, equation (14) is not a statistical expression, but holds for a specific dielectric slab waveguide. We entered the power spectrum in the combination $L|\varphi|^2$ in equation (14) since this combination is independent of L for a randomly varying function $f(z) - d$.

Equation (14) allows us immediately to determine the range of mechanical frequencies θ which contribute to the radiation loss. The integral in equation (14) is extended from $-k$ to k, the β range of continuous radiation modes. The range of mechanical frequencies contributing to the scattering loss is therefore given by

$$\beta_0 - k < \theta < \beta_0 + k. \tag{16}$$

This is an important result since it states that those parts of the power spectrum which lie outside of the range, equation (16), do not contribute to radiation loss.

This last statement must not be misconstrued to mean that a waveguide with a sinusoidal wall distortion extending over length L

$$f(z) = d + a \sin \theta' z \qquad 0 \leqq z \leqq L \tag{17}$$

with θ' lying outside the range of equation (16) does not lose power by radiation. The power spectrum of equation (17) is

$$|\varphi(\theta)|^2 = \left[\frac{a}{L}\frac{\sin(\theta' - \theta)\frac{L}{2}}{\theta' - \theta}\right]^2. \tag{18}$$

A term with $\theta' + \theta$ in the denominator has been neglected in equation (18). The accuracy of this approximation improves with increasing values of L.

It is apparent from equation (18) that $|\varphi(\theta)|^2$ has non-vanishing values for $\theta \neq \theta'$ so that there is some small contribution to radiation loss even if θ' lies outside of the range of equation (16).

However, if we consider the limit $L \rightarrow \infty$ we can approximate the power spectrum, equation (18), by a δ-function:

$$\lim_{L\rightarrow\infty} |\varphi(\theta)|^2 = \frac{\pi a^2}{2L}\,\delta(\theta - \theta'). \tag{19}$$

In this special case the expression (14a) for the scattered power becomes

$$\frac{\mathrm{d}}{\mathrm{L}} \frac{\Delta P}{P} = \frac{\pi}{2} \left(\frac{a}{d}\right)^2 I(\beta_0 - \theta'). \tag{20}$$

The scattering from a dielectric waveguide with a wall distortion function whose power spectrum is a δ-function is proportional to $I(\beta_0 - \theta')$.

The function $I(\beta)$ is plotted in Fig. 1 for $n_g = 1.01$, $kd = 8.0$, and $\beta_0 d = 8.041$. The scattering caused by a wall distortion with a δ-function spectrum (a sinusoidal wall distortion of infinite length) is nearly independent of the value of $\beta = \beta_0 - \theta'$ over most of the β-range. There are two sharp peaks at $\beta \approx k$ and $\beta \approx -k$. The physical reasons for the sharp increase in loss at these values is easy to understand if we consider the direction of the radiation pattern as a function of θ'. We show in Section III [equation (35)] that the angle α between the waveguide and the main radiation lobe is given by

$$\cos \alpha = \frac{\beta}{k} = \frac{\beta_0 - \theta'}{k}. \tag{21}$$

The two peaks of the function $I(\beta)$, or correspondingly of the radiation loss, are associated with

$$\alpha \approx 0 \quad \text{and} \quad \alpha \approx \pi. \tag{22}$$

This shows that the radiation loss is high when the radiation pattern is

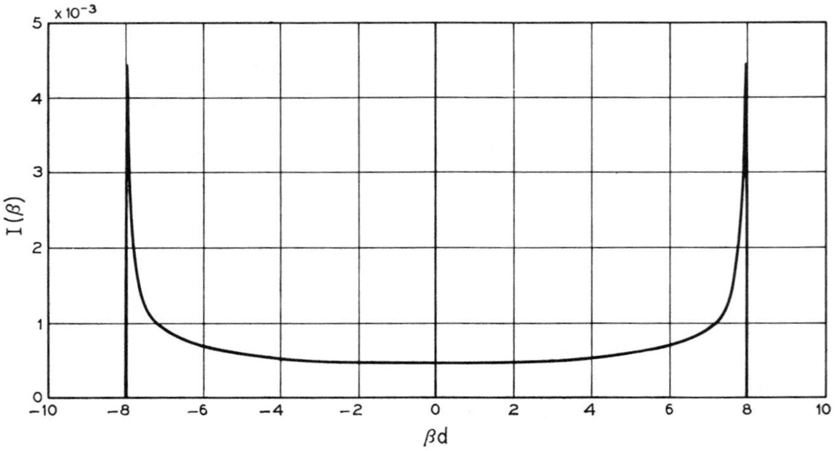

Fig. 1 — Graphical representation of the function $I(\beta)$ [eq. (14b)]. $n_g = 1.01$, $kd = 8.0$, $\beta_0 d = 8.041$.

directed very nearly parallel to the surface of the waveguide. The radiation modes gain more power if the guided mode can interact with them over a longer distance. An observation of this loss peak is reported in Ref. 2.

A power spectrum with sharp peaks much like that of equation (18) or (19) is not likely to occur for dielectric waveguides with random imperfections of the dielectric interface. It is much more reasonable to expect that such waveguides may have spectral distributions which are nearly independent of θ over a certain range of θ values. In the limit of a "white" spectrum,

$$|\varphi(\theta)|^2 = \text{constant}, \tag{23}$$

the scattering loss is proportional to the integral over the function $I(\beta)$ shown in Fig. 1. The two peaks contribute very little to this integral. Numerical integration of $I(\beta)$ of Fig. 1 including and excluding the peaks resulted in the values:

$$\int_{-8}^{8} I(\beta)\, d\beta = 0.011, \int_{-7.8}^{7.8} I(\beta)\, d\beta = 0.0096,$$

$$\text{and} \quad \int_{-7.5}^{7.5} I(\beta)\, d\beta = 0.0087.$$

This result is reassuring for the use of the perturbation theory which was used to derive equation (14). The perturbation theory is based on the assumption that power is converted from the guided mode to the radiation field but that no power is converted back from the radiation field to the guided mode. This approximation is certain to yield better results if the radiation pattern is directed away from the rod. In other words, the perturbation theory will work poorest in the region of the peaks of Fig. 1. However, for spectra that do not particularly favor the regions of these peaks, the contribution of those regions (which at the same time give the least reliable results) to the total radiation loss is only slight.

III. THE FAR FIELD RADIATION PATTERN

The far field pattern of the radiation field (that is excited by the lowest order even guided mode traveling in the dielectric slab with sinusoidal perturbation of one wall) can easily be calculated from equation (10). The even and odd radiation modes were given in Ref. 1 (for $|x| > d$)

$$\mathcal{E}_y^{(e)} = \left[\frac{2\omega\mu P}{\pi\beta(\rho^2\cos^2\sigma d + \sigma^2\sin^2\sigma d)}\right]^{\frac{1}{2}}$$
$$\times [\rho\cos\rho(|x| - d)\cos\sigma d - \sigma\sin\rho(|x| - d)\sin\sigma d]e^{i(\omega t - \beta z)} \tag{24}$$
$$\mathcal{E}_y^{(0)} = \frac{x}{|x|}\left[\frac{2\omega\mu P}{\pi\beta(\rho^2\sin^2\sigma d + \sigma^2\cos^2\sigma d)}\right]^{\frac{1}{2}}$$
$$\times [\rho\cos\rho(|x| - d)\sin\sigma d + \sigma\sin\rho(|x| - d)\cos\sigma d]e^{i(\omega t - \beta z)} \tag{25}$$

With $\psi(\theta) = 0$ and

$$\varphi(\theta) \approx \frac{a}{iL}\exp\left[i(\theta' - \theta)\frac{L}{2}\right]\frac{\sin(\theta' - \theta)\frac{L}{2}}{\theta' - \theta} \tag{26}$$

and with the help of equations (1) and (2) we get from equation (10)

$$E_y = -\frac{ak^2}{(2)^{\frac{1}{2}}\pi}(\omega\mu P)^{\frac{1}{2}}(n_g^2 - 1)\frac{\cos\kappa_0 d}{\left(\beta_0 d + \frac{\beta_0}{\gamma_0}\right)^{\frac{1}{2}}}$$
$$\times \int_0^\infty \frac{\rho}{\beta}\left\{\frac{\cos\sigma d[\rho\cos\rho(x - d)\cos\sigma d - \sigma\sin\rho(x - d)\sin\sigma d]}{\rho^2\cos^2\sigma d + \sigma^2\sin^2\sigma d}\right.$$
$$\left. + \frac{\sin\sigma d[\rho\cos\rho(x - d)\sin\sigma d + \sigma\sin\rho(x - d)\cos\sigma d]}{\rho^2\sin^2\sigma d + \sigma^2\cos^2\sigma d}\right\}$$
$$\times \exp\left[i(\theta' - \theta)\frac{L}{2}\right]\frac{\sin(\theta' - \theta)\frac{L}{2}}{\theta' - \theta} \times e^{i(\omega t - \beta z)}\,d\rho. \tag{27}$$

In the far field with $x \to \infty$ and $z \to \infty$ (but L finite) we can obtain an approximate solution of the integral in equation (27) by the method of stationary phase.[3] The sine and cosine functions of argument $\rho(x - d)$ can be expressed as sums of exponential functions. The most important terms of the integrand of equation (27) are, therefore, of the form

$$\exp[-i(\beta z \pm \rho x)]. \tag{28}$$

This exponential term is an extremely rapidly varying function of ρ as $x \to \infty$ and $z \to \infty$. All other terms in the integrand vary slowly by comparison. According to the method of stationary phase the contribution to the integral comes predominantly from a region that is determined by

$$\frac{\partial}{\partial\rho}(\beta z \pm \rho x) = 0. \tag{29}$$

With the help of equation (6), equation (29) leads to the condition

$$\frac{x}{z} = \pm\frac{\rho_0}{\beta} \tag{30}$$

or

$$\rho_0 = k \sin \alpha \tag{31a}$$

$$\beta = k \cos \alpha \tag{31b}$$

with

$$\cos \alpha = \frac{z}{(x^2 + z^2)^{\frac{1}{2}}} = \frac{z}{r}. \tag{32}$$

For $x > 0$ and $z > 0$ only the $+$ sign in equation (30) is possible. This is an important point. It shows that even though the radiation modes, equations (24) and (25), represent standing wave patterns in x-direction only, the outward traveling part of the decomposition of the standing wave into traveling waves makes a contribution to the radiation field, equation (27).

All terms of the integrand with the exception of equation (28) can be taken out of the integral. The remaining integration can be carried out using the expansion

$$\beta z + \rho x = k(x \sin \alpha + z \cos \alpha) - \frac{1}{2}\frac{z}{k \cos^3 \alpha}(\rho - \rho_0)^2 + \cdots$$

$$\cdot \int_0^\infty e^{-i(\beta z + \rho x)}\, d\rho = (1 + i)(\pi)^{\frac{1}{2}} \frac{(k)^{\frac{1}{2}} \cos \alpha}{(r)^{\frac{1}{2}}} e^{-ik(x \sin \alpha + z \cos \alpha)}. \tag{33}$$

The far field is therefore obtained in the form

$$E_y = \frac{1}{(\pi)^{\frac{1}{2}}} \exp\left(i\frac{\pi}{4}\right) ak^{\frac{3}{2}}(\omega\mu P)^{\frac{1}{2}}(n_g^2 - 1) \frac{\cos \kappa_0 d}{\left(\beta_0 d + \dfrac{\beta_0}{\gamma_0}\right)^{\frac{1}{2}}}$$

$$\cdot \frac{\rho_0^2 \sin 2\sigma_0 d - i\rho_0\sigma_0 \cos 2\sigma_0 d}{(\rho_0^2 + \sigma_0^2) \sin 2\sigma_0 d - 2i\rho_0\sigma_0 \cos 2\sigma_0 d} \frac{\sin (\theta' - \theta)\dfrac{L}{2}}{\theta' - \theta}$$

$$\cdot \exp\left[i(\theta' - \theta)\frac{L}{2}\right] e^{i\rho_0 d} \frac{1}{(r)^{\frac{1}{2}}} e^{i[\omega t - k(x \sin \alpha + z \cos \alpha)]}. \tag{34}$$

The index zero was added to σ to indicate that it must be evaluated from equations (7) and (8) using ρ_0 of equation (31a).

Equation (34) reveals several important features of the far field of radiation. This field is essentially a plane wave traveling in the direction of α ($\tan \alpha = x/z$, and x and z are the coordinates of the point of observation).

The field intensity is inversely proportional to the square root of the distance r from (the sinusoidally distorted) waveguide section. The dependence on distance is inversely proportional to $(r)^{\frac{1}{2}}$ rather than r because the waveguide is infinitely extended in y-direction (see Ref. 1).

The main radiation lobe occurs at the maximum value of $[\sin (\theta' - \theta)L/2]/(\theta' - \theta)$ that is at $\theta = \theta'$ or from equations (5) and (31b) at

$$\cos \alpha_m = \frac{\beta_0 - \theta'}{k} \tag{35}$$

(β_0 = propagation constant of guided mode).

The width of the main lobe depends on the length L of the sinusoidally distorted waveguide section. The difference in angle between the peak of the lobe and the first null determines the half width of the main lobe

$$\Delta\alpha = \frac{2\pi}{Lk \sin \alpha} \text{ for } \alpha \neq 0. \tag{36a}$$

The width of the main radiation lobe is inversely proportional to L. The lobe is narrowest for $\alpha = \pi/2$ and becomes wider as α decreases toward zero. If the peak of the main lobe is at $\alpha = 0$, we obtain

$$\Delta\alpha = \left(\frac{4\pi}{Lk}\right)^{\frac{1}{2}} \text{ for } \alpha = 0. \tag{36b}$$

The peak amplitude of the main radiation lobe is not strongly dependent on α. The increase in radiated power in forward direction ($\alpha = 0$) which is apparent from Fig. 1 is caused by the broadening of the radiation lobe with decreasing angle.

IV. CONCLUSION

The radiation loss of dielectric waveguides caused by deviations from perfect straightness of the waveguide walls depends on the "power spectrum" of the wall deviation function. A sinusoidal wall perturbation gives rise to radiation into a particular direction in space. Each Fourier component of the Fourier expansion of the wall distortion function is responsible for radiation into a particular direction. The width of the radiation lobes is wide for scattering directions parallel to the rod so that those Fourier components responsible for forward and backward scattering contribute more to the radiation loss than those causing scattering in other directions. However, this preferential loss behavior is not very pronounced, so that the Fourier components responsible for forward and backward scattering contribute only a small amount of the total radiation loss caused by a broad power spectrum.

The coupling between two guided modes of the dielectric waveguide is also governed by equation (5). Only one component of the power spectrum of the wall distortion function influences the coupling between two guided modes, while the entire range of mechanical frequencies, equation (16), determines the radiation loss.

The general predictions of this theory have been experimentally verified. Microwave experiments on a periodically corrugated teflon rod have shown that the radiation losses are negligibly small if the period of the corrugation is such that θ lies outside of the interval indicated by equation (16).[2] However, if θ falls inside of the interval, equation (16), considerable radiation losses do occur. The peak of the radiation losses shown in Fig. 1 and the direction and width of the radiation lobes have also been observed in agreement with this theory.

REFERENCES

1. Marcuse, D., "Mode Conversion Caused by Surface Imperfections of a Dielectric Slab Waveguide," B.S.T.J., this issue, pp. 3187–3215.
2. Marcuse, D., and Derosier, R. M., "Mode Conversion Caused by Diameter Changes of a Round Dielectric Waveguide," B.S.T.J., this issue, pp. 3217–3232.
3. Mathews, J., and Walker, R. L., "Mathematical Methods of Physics," New York: W. A. Benjamin, 1965, pp. 85–86.

Part 6
Integrated Optics Lasers

JUNCTION LASERS WHICH OPERATE CONTINUOUSLY AT ROOM TEMPERATURE

I. Hayashi, M.B. Panish, P.W. Foy, and S. Sumski

Bell Telephone Laboratories, Murray Hill, New Jersey 07974

(Received 8 June 1970)

Double-heterostructure GaAs-$Al_xGa_{1-x}As$ injection lasers which operate continuously at heat-sink temperatures as high as 311 °K have been fabricated by liquid-phase epitaxy. Thresholds for square diodes as low as 100 A/cm^2 and for Fabry-Perot diodes as low as 1600 A/cm^2 have been obtained. Some details of preparation and properties are given.

We have reported studies of heterostructure injection lasers[1,2] which were aimed at the achievement of low current thresholds at 300 °K [J_{th} (300 °K)]. The lowest thresholds achieved at this temperature were 2300 A/cm^2. Since then we have modified our method of growth of the structures and the doping of the various layers so as to achieve further reduction of J_{th} (300 °K). These modifications have enabled us to grow by liquid-phase epitaxy a structure consisting of alternate layers of GaAs and $Al_xGa_{1-x}As$, appropriately doped, with which J_{th} (300 °K) as low as 1000 A/cm^2 has been observed in square fully internally reflecting diodes, and as low as ~1600 A/cm^2 in uncoated Fabry-Perot diodes (560 μm long). We believe that the further reduction in threshold results primarily from the attainment of a narrower active region than previously.

The diodes were four-layer structures similar to those illustrated in Ref. 2 except that an additional layer (layer 4) of GaAs was used to provide a better contact to the *p* side of the diode than with $Al_xGa_{1-x}As$. The layers were grown on polished and etched (111) and (100) faces of *n*-type GaAs in the apparatus shown in Fig. 1. After cooling rates of 1–3 °C/min had been established, the solutions were successively brought into contact with the seed. Layer 1 was grown from 840 to 830 °C, and layers 2, 3, and 4 by allowing solutions 2, 3, and 4 to remain on the seed for 15 sec each while cooling. The solution compositions and layer thicknesses for a typical run are shown in Table I.

Dopant concentrations were selected on the basis of the literature and our unpublished data on doping to be expected with liquid-phase epitaxy with Si,[3,4] Zn,[5,6] Ge,[7] and Sn.[8,9] Al concentrations

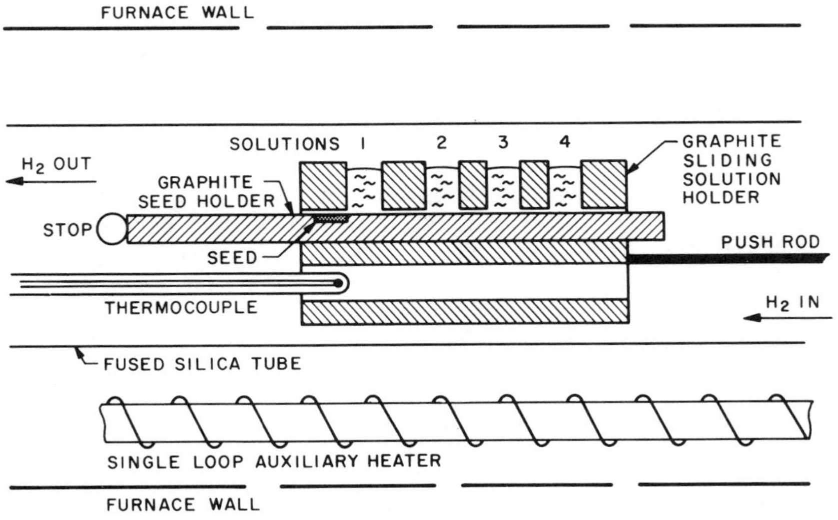

FIG. 1. Apparatus used for the liquid-phase epitaxial growth of four-layer double-heterostructure wafers.

Reprinted with permission from *Appl. Phys. Lett.*, vol. 17, pp. 109–111, Aug. 1, 1970.

TABLE I. Solution compositions and layer thickness for a typical run.

Solution	Ga	Al	Atom % As	Dopant	Layer Thickness (μm)
1	92.0	0.5	3.0	Sn/4.5	3–4
2	94.0	...	4.0	Si/2.0	$\lesssim 1$
3	95.9	0.8	2.8	Zn/0.5	~1.5
4	95.7	...	4.0	Ge/0.3	~1.5

for the liquid were chosen to yield $x \approx 0.2-0.4$ in the $Al_xGa_{1-x}As$. We have not determined concentrations in the layers, but the liquid concentrations were selected to yield for layer 1, $n \approx (1-4) \times 10^{18}$; layer 2, $p \approx (1-4) \times 10^{18}$ (compensated); layer 3, $p \approx (5-8) \times 10^{18}$; and layer 4, $p \approx (3-5) \times 10^{18}$ cm^{-3}. A small excess of GaAs was used in each solution to ensure saturation. We have prepared lasers with a range of Si concentrations (1–15 mg/g Ga). Within the scatter of our data this factor does not appear to be the major one effecting laser performance. The most important factor is that the second layer be thin and uniform although we have not, as yet, optimized the width at less than 1 μm. We have not yet attempted to optimize other dopant concentrations.

The layer thicknesses indicated in Table I are typical of those obtained. In a given wafer the thicknesses may vary by a factor of 2 across the crystal. It is desirable to have layers 1, 3, and 4 as thin as possible for efficient heat sinking. For continuous operation the as-grown wafer was diffused with Zn to form a thin (~0.2 μm) p^+ region. Contacts were vapor plated onto the wafer, Cr then Au on the p side, and a Sn-Pt-Sn sandwich on the n side. The Au sides of the units were bonded to a diamond upon which layers of Cr and then Sn had been vapor deposited, and an In-coated contact was pressed on the n side. The diamond was bonded to a Cu block. Reported heat-sink temperatures are those of this Cu block. This technique is a modification of that used by Dyment and D'Asaro with homostructure diodes.[10]

To determine heat-sinking characteristics the intensity of light output was plotted as a function of current as shown at 297 °K in Fig. 2 for dc and pulsed operation at a 10% duty cycle. Lasing was observed by watching the "turn on" of several spots with an image converter microscope or spectrally, Fig. 3. In initial experiments, a number of diodes which lased continuously with heat-sink temperatures as high as 270–311 °K were studied. The data of Figs. 2 and 3 are for an uncoated Fabry-Perot double heterostructure diode

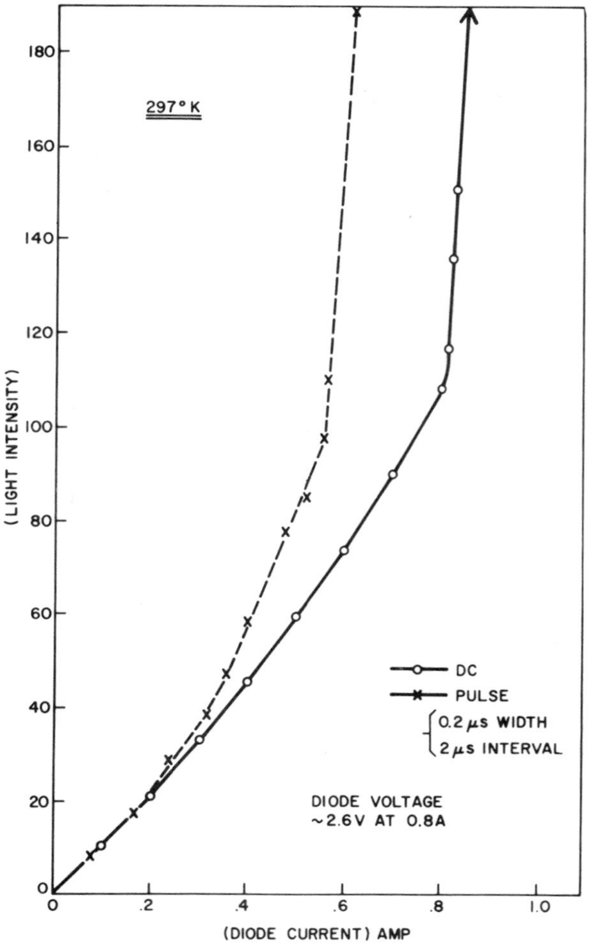

FIG. 2. Light output versus current for a heat-sinked Fabry-Perot double-heterostructure diode in pulsed and dc operation at 297 °K.

370 μm long and about 80 μm wide. With the heat sink at 311 °K and in continuous operation, the threshold current density $J_{th} \approx 3.5$ kA/cm^2 and in pulsed operation, $J_{th} \approx 2.0$ kA/cm^2. For the diode of Fig. 2 the temperature of the junction is estimated to have been about 25 °K above the heat-sink temperature when the latter was at 297 °K. This temperature rise would be expected with a thickness of $Al_xGa_{1-x}As$ of ~1.8 μm in layer 3 if the heat conductivity of the $Al_xGa_{1-x}As$ were about 15% that of GaAs. The actual heat conductivity of $Al_xGa_{1-x}As$ is not known. The power output of this diode on its heat sink at 297 °K was at least 20 mW at 30% above threshold. The emission energy of this diode was 80 meV higher than expected for GaAs. We believe therefore that a small amount of Al was transferred from solution 1 to solution 2 during the growth procedure.

We will report upon these and similar diodes in more detail in other papers now being prepared. Techniques which provide for thinner layers, bet-

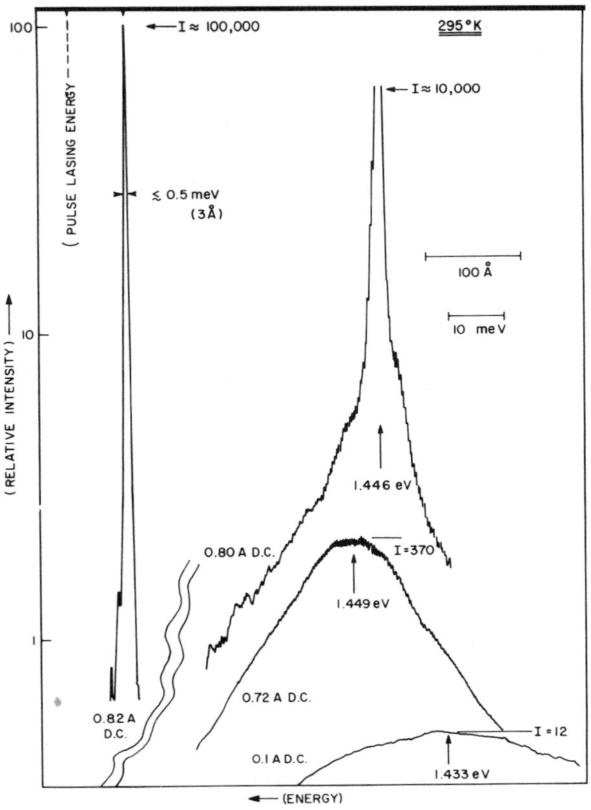

FIG. 3. Spectra for the diode of Fig. 3 at 295°K at several current densities. The numbers (I) are the relative intensities. The peak at 0.8 A is incomplete but is included to illustrate the shape of the base.

ter heat sinking and mirror coatings on the diodes are expected to increase considerably the maximum lasing temperature of double-heterostructure diodes.

We thank J.C. Dyment for some experimental help and for useful discussions, and J.K. Galt for discussions and encouragement throughout the course of this work.

[1]I. Hayashi and M.B. Panish, J. Appl. Phys. 41, 150 (1970).

[2]M.B. Panish, I. Hayashi, and S. Sumski, Appl. Phys. Letters 16, 326 (1970).

[3]W.G. Spitzer and M.B. Panish, J. Appl. Phys. 40, 4200 (1969).

[4]M.B. Panish and S. Sumski, J. Appl. Phys. 41, 3195 (1970).

[5]M.B. Panish and H.C. Casey, J. Phys. Chem. Solids 28, 1673 (1967).

[6]A.S. Jordan (unpublished).

[7]F.E. Rosztoczy, F. Ermanis, I. Hayashi, and B. Schwartz, J. Appl. Phys. 41, 264 (1970).

[8]R. Solomon, in *Proceedings of the Second International Symposium on GaAs* (The Institute of Physics and The Physical Society, London, 1968), p. 11.

[9]J. Kinoshita, W.W. Stein, G.F. Day, and J.B. Mooney, in *Proceedings of the Second International Symposium on GaAs* (The Institute of Physics and The Physical Society, London, 1968), p. 22.

[10]J.C. Dyment and L.A. D'Asaro, Appl. Phys. Letters 11, 292 (1967).

Surface Lasers

F. Varsanyi
Department of Physics, Stanford University, Stanford, California 94305
and
Isoray International Corporation, Palo Alto, California 94304
(Received 29 April 1971; in final form 18 June 1971)

Room-temperature visible laser action is reported in $PrCl_3$ and $PrBr_3$. Optical pumping directly into the upper laser level is accomplished by a wavelength-tunable dye laser. Threshold is less than 1 μJ. Superradiant laser action within physical dimensions of a few microns is generated by focusing the excitation light on the cleaved crystal surface. Use as an active element in integrated optics networks is indicated with over 10^6 individual lasers per square inch.

A new kind of solid-state laser is reported. The first observations were made on 100% $PrCl_3$ single crystals. Illuminating the cleaved surface of the bulk crystal with monochromatic light[1] of sufficient intensity and of such color that direct excitation into the upper laser level occurs, superradiant laser action is generated within the first few ionic layers near the surface, and the beam freely travels to the edge of the crystal. The high density of active ions self-generates this geometry, the penetration depth of the exciting radiation being in the order of 1 μ. The combination of the high-ion density with the essentially 100% radiative quantum efficiency[2] creates an extremely large gain per unit length and makes the device very compact. If the excitation light is projected on the crystal surface in the form of a small elongated slit, threshold is first reached along the longest dimension and the laser action occurs along this axis. The amplification is so large that a strong directional laser beam emerges from an optically pumped length of just a few microns along the surface. The threshold is less than 1 μJ of pump light. Upon further increase of the excitation energy density, a second threshold is reached, whereby the incoming monochromatic light saturates layer after layer of active ions until the dimension of this excitation pocket becomes larger in the direction perpendicular to the surface than along the surface. At this point the surface-laser action abruptly ceases, and a strong laser beam of the same wavelength as the surface one exits the crystal both in the forward and in the backward direction.

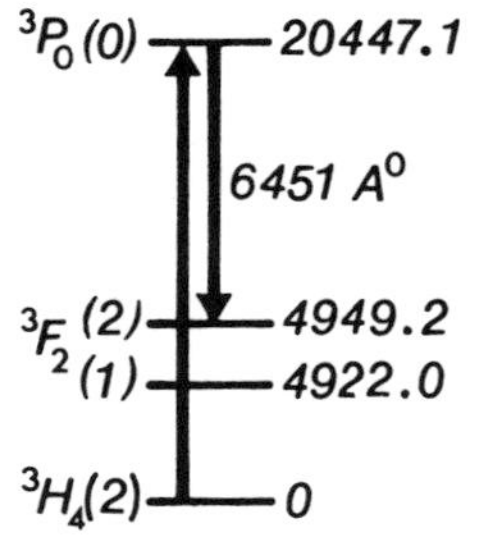

FIG. 1. Relevant energy levels of the Pr^{3+} spectrum. Laser output is frequency tunable with magnetic field due to the $S_1 = 2.74$ splitting factor of the $^3F_2(2)$ terminal state.

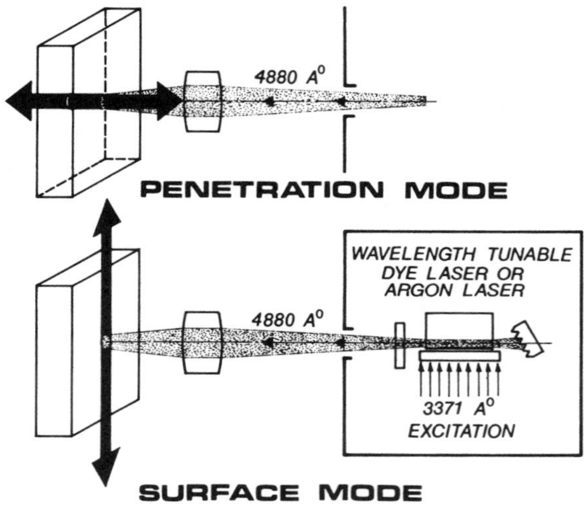

FIG. 2. Operation of the room-temperature 6451-Å $PrCl_3$ laser in the "surface mode" and in the "penetration mode". Excitation threshold with 4880-Å illumination is less than 1 μJ.

The relevant levels of the well-known[3] Pr^{3+} spectrum are reproduced on Fig. 1. The $^3P_0$ level is nondegenerate at 20 447.1 cm^{-1} and is reached directly with the 4880-Å excitation from the $^3H_4(2)$ ground state. Laser action originates directly from $^3P_0$ and terminates on $^3F_2(2)$ at 4949.2 cm^{-1} giving 15 497.9 cm^{-1} (vacuum) or 6451 Å (air) for the laser output. Terminating so high above the ground state makes the efficient room-temperature operation possible. The terminal state has a spectroscopic splitting factor $S_1 = 2.74$ which allows considerable frequency tuning by a magnetic field.

Examination with a Fabry-Perot etalon shows that the output is a single component with a spectral width of about 0.01 cm^{-1}. The output frequency is very stable, with no noticeable shift within the 0.01-cm^{-1} instrumental resolution in several hours.

The experimental details are shown in Fig. 2. The $PrCl_3$ crystal is held at room temperature in a quartz capsule to protect it from atmospheric moisture and dust, has about 5×10-mm cleaved surface on two opposing sides, and is about 3 mm thick. It is as clear and transparent as high-quality optical glass with the characteristic pale-green color of the praseodymium ionic absorption. $PrCl_3$ is hexagonal and cleaves parallel to the C_3 axis. Cleavage is so perfect that at the proper angle of orientation we observe rings of light, resembling a Fabry-Perot pattern, generated by the "perpendicular" laser beam

Reprinted with permission from *Appl. Phys. Lett.*, vol. 19, pp. 169–171, Sept. 15, 1971.

and the two opposing cleaved surfaces. It has to be noted, however, that neither the "surface" nor the "penetration" mode of operation depends on outside reflectors. This becomes obvious when the crystal is tilted off perpendicular with respect to the excitation beam. A strong laser output is still observed, but now the material acts as its own wavelength selector prism, with the excitation and generated laser beams clearly separated, and nonperpendicular to the crystal face.

An ordinary 50-mm focal-length photographic lens focuses the exciting 4880-Å radiation on the $PrCl_3$ crystal surface from a wavelength-tunable dye laser, which in turn is excited by a repetitively pulsed nitrogen laser. The active dye in our experiments was 4-methyl-umbelliferone in a 5×10^{-3}-mole/liter ethanol solution with 20% of 1*M* HCl-water solution added. The nitrogen laser emits at 3371 Å and is focused into the dye cell with a simple 6-mm-diam quartz rod acting as a cylindrical lens. The use of a flow dye cell allowed pulse-repetition rates of up to 100/sec, and we observed no saturation, heating, or other adverse effects on the $PrCl_3$ laser output up to this pulse rate. The 4880-Å excitation pulse is about 5 nsec long. Examination with a Tektronix 519 oscilloscope showed the 6451-Å output from the $PrCl_3$ laser to closely follow the time behavior of the excitation. The photographic lens is mounted on a micrometer screw allowing precise movement along the light beam, and the excitation energy density was changed on the crystal face by focusing or defocusing as an alternate to changing the nitrogen laser output.

It is interesting to note that the state $^3F_2$ has its lowest component at 4922 cm^{-1}, some 27 cm^{-1} below the 4949-Å terminal laser level. Rapid phonon-assisted relaxation is known[3] to occur in a like situation, thus copious quantities of monoenergetic 27-cm^{-1} phonons are generated. We believe to have observed some evidence of interaction between the 6451-Å laser beam and the possibly coherent phonon field. We are now further investigating this potentially important aspect of the laser.

Under the present experimental conditions (that is, a fast 5-nsec excitation pulse) we observe no evidence of self-termination of the $PrCl_3$ laser output. This fact and the known high-infrared radiative quantum efficiency[4] make it possible that in a cascadelike process an infrared laser is also generated along with the visible beam. We, however, have not investigated this experimentally.

Extending the above observations to other crystals is a natural course to follow. We have already observed surface-laser action in a second Pr^{3+} salt, praseodymium tribromide. The behavior is very similar to $PrCl_3$. This is expected in view of earlier observations[5] which showed the strong analogy in optical and fluorescence behavior.

There are other high-transition probability lines[6] originating on the $^3P_0$ laser level, and with a frequency selective cavity arrangement, operation in the yellow, green, and blue seems certain. Coincidence between the 4880-Å strong argon laser line and the 4880-Å $^3P_0$ level of Pr^{3+} was noted[7] earlier years ago, and efforts to extend the operation of the surface laser to the room-temperature cw mode are under way. The presently reported observations already make pulsed argon laser excitation at room temperature a certainty, and the compactness and reliability of an argon-excited $PrCl_3$ surface laser is attractive in systems applications.

An obvious role for a surface laser would be as the active element in integrated optics networks. This would be facilitated by its extremely small size. Packing densities of well over 10^6 independent lasers on 1 $in.^2$ of crystal surface seem entirely reasonable. Also important is the steerability of the beam direction. It was mentioned earlier that the surface-laser action is generated along the longest dimension of the excitation beam on the crystal face; thus shifting or rotating the excitation image causes a correlated change in the output-beam position and/or direction. Pumping a spot on the crystal face below threshold with 4880-Å radiation creates a memory cell with a $1/e$ decay time of about 10 μsec. The arrival of an additional 5-nsec excitation pulse of proper intensity within a specified time would then either replenish this memory, make it emit 6451-Å laser radiation along the surface, or make it "read out" by reaching the second threshold of the penetration mode.

The active area of the surface laser is in the order of a few square microns. This made possible some visually spectacular experiments, whereby the excitation light was focused on a tiny $PrCl_3$ or $PrBr_3$ crystallite which was hardly visible to the naked eye, and upon reaching the threshold energy density the little "dust" particle lit up with the brilliant red glow of the laser output. It is not unlikely that this "powder laser" dispersed in some appropriate carrier (liquid, gas, or solid) opens up new possibilities in display technology.

In conclusion, it is worthwhile to note that both $PrCl_3$ and $PrBr_3$ can easily be vacuum deposited (about 2-mm vapor pressure at 1000 °C)[8]; thus the well-developed masking techniques of integrated electrical-circuit technology can be utilized in connection with integrated-optics-network applications.

The author would like to thank Professor Arthur L. Schawlow for the critical reading of the manuscript and for his stimulating interest in this work.

---

[1]F. Varsanyi and G.H. Dieke, J. Chem. Phys. 31, 1066 (1959).

[2]F. Varsanyi and G.H. Dieke, Phys. Rev. Letters 7, 442 (1961).

[3]Gerhard H. Dieke, *Spectra and Energy Levels of Rare Earth Ions in Crystals* (Wiley, New York, 1968).

[4]F. Varsanyi, Phys. Rev. Letters 11, 314 (1963).

[5]B. Toth and F. Varsanyi, Bull. Am. Phys. Soc. 12, 293 (1967).

[6]E. Dorman, J. Chem. Phys. 44, 2910 (1966).

[7]T.C. Damen, S.P.S. Porto, and F. Varsanyi, Bull. Am. Phys. Soc. 12, 273 (1967).

[8]F.H. Spedding and A.H. Daane, *The Rare Earths* (Wiley, New York, 1961).

STIMULATED EMISSION IN A PERIODIC STRUCTURE

H. Kogelnik and C. V. Shank
Bell Telephone Laboratories, Holmdel, New Jersey 07733
(Received 23 November 1970)

We have investigated laser oscillation in periodic structures in which feedback is provided by backward Bragg scattering. These new laser devices are very compact and stable as the feedback mechanism is distributed throughout and integrated with the gain medium. Intrinsic to these structures is also a gratinglike spectral filtering action. We discuss periodic variations of the refractive index and of the gain and give the expression for threshold and bandwidth. Experimentally we have induced index periodicities in gelatin films into which rhodamine 6G was dissolved. The observed characteristics of laser action in these devices near 0.63 μm are reported.

Laser oscillators consist of a laser medium which provides gain and a resonator structure which provides the feedback necessary for the build-up of oscillation. The resonator is commonly formed by two (or more) end mirrors terminating the laser medium. In this paper we propose mirrorless laser devices in which the feedback mechanism is distributed throughout and integrated with the gain medium. In particular, the feedback mechanism is provided by Bragg scattering from a periodic spatial variation of the refractive index of the gain medium, or of the gain itself. Distributed feedback (DFB) lasers are very compact and have a mechanical stability which is intrinsic to integrated optical devices.[1,2] In addition, the gratinglike nature of the device provides a filter mechanism which restricts the oscillation to a narrow spectral range. The DFB approach appears applicable to such lasers as dye lasers, semiconductor lasers, and possibly parametric oscillators. In

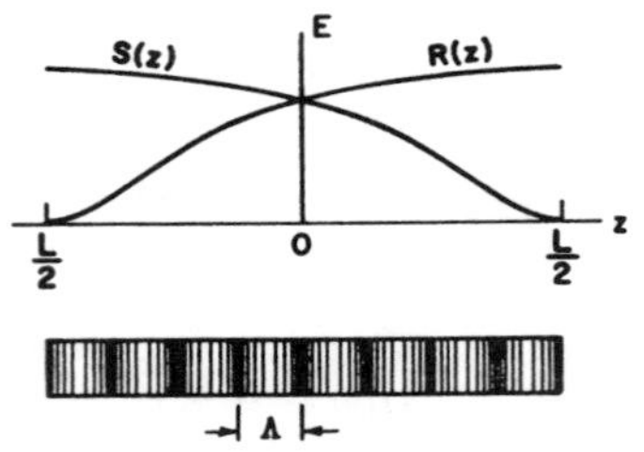

FIG. 1. Spatial dependence of the amplitudes of the two counterrunning waves $R(z)$ and $S(z)$.

Reprinted with permission from *Appl. Phys. Lett.*, vol. 18, pp. 152–154, Feb. 15, 1971.

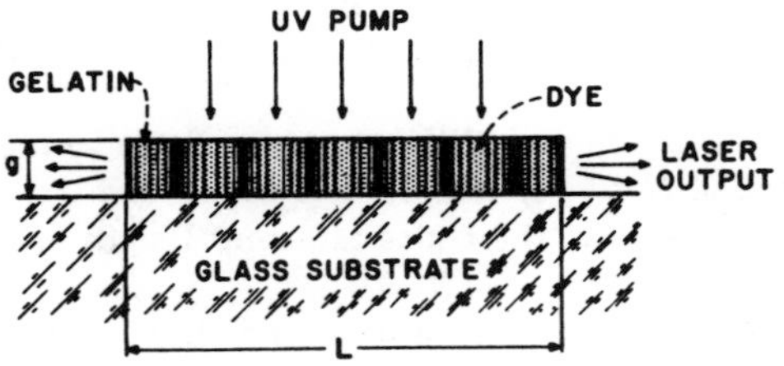

FIG. 2. Cross section of distributed feedback device consisting of dyed gelatin on a glass substrate.

the following we will first discuss the general properties of DFB lasers including the conditions for threshold, and then describe experiments on DFB dye lasers which indicate a confirmation of our theoretical ideas.

One can produce a DFB structure by inducing a periodic spatial variation of the refractive index n or of the gain constant α of the laser medium such as

$$n(z) = n + n_1 \cos Kz \tag{1}$$

or

$$\alpha(z) = \alpha + \alpha_1 \cos Kz, \tag{2}$$

where z is measured along the optic axis and $K = 2\pi/\Lambda$. Here Λ is the period (or "fringe spacing") of the spatial modulation, and n_1 and α_1 are its amplitudes. A DFB structure of this kind will oscillate in the vicinity of a wavelength λ_0 given by the Bragg condition

$$\lambda_0/2n = \Lambda. \tag{3}$$

We have derived[3] expressions for the threshold and the spectral bandwidth of stimulated emission in such periodic structures from a simple coupled-wave analysis which is similar to that discussed in Ref. 4. The coupled-wave picture assumes that the field E in the device is of the form

$$E = R(z)e^{-jKz/2} + S(z)e^{jKz/2}, \tag{4}$$

consisting of two counterrunning waves with the complex amplitudes R and S. As indicated in Fig. 1 these waves grow in the presence of gain, and they feed energy into each other due to the spatial modulation of n or α. The boundary conditions for the wave amplitudes are

$$R(-L/2) = S(L/2) = 0, \tag{5}$$

where L is the length of the DFB laser. At the endpoints of the device a wave starts with zero amplitude receiving its initial energy through feedback from the other wave.

While nonlinear calculations are necessary to determine the final oscillation amplitudes, we can obtain the threshold conditions from a linear analysis. We give its results in a simplified form, which is valid for large gain factors $G = \exp(2\alpha L)$. The start oscillation condition is

$$4\alpha^2 G = (\pi n_1/\lambda)^2 + \alpha_1^2/4. \tag{6}$$

If only the refractive index is modulated, the threshold condition becomes

$$n_1 = (\lambda_0/L)[\ln G/\pi(G)^{1/2}] \tag{7}$$

and for pure gain modulation we obtain

$$\alpha_1/\alpha = 4G^{-1/2}. \tag{8}$$

When the gain G exceeds the threshold value at center frequency by a factor of 2, the threshold is exceeded over a spectral bandwidth $\Delta\lambda$ approximately given by

$$\Delta\lambda/\lambda_0 = (\lambda_0/4\pi n L) \ln G. \tag{9}$$

Nonlinear effects will tend to further narrow the spectral width of the oscillator output. To illustrate the above relations let us assume a device with a length of $L = 10$ mm and a gain[5] of $G = 100$ operating at $\lambda_0 = 0.63$ μm. Equation (7) indicates that oscillation will occur if $n_1 \gtrsim 10^{-5}$. This value is relatively easy to achieve by various methods. For the bandwidth of the device we obtain $\Delta\lambda \approx 0.1$ Å.

As a first test of the above ideas we have implemented and tested a DFB dye laser in a gelatin film. The laser was $L = 10$ mm long and about 0.1 mm wide. The film was deposited on a glass substrate as shown in Fig. 2. The gelatin was di-

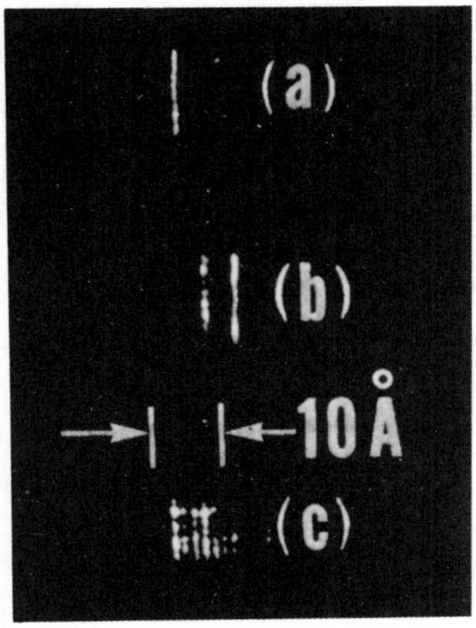

FIG. 3. Output spectra of (a) single mode of a 14-μ-thick DFB laser pumped just above threshold (wavelength is 6275 Å); (b) multiple film modes of a 14-μ-thick DFB laser pumped well above threshold; and (c) multiple film modes of a 45-μ-thick DFB laser.

chromated and exposed to the interference pattern produced by two coherent uv beams from a He-Cd laser. The fringe spacing (in the gelatin) was about 0.3 μm. After exposure the gelatin was developed using techniques which are well known in holography[6,7] and which result in a spatial modulation of the refractive index. The developed gelatin was soaked in a solution of rhodamine 6G to make the dye penetrate into the porous gelatin layer. After drying, the resulting DFB structure was transversely pumped with the uv radiation from a nitrogen laser as indicated in Fig. 2. At pump densities above 10^6 W/cm^2 we observed laser oscillations at a wavelength of about 0.63 μm. We analyzed the spectrum of the laser output using a spectrometer. We measured an output linewidth of less than 0.5 Å, which was the resolution limit of our instrumental technique. This should be compared to the stimulated fluorescence linewidth of rhodamine 6G in uniform gelatin under the same pumping conditions, which is approximately 50 Å and centered about 0.59 μm.[8] Considerable narrowing due to the distributed feedback is apparent.

We investigated devices of various gelatin thicknesses at various pumping levels. While it is possible to obtain single-line operation we have also observed simultaneous oscillation at several wavelengths. We have illustrated this in Figs. 3(a) and 3(b), which show the output spectrum obtained from a DFB laser with a gelatin thickness of $g = 14$ μm, and in Fig. 3(c), which shows another DFB structure with $g = 45$ μm. The line spacing of the multiline output of Fig. 3(b) is about 5 Å. We attribute the multiline nature of the output to oscillations of several modes of the gelatin film. These modes correspond to plane waves propagating in the film at discrete grazing angles which are reflected from the gelatin-air and gelatin-substrate interfaces. Their reflection losses are balanced by the gain in the medium. A film mode of mode number N will oscillate at a wavelength λ_N such that its propagation constant β_N obeys

$$K = 2\beta_N . \tag{10}$$

If $\lambda_0 \ll 2gn$ we can derive an approximate expression for the mode spacing $\lambda_N - \lambda_{N-1}$ which is

$$\frac{\lambda_N - \lambda_{N-1}}{\lambda_0} = -\frac{2N-1}{2}\left(\frac{\lambda_0}{2gn}\right)^2 . \tag{11}$$

The observed mode splitting corresponds to mode numbers of $N = 3$, 4, and 5. The lower orders are suppressed due to the finite (10 mm) length of the film.

The above observations clearly indicate DFB laser action which is due to the refractive-index modulation in the gelatin film. While this device was relatively easy to implement there are several other DFB structures which are promising possibilities. For example, one can employ the uv-induced index variations in poly(methylmethacrylate)[9] or the index damage in ferroelectrics which can be produced with a high spatial resolution.[10] Etched periodic deformations or loading of thin films offer further possibilities. Spatial variations of the current densities in semiconductor lasers are yet another means to achieve variations in the gain and the refractive index. The use of acoustic waves offers easy tunability. However, to satisfy the Bragg condition of Eq. (3) in the visible one needs acoustic frequencies of the order of 10 GHz, which poses technical problems. Finally, one can spatially modulate the pump intensity, and thereby the gain, by interfering two coherent pump beams at the proper angle. A change of the angle will then tune the oscillating wavelength. Further work is necessary to determine the advantages and disadvantages of the various DFB structures.

We would like to acknowledge, with thanks, the technical assistance of M.J. Madden.

---

[1]P.K. Tien, R. Ulrich, and R.J. Martin, Appl. Phys. Letters 14, 291 (1969).

[2]S.E. Miller, Bell System Tech. J. 48, 2059 (1969).

[3]H. Kogelnik and C.V. Shank (unpublished).

[4]H. Kogelnik, Bell System Tech. J. 48, 2909 (1969).

[5]C.V. Shank, A. Dienes, and W.T. Silfvast, Appl. Phys. Letters 17, 307 (1970).

[6]T.A. Shankoff, Appl. Opt. 7, 2101 (1968).

[7]L.H. Lin, Appl. Opt. 8, 963 (1969).

[8]T.W. Hansch, M. Pernier, and A.L. Schawlow, IEEE J. Quantum Electron. (to be published).

[9]W.J. Tomlinson, I.P. Kaminow, E. Chandross, R.L. Fork, and W.T. Silfvast, Appl. Phys. Letters 16, 486 (1970).

[10]F.S. Chen, J.T. LaMacchia, and D.B. Fraser, Appl. Phys. Letters 12, 223 (1968).

A Thin-Film Ring Laser

H. P. Weber and R. Ulrich
Bell Telephone Laboratories, Holmdel, New Jersey 07733
(Received 25 March 1971; in final form 17 May 1971)

We report the operation and characteristics of a ring laser formed by a single-mode light-guiding thin film. The rhodamine 6G doped polyurethane film is coated on the surface of a cylindrical glass rod. This geometry establishes feedback for laser oscillation around the circumference of the rod. A nitrogen laser serves as pump source.

The observation of stimulated amplification of light in thin-film light guides[1] offers the possibility of constructing thin-film lasers. Gains as high as 100 dB/cm had been observed in single-mode light-guiding films of polyurethane,[2] doped with rhodamine 6 G (Rh 6 G) and pumped with a N_2 laser. These gains were determined by measuting the amplification of the spontaneous emission[3] as described by Shaklee and Leheny.[4]

In order to provide the feedback necessary for laser action, we applied the doped light-guiding film on the surface of a cylindrical glass rod. In this way, a closed optical path is established along any circumference of the rod.[5] Laser action will then take place if the gain along the circumference exceeds the round-trip loss. A N_2 laser ($\lambda = 3371$ Å) emitting a sheetlike beam is used to pump a narrow circumferential strip of the film (Fig. 1).

The light generated in the film is coupled out by a prism-film coupler.[6,7] Because of the directional nature of this coupler we obtain two output beams, corresponding to clockwise and counterclockwise oscillation of the laser. The beams are well collimated in the vertical direction due to the narrow height (approximately 0.2 mm) of the pumped region. In the horizontal direction, they are spread over an angle of approximately 1°. This is due to the finite spectral width of the laser light and the curvature of the film. The latter results in a very short coupling length[7] and, therefore, in angular spread of the output beam. The gap width between prism and film, and hence the degree of output coupling, could be optimized by adjusting the pressure between prism and laser rod.

It may appear that there should also be radiation loss caused by the curvature of the thin-film guide.

Reprinted with permission from *Appl. Phys. Lett.*, vol. 19, pp. 38–40, July 15, 1971.

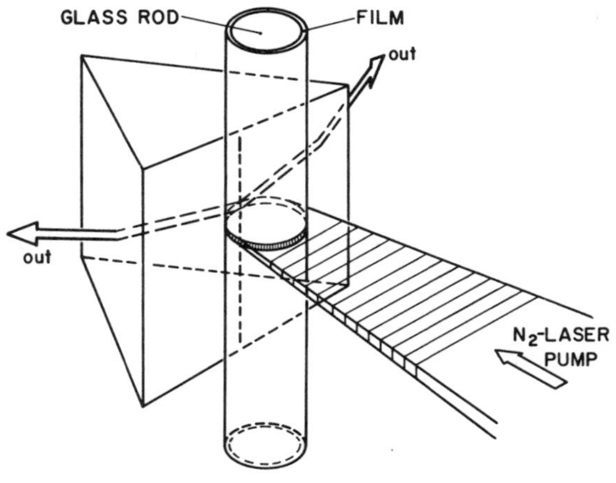

FIG. 1. Ring-laser arrangement. A 0.8-μ-thick film of polyurethane, doped with Rh 6 G, is coated on the outside of a glass rod of 5 mm diam. The beam of a N_2 laser is focused into a line, pumping a narrow section of the rod. The resulting laser light travels around the rod. It is coupled out by a prism, which is in loose contact with the rod. The two output beams correspond to clockwise and counterclockwise oscillation of the ring laser.

Such radiation would leave the laser in a tangential direction and would be highly collimated in the plane of the ring. A quantitative estimate following the theories of Marcatili and Marcuse[8] shows, however, that this output coupling mechanism is entirely negligible (far less than 10^{-10} per round trip) for the radii of curvature use here ($r \geq 0.5$ mm). Our observations are in agreement with this prediction.

For the preparation of the light-guiding polyurethane films we mixed the two compounds of the material in the ratio 1 : 1 as recommended by the manufacturer.[2] The mixture was diluted five to ten times with an organic solvent[9] in order to reduce its solid content and, thus, the thickness of the resulting film. The rhodamine, in the form of a 10-g/liter solution in ethanol, had been added to the organic solvent. The glass rod was dipped into the mixture and was air dried in a vertical position for 30 min. Heating the rod in an oven to 60 °C for 1 h cured the polyurethane. The Rh 6G concentration in the film was typically 8×10^{-3} mol/liter, as calculated from the preparation procedure. The film has a refractive index of $n_{film} = 1.55 \pm 0.01$, depending on the exact mixing ratio, and a thickness of typically 0.8 μ. When coated on the Pyrex ($n_{glass} = 1.47$) glass substrate, it can support only the fundamental modes of both polarizations, i.e., the TE_0 and TM_0 mode. We observed no gross differences between the laser action in either one of the modes.

The pump light was focused into a line of approximately 0.2 mm height and 5 mm width, matching the diameter (5 mm) of the coated rod. We measured the output power as a function of the pump power, and observed a well defined threshold. At peak intensity, a power of approximately 15 kW in 10-nsec pulses was incident on the laser rod. This corresponds to a pump intensity of 1.5 MW/cm^2. Only a small fraction of this intensity is actually used for pumping because a single-mode film is so thin that it absorbs very weakly. In addition, at both ends of the pumped region the angle of incidence of the pump light is large so that a considerable fraction of the pump radiation is reflected. We estimate that about 1 kW of the pump light is absorbed. The measured output intensity was about 100 W in each of the two emerging beams, so that the net efficiency of this laser must be quite high. It is apparent that the over-all efficiency of the laser could be considerably improved (i.e., the pump-power requirements reduced) by modifying the pump geometry so that the pump light passes repeatedly through the film, or by pumping in the green absorption band of the Rh 6G where its absorption cross section is more than an order of magnitude higher than that at 3371 Å.

The output spectrum of the ring laser peaks at 6200 Å [see Fig. 2(a)]. This relatively long wavelength is a result of the heavy doping. The bandwidth at half-intensity is 110 Å. This is much narrower than the width of the spontaneous emission of several hundred angstroms, measured in the same host material.[1] Such a narrowing of the output spectrum is also characteristic of superradiant emission. However, a direct proof of a feedback around the rod is obtained from our observation [Fig. 2(b)] of the individual modes of the ring-laser resonator.

These modes may be viewed in analogy with the modes of a plane-parallel Fabry-Perot resonator, whose mirror spacing is equal to half the circumference of the rod, and which is filled with an appropriate dielectric medium. The tangential (circumferential) modes of the ring resonator are analogous to the longitudinal modes of the Fabry-Perot resonator. Their wavelength spacing is $\Delta\lambda = \lambda^2/\pi d N_g$, where d is the diameter of the rod and $N_g = c/v_{group}$ is the effective group velocity index of the guide. Corresponding to the transverse modes of the Fabry-Perot resonator are the radial and axial modes of the thin film on the outside of the rod. In this classification, the radial modes are the well-known TE_m and TM_m modes of a thin-film guide.[6-8] Our films permitted propagation of only the fundamental ($m=0$) modes in the radial direction. In the axial direction, there are no relevant boundaries to define axial modes in the absence of pumping. However, in the pumped rod the strip of high gain is bordered at both sides by unpumped slightly lossy regions of the film defining the modes in axial direction, the fundamental one having the lowest loss.

From the uniform distribution of the light intensity across the output beams and from the diffraction-limited divergence, we conclude that we did obtain laser oscillation in the fundamental transverse (radially as well as axially) mode of the ring laser. Any nonuniformity of the thin-film guide, e.g., a dust

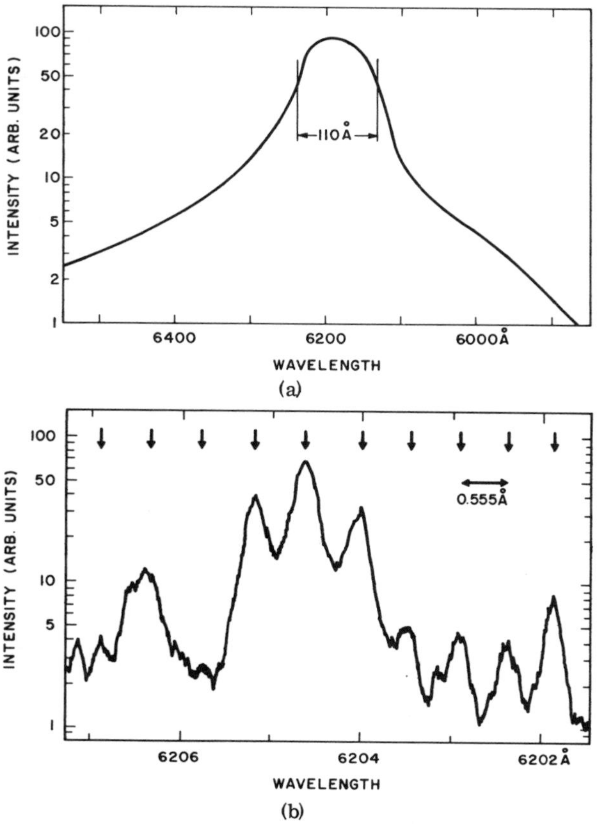

FIG. 2. Output spectrum of the laser: (a) total spectrum of the 5-mm laser rod; (b) spectrum with high resolution showing the individual longitudinal modes for a rod of 1.397 mm diam.

particle, will spoil the ideal modes described above. When pumped near such nonuniformity, the laser output beams broke up into a series of several horizontal lines. They covered a wider vertical angle than the fundamental mode. This indicates oscillation of higher-order axial modes.

From the effective index N of the guide[6,7] we compute the group velocity index as $N_{group} = 1.583 \pm 0.010$ for a wide range of film thicknesses around 0.8 μ. Assuming that the group velocity of the optically pumped system does not differ significantly from that one of the unpumped system, we calculate for the longitudinal modes at $\lambda = 6200$ Å a spacing $\Delta\lambda = 0.155 \pm 0.001$ Å for a 5-mm rod. This is difficult to resolve with a spectrometer. In order to increase the mode spacing, we pumped a thinner rod of 1.397 mm in diameter having a theoretical mode spacing of $\Delta\lambda = 0.553 \pm 0.003$ Å. The resolved output of this rod is shown in Fig. 2(b). The individual modes are clearly separated. Their measured spacing is 0.555 ± 0.005 Å, in good agreement with the theoretically expected value, In the thin rod the axial-mode structure was not as clean as in the 5-mm-thick rod. We believe that this is the reason for the weak secondary set of lines observed in Fig. 2(b). In order to prove that we have seen the longitudinal modes of the ring resonator and not the spectrum due to any structure of the Rh 6G molecule, we pumped another thin rod of 1.05 mm in diameter. This one was doped less heavily and had its peak output at 6050 Å. Its theoretical mode spacing is $\Delta\lambda = 0.70$ Å; we measured 0.71 ± 0.02 Å.

As in other host materials,[10,11] the Rh 6G showed irreversible bleaching, resulting from the pump radiation. This effect limited the laser operation to about 10^3–10^4 shots at a given spot on the laser rod. Laser action started again when the pump beam was moved to a fresh spot on the rod.

The authors are indebted to K. L. Shaklee and R. F. Leheny for making available to us their pump and detection system for this study. We also thank R. P. Reeves for his excellent technical assistance during the measurements and F. A. Dunn for constructing the prism coupling system.

[1]R. Ulrich and H. P. Weber, J. Opt. Soc. Am. 61, 1129 (1971).

[2]Epoxylite Type 9653, Epoxylite Corp., South El Monte, Calif.

[3]C. V. Shank, A. Dienes, and W. T. Silfvast, Appl. Phys. Letters 17, 309 (1970).

[4]K. L. Shaklee and R. F. Leheny, Appl. Phys. Letters 18, 475 (1971).

[5]This is one of a variety of possible thin-film configurations for ring lasers which were independently conceived by R. Rosenberg.

[6]P. K. Tien, R. Ulrich, and R. J. Martin, Appl. Phys. Letters 14, 291 (1969).

[7]R. Ulrich, J. Opt. Soc. Am. 60, 1337 (1970).

[8]E. A. J. Marcatili, Bell System Tech. J. 48, 2103 (1969); D. Marcuse, *ibid.* (to be published). We acknowledge a fruitful discussion with the two authors concerning the radiation loss.

[9]Laminar X-500 Reducer 66-C-20, produced by Midland Division, Dexter Corporation, Hayward, Calif. 94544.

[10]E. P. Ippen, C. V. Shank, and A. Dienes, IEEE J. Quantum Electron. QE-7, 178 (1971).

[11]I. P. Kaminow, H. P. Weber, and E. A. Chandross, Appl. Phys. Letters 18, 497 (1971).

Part 7
Modulators and Light Deflectors

DEFLECTION OF AN OPTICAL GUIDED WAVE BY A SURFACE ACOUSTIC WAVE

L. Kuhn, M. L. Dakss, P. F. Heidrich, and B. A. Scott
IBM T. J. Watson Research Center, Yorktown Heights, New York 10598
(Received 23 July 1970)

The experimental demonstration of deflection of an optical film-guided wave by a surface acoustic wave is reported. When Bragg conditions are satisfied, 0.18 W acoustic power gives rise to 66% deflection efficiency as measured by the depletion of the incident optical guided wave.

In this letter, we report the experimental demonstration of deflection of an optical film-guided wave by a surface acoustic wave. The layout of the experiment is shown schematically in Fig. 1, where a surface acoustic wave was propagated along an α-quartz crystal. A thin glass film deposited on the crystal surface served as an optical waveguide. The acoustic wave was not perturbed by the glass film since the film thickness was much less than an acoustic wavelength. However, the acoustic wave imposed a periodic strain on the film giving rise to deflection of the optical guided wave in the film.

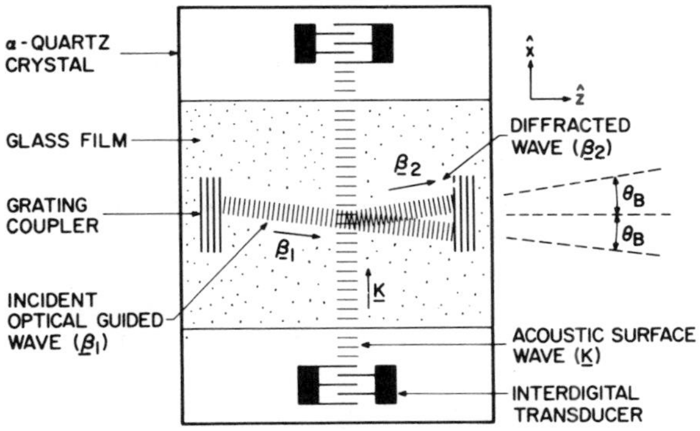

FIG. 1. Schematic layout of the experiment.

The optical waveguide was a high index ($n \simeq 1.73$) glass film[1] (thickness $d \simeq 0.8\,\mu$), which was deposited onto an α-quartz crystal by rf sputtering. A 6328-Å laser beam was coupled into the glass film by a grating coupler.[1] The grating can be thought of as diffracting the incident laser beam into the film at an angle which coincides with the characteristic angle of one of the modes of the film. The optical guided wave, thus excited, propagated approximately parallel to the z axis with a propagation factor $\exp(i\beta z)$. Using the observed coupling angle and the procedure described in Ref. 1, an experimental value $\beta = 1.67 \times 10^5$ cm^{-1} was determined. A second grating coupler was used to couple the deflected and undeflected waves out of the film. All measurements were made on the output coupled beams.

The surface acoustic wave (frequency $f \simeq 191$ MHz) propagated in the x direction on a y face of the α-quartz crystal substrate. The wave was launched and detected by interdigital transducers which had periodicity $\Lambda = 16\,\mu$, aperture $w = 0.035$ in., and ten finger pairs. The total electrical insertion loss of the acoustic channel was 23 dB. To within the accuracy of the experiment this insertion loss was found to be independent of the propagation length on the surface and was unaffected by the presence of the glass film. Series inductors were used to resonate the transducers at the operating frequency.

The deflection efficiency was measured as the fractional decrease in the output from the undeflected guided wave. Although the output from the deflected guided wave was easily visible, a measurement on the undeflected wave was found to be more accurate and convenient because of the partial overlap of the two beams. In Fig. 2 we plot the measured value of the deflection efficiency η as a function of the relative orientation θ of the optical guided wave with respect to the acoustic wave. Variation in θ was accomplished by changing the angle of incidence of the input laser beam. As θ was varied, the deflected beam was observed first on one side of the undeflected beam [region (a) in Fig. 2], then on both sides [region (c)], and finally on the opposite side [region (b)]. The maximum intensity of the deflected beam coincided with maximum depletion of the undeflected beam. We believe that the peaks (a) and (b) correspond to satisfying the Bragg conditions, as indicated in the upper part of Fig. 2. According to this interpretation, the angular separation of the two peaks 1.36°, is twice the Bragg angle, $2\theta_B$, and is equal to the angular de-

Reprinted with permission from *Appl. Phys. Lett.*, vol. 17, pp. 265–267, Sept. 15, 1970.

flection in the film. The Bragg angle is given theoretically by[2] $\theta_B = \arcsin(\frac{1}{2}\lambda_g/\Lambda)$, where $\lambda_g = 2\pi/\beta$ is the wavelength of the optical guided wave and Λ is the surface acoustic wavelength. Using the experimental value for β and $\Lambda = 16\,\mu$, we find $2\theta_B = 1.32°$, in good agreement with the observed angular separation. It is to be noted that for this experiment $w\lambda_g/\Lambda^2 \simeq 1.3$, which indicates[2] that the scattering is slightly within the Bragg regime $w\lambda_g/\Lambda^2 > 1$; this is consistent with the observations discussed here.

The efficiency when Bragg conditions are satisfied (η_B) was measured as a function of electrical input power P_{el}. The results are indicated in Fig. 3. The maximum observed efficiency was 66%; this occurred for an electrical input power of 2.5 W, or an acoustic power of 0.18 W. At higher power levels the transducer was destroyed by arcing between the fingers. Higher values of efficiency for smaller electrical power should be achievable with a better transducer design.

The electrical input power is only an indirect measure of the amplitude of the acoustic wave. Using the conventional optical probing technique,[3,4] the acoustic strain in the interaction region was measured directly. In this technique, a laser beam reflected from the surface is diffracted by the rippling of the surface associated with the acoustic wave. The probe efficiency η_P, defined as the ratio of the intensity of either of the first-order diffracted beams (in reflection) to that of the direct reflected beam, is a measure of the normal component of the surface displacement. Any component of the acoustic strain can be determined using Ref. 5. In Fig. 3 the probe efficiency η_P resulting from a measurement in the interaction region is plotted as a function of electrical input power P_{el}. The fact that η_P varies linearly with P_{el} indicates that no nonlinear acoustic effects, such as harmonic generation, are responsible for the saturation of η_B seen in Fig. 3.

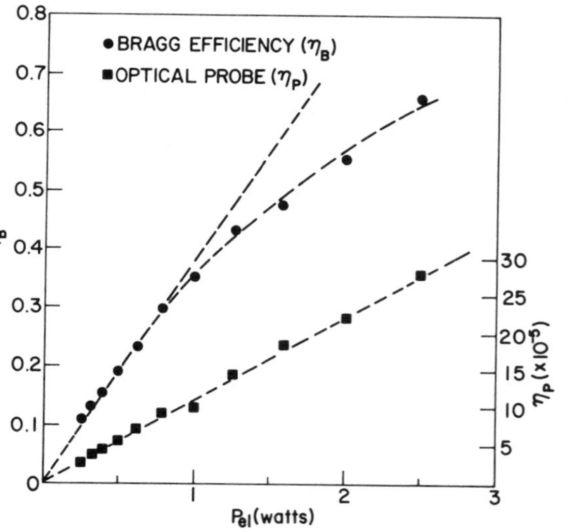

FIG. 3. Dependence of the Bragg deflection efficiency η_B and the optical probe efficiency η_P on electrical input power P_{el}.

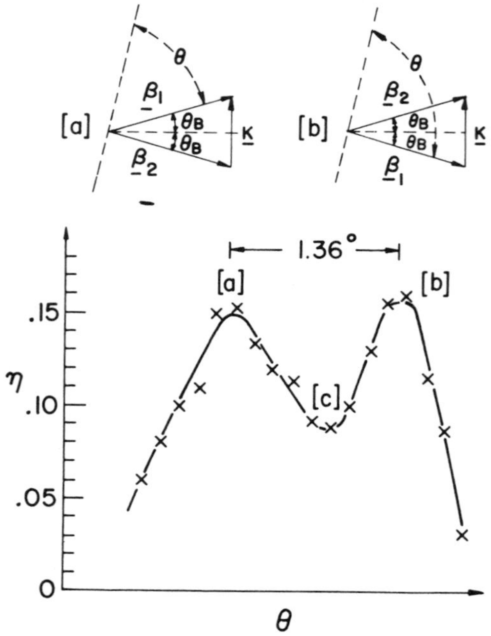

FIG. 2. Dependence of the deflection efficiency η on the relative orientation of the optical guided wave ($\vec{\beta}_1$) with respect to the fixed acoustic surface wave ($\vec{K}$). Electrical input power was $P_{el} = 0.4$ W; θ was measured from an arbitrary axis (dotted line in inset).

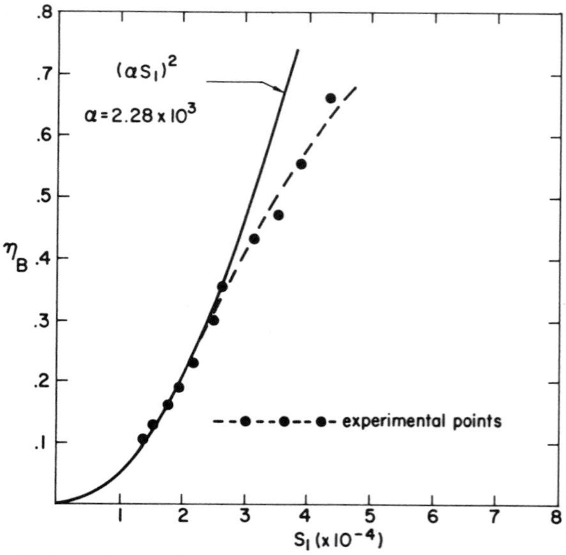

FIG. 4. Bragg deflection efficiency plotted as a function of acoustic strain S_1 as determined from optical probe measurements. The experimental points at the higher strain levels fit equally well to the theory of Refs. 2 and 6.

The optical guided-wave deflection efficiency can be estimated theoretically. Assuming that the compressive strain S_1, measured on the glass surface, is uniform throughout the film, we can treat the acoustic strain imposed on the film as a plane wave in the film. This type of interaction is treated theoretically in Refs. 2 and 6. Although the results described in those references are somewhat dissimilar, for small acoustic strain both yield a deflection efficiency $\eta_B = (\alpha S_1)^2$ where α is given by $\alpha = \frac{1}{4} n^2 P_{11} w \beta / \cos\theta_B$ and P_{11} is a photoelastic constant for the glass.

In Fig. 4 we have plotted the Bragg efficiency η_B as a function of the strain S_1 obtained from the optical probe data of Fig. 3. For low efficiencies the points fit well to a theoretical curve $\eta_B = (\alpha S_1)^2$ with $\alpha = 2.28 \times 10^3$. The theoretical and experimental values of α are in agreement for $P_{11} = 0.2$, a value which is consistent with various measurements and models for the photoelectric constants of glasses.[7] The saturation at high efficiencies is described theoretically in Refs. 2 and 6. Our data do not go to high enough strain levels to show which model would be most appropriate for describing our experiment.

It is expected that with further development, the type of interaction studied here may be utilized in practical integrated acousto-optic devices for a variety of applications, such as switching (deflection) and analog data processing.[8]

The authors are grateful to E. G. H. Lean for many helpful suggestions and discussions.

---

[1]M. L. Dakss, L. Kuhn, P. F. Heidrich, and B. A. Scott, Appl. Phys. Letters 16, 523 (1970).

[2]C. F. Quate, C. D. W. Wilkinson, and D. K. Winslow, Proc. IEEE 53, 1604 (1965).

[3]E. G. H. Lean, C. C. Tseng, and C. G. Powell, Appl. Phys. Letters 16, 32 (1970).

[4]E. G. H. Lean and C. G. Powell, Proc. IEEE (to be published).

[5]K. A. Ingebrigsten and A. Tonning, Appl. Phys. Letters 9, 16 (1966).

[6]H. Kogelnik, Bell System Tech. J. 48, 2909 (1969).

[7]D. A. Pinnow, IEEE J. Quantum Electron. QE-6, 223 (1970).

[8]R. Shubert and J. H. Harris, IEEE Trans. Microwave Theory Tech. MTT-16, 1048 (1968).

Voltage-Induced Optical Waveguide

D. J. Channin
RCA Laboratories, Princeton, New Jersey 08540
(Received 21 May 1971)

An optical waveguide based on electro-optic index changes in $LiNbO_3$ has been constructed. Voltage and polarization dependence of the waveguide mode structure has been observed and is described qualitatively with a simple theory.

Optical waveguide techniques[1] including modulation and switching techniques[2,3] have attracted considerable interest because of their applicability to optical communication and logic. We have demonstrated a new means for electrically generating a controllable low-loss optical waveguide structure within an electro-optic crystal. The observed mode structure and polarization phenomena may be qualitatively understood with a simple dielectric waveguide analysis.

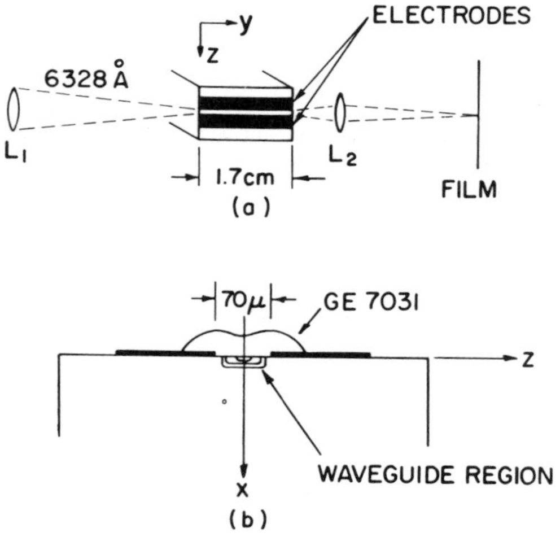

FIG. 1. (a) Experimental arrangement. (b) Cross section of optical waveguide. GE 7031 varnish prevents arcing.

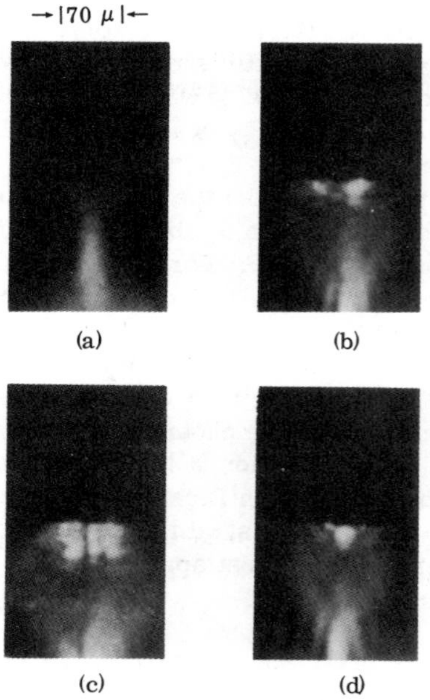

FIG. 2. (a) 100 V, z-polarized light, no waveguide modes. (b) 300 V, z-polarized light, onset of waveguide modes. (c) 450 V, z-polarized light, high-order waveguide modes. (d) 700 V, x-polarized light, low-order waveguide modes.

The experiment is shown in Fig. 1. Two parallel electrodes are deposited on the optically polished x-cut surface of lithium niobate. A voltage difference applied to the electrodes produces a high electric field concentration in the crystal directly below the 0.007-cm gap separating the electrodes. (The voltage is applied as a 60-cycle sine wave to prevent permanent refractive index changes.[4] Laser light at 6328 Å is focused by lens L_1 into one end of this high-field region and passed out the other through polished ends of the crystal. The near-field pattern is focused by lens L_2 onto a photographic plate.

The optical waveguide is formed in the localized region of modified refractive index produced by the electric field through the electro-optic effect. On the half-cycle for which the index change is positive the high index region is bounded by lower index material outside the high electric field. Provided the index change is sufficiently large, the high index region will support waveguide modes. The depth to which the waveguide extends into the material from the surface should increase with the electrode voltage.

The simplest description of such a waveguide is the anisotropic planar slab model. In applying this model to our experiment, we neglect completely electric field variation in the z direction. Along the x axis centered in the middle of the gap, only $E_z(x)$ is nonzero and is approximated by

$$E_z(x) = V/2aK_z, \quad 0 \le x \le d, \qquad = 0, \quad x > d, \tag{1}$$

where V is the electrode voltage, $2a = 0.007$ cm is the electrode gap, and d is the observed depth the waveguide modes extend below the surface. The factor $1/K_z$ accounts for the dielectric constant change at the crystal surface. The electro-optic index changes for light polarized parallel to the z and x axes are

$$\Delta n_z = \tfrac{1}{2}(n_z^3 r_{33}) E_z(x),$$

Reprinted with permission from *Appl. Phys. Lett.*, vol. 19, pp. 128–130, Sept. 1, 1971.

$$\Delta n_x = \tfrac{1}{2}(n_x^3 r_{13}) E_z(x). \tag{2}$$

In $LiNbO_3$, $r_{33} = 30.0\times10^{-12}$m/V, $r_{13} = 10\times10^{-12}$m/V, $n_x = 2.30$, and $n_z = 2.21$.

In this planar slab model for optical waveguides, z-polarized light is guided as TE modes and x-polarized light as TM modes. For a highly anisotropic guide such as ours, the traverse propagation constants q_x (describing the spatial variation of the guided waves along the x axis) for both nonradiating TE and TM modes are degenerate and satisfy[5]

$$\tan(q_x d) = -\frac{q_x/\beta_0(2n\Delta n)^{1/2}}{(1 - q_x^2/\beta_0^2 2n\Delta n)^{1/2}} \tag{3}$$

and

$$q_x/\beta_0 < 2n\Delta n, \tag{4}$$

where β_0 is the propagation constant for free space, n and Δn are the appropriate index and electro-optic index change, and d is the depth the guide extends into the crystal. The number of solutions of (3) and (4) at a given voltage represents the number of lobes one should see on the near-field pattern.

Near-field patterns for various voltages and polarizations are shown in Fig. 2. With the light polarized in the z direction and the voltage increasing from 0 to 200 V, a pattern due to electro-optic lensing is seen [Fig. 2(a)]. Above 200 V, a pattern of two bright and discrete spots appears just below the crystal surface [Fig. 2(b)]. With increasing voltage, the number of spots increases and the pattern extends deeper into the crystal [Figs. 2(c) and 3]. When the light is polarized in the x direction, the pattern behaves similarly with voltage except that its onset is at nearly 600 V [Fig. 2(d)]. That these spot patterns represent waveguide modes localized within the crystal may be demonstrated by translating the incident light beam across the entrance face. The lensing pattern shifts while the waveguide pattern remains fixed in position. The calculated number of lobes is shown in Fig. 3 and is in approximate agreement with the observed near-field mode patterns. The reason that the onset of x-polarized modes occurs at nearly three times the voltage for the appearance of the z-polarized modes is apparent from Eq. (1), since this ratio is that of the appropriate electro-optic coefficients r_{13} and r_{33}. Effectively, the strength of the guide changes with polarization.

Finally, there is no apparent scattering loss in the form of a bright streak of light scattered from the sides of the guide, as is usually seen in waveguides made of deposited dielectric films. This is quite

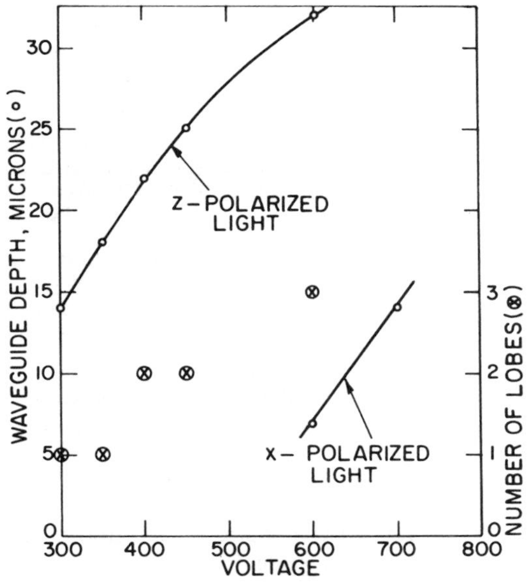

FIG. 3. Dots are observed waveguide depth along x axis. Crosses are calculated number of lobes for z-polarized light and the observed depth values.

understandable since in the electro-optic guide, light is transmitted through an optical quality crystal. The serious materials problem of producing optical waveguides free of light-scattering inhomogeneities is thereby reduced.

In conclusion, we note that this electrically controlled waveguide was produced very simply by depositing electrodes onto bulk electro-optic material. Therefore, complex optical switching, gating, and modulating systems should be readily constructed using standard photolithography. High-efficiency holographic[6] or prism[7] couplers should be directly applicable to this type of guide, in addition to the end-fire technique used here.

The author wishes to acknowledge valuable discussions with J.M. Hammer and the technical assistance of J. Minville.

[1]James E. Goell and Robert D. Standley, Proc. IEEE 58, 1504 (1970).
[2]David Hall, Amnon Yariv, and Elsa Garmire, Appl. Phys. Letters 17, 127 (1970).
[3]J.M. Hammer, Appl. Phys. Letters 18, 147 (1971).
[4]F.S. Chen, J. Appl. Phys. 38, 3418 (1967).
[5]W. E. Anderson, IEEE J. Quantum Electron. QE-1, 228 (1965).
[6]L. Kuhn, M.L. Dakss, P.F. Heidrich, and B.A. Scott, Appl. Phys. Letters 17, 265 (1970).
[7]P.K. Tien, R. Ulrich, and R.J. Martin, Appl. Phys. Letters 14, 291 (1969).

OBSERVATION OF PROPAGATION CUTOFF AND ITS CONTROL IN THIN OPTICAL WAVEGUIDES*

David Hall, Amnon Yariv, and Elsa Garmire

Division of Engineering and Applied Science, California Institute of Technology, Pasadena, California 91109
(Received 8 June 1970; in final form 15 June 1970)

The first observation of optical cutoff in thin-film waveguides is reported. The waveguides consist of thin ($\sim 10\mu$) epitaxial layers of high-resistivity GaAs deposited on lower-resistivity GaAs substrates. The optical cutoff is controlled through the electro-optic effect by applying an electric field across the epitaxial layer.

Guiding and electro-optic modulation of light in thin epitaxial semiconductor films has recently been demonstrated.[1] In this paper we report the first observation and control of optical cutoff in such waveguides.

The optical waveguide consists of a GaAs high-resistivity epitaxial layer ($\sim 12\ \mu$) sandwiched between a metal film and lower-resistivity GaAs substrate as shown in Fig. 1. The existence of confined modes is due to a discontinuity Δn of the index of refraction at the epitaxial layer–substrate interface.[1]

The theory describing the propagation of modes in this structure can be adapted from that of the symmetric dielectric waveguide.[2] The symmetric guide can support, in general, two types of modes: *TE* waves where $\vec{E}$ is parallel to y, and *TM* waves in which $\vec{E}$ is parallel to x. The existence of a conducting plane (metal film) at $x = 0$, however, limits the *TE* and *TM* modes in our case, to those possessing odd symmetry about $x = 0$. These can be written as

$$E_y(TE) \propto \sin(hx)\exp(-i\beta z),$$
$$H_y(TM) \propto \sin(hx)\exp(-i\beta z), \quad |x| < t \tag{1}$$

$$E_y(TE) \propto \exp[-p(x-t) - i\beta z],$$
$$H_y(TM) \propto \exp[-p(x-t) - i\beta z], \quad |x| > t. \tag{2}$$

The lowest-order TE_1 and TM_1 (the numerical subscript gives the number of zero crossings in the interval $|x| < t$) modes can exist only if the condition

$$(2n_0\Delta n)(2\pi t/\lambda_0)^2 > (\pi/2)^2 \tag{3}$$

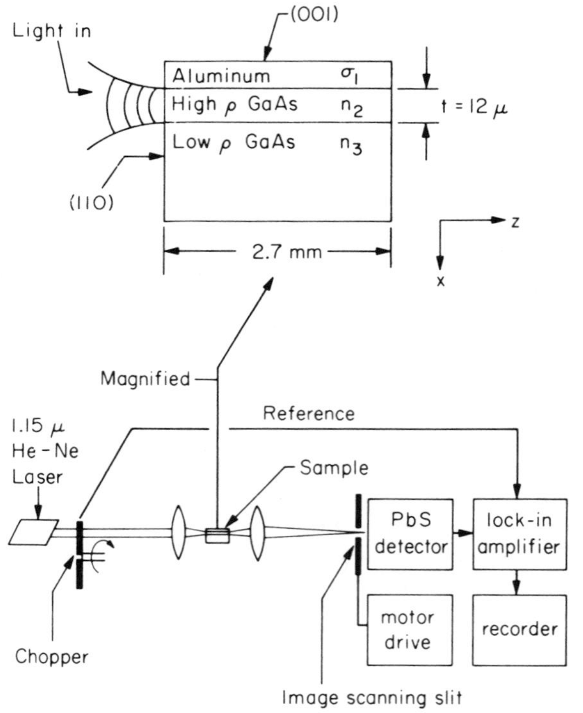

FIG. 1. The experimental setup.

Reprinted with permission from *Appl. Phys. Lett.*, vol. 17, pp. 127–129, Aug. 1, 1970.

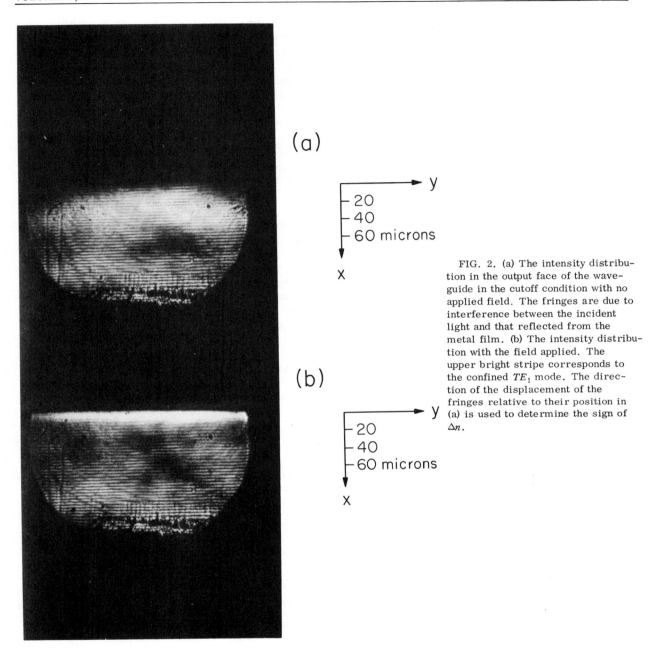

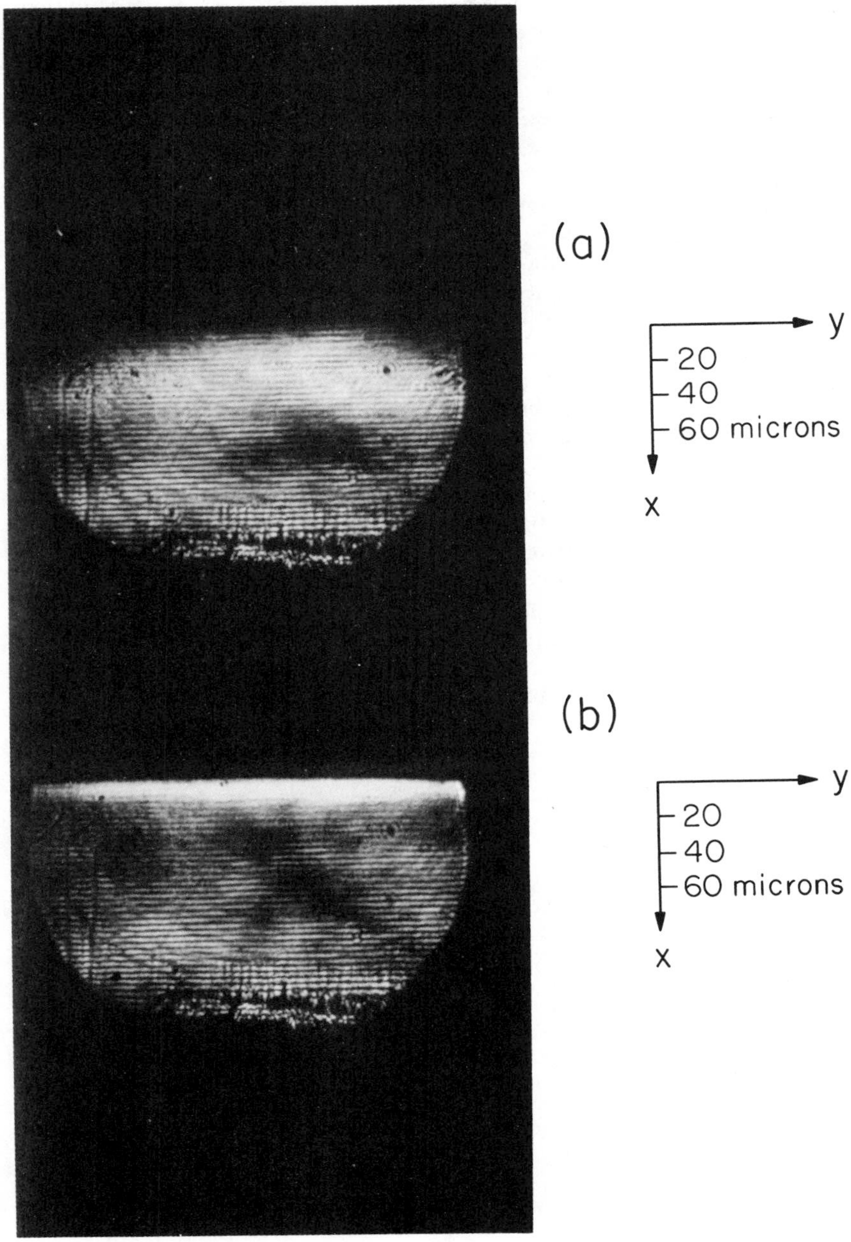

FIG. 2. (a) The intensity distribution in the output face of the waveguide in the cutoff condition with no applied field. The fringes are due to interference between the incident light and that reflected from the metal film. (b) The intensity distribution with the field applied. The upper bright stripe corresponds to the confined TE_1 mode. The direction of the displacement of the fringes relative to their position in (a) is used to determine the sign of Δn.

is satisfied. When the sign of the inequality in (3) is reversed, the field intensity increases with x so that confined propagation does not exist. This condition is referred to as cutoff.

The index discontinuity Δn at the interface $x=t$ is due, in our experiment, to two mechanisms. The first is the dependence of the index on the doping level. This discontinuity, which we denote as $\Delta n_{\text{chemical}}$, is known to lead to mode confinement in p-n junctions.[3,4] In our guide $\Delta N \approx 2\times 10^{16}$ cm^{-3} and we estimate $\Delta n_{\text{chemical}} \sim 10^{-4}$. The second contribution is due to the linear electro-optic effect in GaAs and is proportional to the reverse bias V applied to the metal-semiconductor junction. For the crystal orientation shown in Fig. 1 the electro-optic contribution to the index of a wave polarized along y is

$$\Delta n = n_0^3 r_{41} E_x/2,$$

and is zero for waves polarized along x.[5] We can consequently, write condition (3) for confined propagation as

$$\Delta n_{\text{chemical}} + \frac{n_0^3 r_{41} V}{2t} > \frac{1}{32 n_0}\left(\frac{\lambda_0}{t}\right)^2, \tag{4}$$

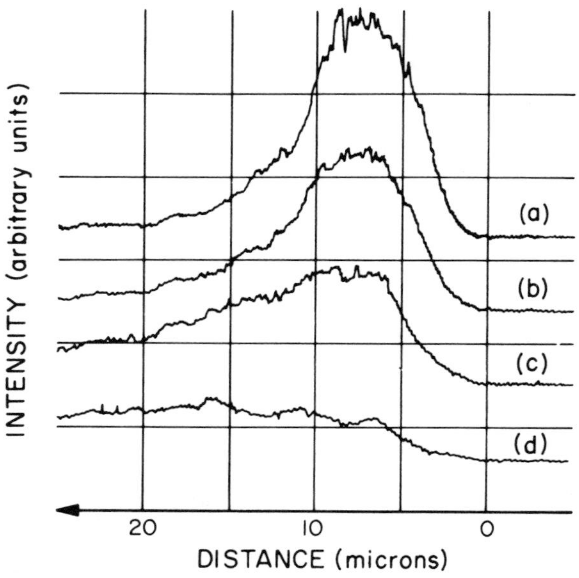

FIG. 3. The intensity distribution of the TE_1 mode measured with applied voltages of (a) 130, (b) 100, (c) 70, and (d) 0 V.

for the TE mode and as

$$\Delta n_{\text{chemical}} > (1/32n_0)(\lambda_0/t)^2, \qquad (5)$$

in the case of the TM wave. r_{41} is the electro-optic coefficient of GaAs and the applied voltage is $V = E_x t$.

The doping level of the substrate ($N \sim 2\times 10^{16}$ cm^{-3}) and the thickness $t = 12\ \mu$ were chosen in our experiment so that at $\lambda_0 = 1.15\ \mu$ condition (5) was not fulfilled and no confined modes can exist. The application of a voltage V increases Δn by adding, as indicated in (4), an electro-optic contribution making it possible for a confined TE wave to exist. A magnified image of the output face of the crystal with and without an applied bias ($V = 130$ V) is shown in Fig. 2. The existence of a confined mode with the voltage on is clearly evident. In addition, we made the following observations: (1) When the optical polarization is rotated by 90° ($\vec{E} \parallel x$) so as to excite a TM wave, no confined mode exists with or without an applied field. This is consistent with the fact that there is no electro-optic contribution to Δn in the case of the TM wave. (2) By changing the crystal orientation we can reverse the sign (see caption under Fig. 2) of the electro-optic contribution to Δn as "seen" by the TE mode from (+) to (−). When this is done the application of a voltage does not lead to a confined mode in agreement with (4).

The gradual onset of confinement with increasing voltage is shown by the intensity profile plots in Fig. 3. The dependence of the guided intensity on the applied voltage can be used for modulation purposes.

In summary: The phenomena of propagation cutoff in thin optical waveguides is demonstrated. A continuous electro-optic control of the cutoff condition is used to demonstrate its effect on the intensity distribution of the dominant TE mode.

The authors wish to thank A. Shuskus and W. Oshinsky of United Aircraft Research Laboratories in East Hartford, Conn. for supplying the GaAs samples.

---

*Research supported by the Office of Naval Research and by the Advanced Research Projects Agency through the Army Research Office, Durham, N.C.

[1]D. Hall, A. Yariv, and E. Garmire, Opt. Commun. May 1970 (to be published).
[2]See, for example, R.E. Collin *Field Theory of Guided Waves* (McGraw-Hill Book Co., New York, 1960).
[3]A. Yariv and R.C.C. Leite, Appl. Phys. Letters 2, 55 (1963). Also W.L. Bond, B.G. Cohen, R.C.C. Leite, and A. Yariv, *ibid* 2, 57 (1963).
[4]D.F. Nelson and F.K. Reinhart, Appl. Phys. Letters 5, 148 (1964).
[5]S. Namba, J. Opt. Soc. Am. 51, 76 (1961).

Efficient GaAs-$Al_xGa_{1-x}As$ Double-Heterostructure Light Modulators

F. K. Reinhart
Bell Telephone Laboratories, Murray Hill, New Jersey 07974

and

B.I. Miller
Bell Telephone Laboratories, Holmdel, New Jersey 07733
(Received 30 September 1971)

Properly designed GaAs-$Al_xGa_{1-x}As$ double heterostructures produce strong optical waveguides. The propagation constants of the waveguide modes can be readily modulated by the linear electro-optic effect. Measurements at a wavelength $\lambda = 1.153$ μm have yielded a phase modulation of 180° with −10 V applied bias to a device only 1 mm long. The power necessary to phase modulate light at $\lambda \approx 1$ μm by 1 rad is of the order of 0.1 mW per 1-MHz bandwidth. The power dissipation is very strongly dependent on wavelength. At present, the high-frequency modulation is limited by the series resistance and capacitance of the device. The highest cutoff frequency determined thus far, ≈ 4 GHz, is considerably lower than that calculated based on the geometry and material properties.

Significant phase modulation of visible light was achieved in reverse-biased GaP *p-n* junctions due to a naturally occurring optical dielectric waveguide and a sizable linear electro-optic or Pockels effect.[1] Similar effects are expected in cw laser-type GaAs-$Al_xGa_{1-x}As$ double heterostructures (DH) because of a very pronounced and controllable optical dielectric waveguide[2,3] coupled with a strong Pockels effect.[4] The optical waveguide of the DH structures may lead to a much tighter coupling of the optical field to the junction field than that characteristic of GaP *p-n* junctions. Since this interaction can be controlled in the DH structures, very strong phase modulation may result in the red and infrared portion of the spectrum.

The *p-n* junction of DH lasers usually occurs at one of the boundaries of the high-optical-dielectric-constant region (see Figs. 1 and 3 of Ref. 2). Under reverse bias the boundaries of the depletion layer move, respectively, further into the regions having the high and low optical dielectric constants, n_1^2 and n_2^2. The penetration depth of the depletion layer essentially depends on the doping profiles. The junction electric field E_j perturbs the optical-dielectric-constant profile via the Pockels effect. This perturbation is very small compared to the fractional-optical-dielectric-constant step, $\Delta = (n_1^2 - n_2^2)/n_1^2$, which is of the order of 0.1 for typical DH lasers.[3] Significant perturbations of the mode parameters will result only if the high-optical-field region coincides with the depletion layer. From this argument it follows that efficient interaction can be achieved by reducing the high-optical-dielectric-constant region having a width $2w$, and by having the depletion layer extend over $2w$. These conditions may also be obtained, in principle, with *p-i-n*, *p-i-p*, or *n-i-n* structures, where the *i* layer is essentially coincident with the high-optical-dielectric-constant layer.

The DH lasers or modulators were grown on a [100] substrate of *n*-GaAs by liquid-phase epitaxy.[5,6] The first and third layer consist of $Al_{0.3}Ga_{0.7}As$ of *n* and *p* type, respectively. Typical doping levels for these layers are $(1-5)\times10^{17}$ cm^{-3}. The second layer is typically *n*- or *p*-type GaAs with doping levels ranging from 10^{16} to 10^{17} cm^{-3} as determined from junction-capacitance measurements. The growth of *i* layers by liquid-phase epitaxy appears quite feasible at present. We have obtained π-ν—type layers instead of *i* layers.

TABLE I. Characteristic data of a DH junction phase modulator.

	Calculated						Measured
	λ (μm)	w (μm)	C (pF)	Orientation [100]	Orientation [110]	Orientation [111]	Orientation [100]
P_O (mW/MHz rad²)	1.153	0.203	14.0	0.22	0.22	0.16	...
(Phase modulation)	0.9	0.152	18.8	0.077	0.077	0.058	...
P_{OP} (mW/MHz rad²)	1.153	0.203	14.0	0.22	0.055	0.073	0.21
	0.9	0.152	18.8	0.077	0.019	0.026	...
(Polarization modulation)							
	1.153	0.203	14.0	182	364	315	173
M_{00} (deg/V cm)	0.9	0.152	18.8	354	708	613	...

Reprinted with permission from *Appl. Phys. Lett.*, vol. 20, pp. 36–38, Jan. 1, 1972.

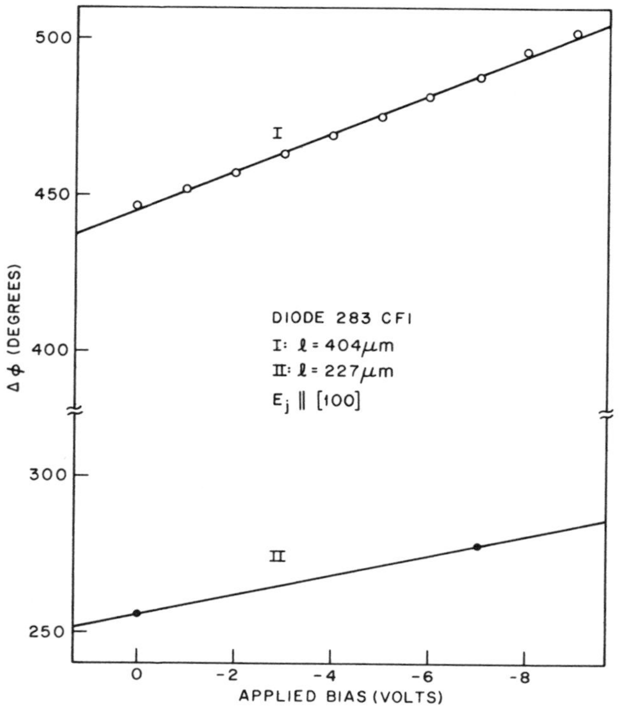

FIG. 1. Phase difference vs applied bias at $\lambda = 1.153$ μm for two lengths of the same diode.

In such structures depletion-layer widths up to 0.6 μm have been observed.

The presence of the junction electric field lowers the symmetry of the crystal, and thus the originally isotropic crystal becomes birefringent in the depletion-layer region and the resulting propagation constant is orientation dependent. The orthogonally polarized modes are excited by means of a focused linearly polarized light beam, the polarization of which is oriented at 45° to the junction electric field. The elliptically polarized light characterized by the phase difference $\Delta\varphi$, emerging from the exit face of the modulator, is collimated by a second lens and passed throught an Ehringhaus or Sénarmont compensator and polarizing filter arrangement. The polarization modulation is measured either visually with the aid of an ir converter or through a slit positioned at the real image of the exit face and a photomultiplier tube. In the latter case, a slit is desirable in order to discriminate against the continuum mode background.[1] The background can be reduced by proper excitation of the waveguide modes also. With the aid of oil immersion, up to 90% of the transmitted light has been excited as discrete waveguide modes. In order to obtain such a high-mode excitation efficiency, it is necessary to carefully match and align the incoming focused beam with respect to the waveguide. The adjustment of position vertical to the junction plane may be critical to about ± 0.1 μm.

In Fig. 1, $\Delta\varphi$ measurements are shown for a unit with $2w = 0.14 \pm 0.02$ μm and E_j parallel to the crystallographic [100] direction and with the light ($\lambda = 1.153$ μm) traveling along the [0$\bar{1}$1] direction. In order to determine $\Delta\varphi$ between the TE and TM modes on an absolute basis, the diode was measured in the same area before and after cleaving, hence for two different lengths (I) and (II). The ratio of the $\Delta\varphi$ values agrees with the length ratio well within the accuracy of measurement (± 5%). The large $\Delta\varphi$ value at $E_j \approx 0$, $\Delta\varphi_0$, is due to the inherent different propagation constants of the TE and TM modes in the waveguide apart from the effects of birefringence. From $\Delta\varphi$, the value present at full forward bias (no "junction field"), a fairly accurate determination of the effective Δ is obtained.[3] For this case a value of $\Delta = 0.094$ is deduced. It is noteworthy that such modulator devices may also be useful as simple static phase plates.

We have also studied the intensity distribution of the light across the real image of the exit face. The following results are quite remarkable: (i) The shape of the planar waveguide modes appears to be independent of the additional phase shifts imposed by the compensator; (ii) very high extinction ratios (up to 20 dB) have been found; and (iii) occasional focusing of the modes within the plane of the junction occurs. This latter fact would be very desirable from a device point of view, if it can be controlled.

It is customary to compare phase modulators by the power they consume in order to obtain 1 rad of phase modulation per 1 MHz of bandwidth.[4,7] Typical projected data for fundamental modes are compiled in Table I for various crystallographic orientations of E_j with λ as a parameter. For the calculation it was assumed that $2w$ is identical with the depletion-layer width. The values shown for phase modulation apply for $E_j \parallel$ [100] for the TE mode, $E_j \parallel$ [111] for the TM mode, and in the case of $E_j \parallel$ [110] to either of the hybrid modes.[1] The calculations were performed to obtain the least power per unit length of light travel similar to Ref. 8 yielding optimum values of w. A length, $l = 1$ mm, a width along the junction plane, $b = 50$ μm, and $\Delta = 0.1$ were assumed. Such dimensions are in the range of what can be readily obtained and handled. In order to use such narrow modulators the exciting beam has to be astigmatic.[9] With astigmatic beams the light interaction can, in principle, be made arbitrarily long. Using arguments similar to the ones in Ref. 4, it can be shown that the characteristic modulation power P_0 decreases with $l^{-1/2}$, if a simple focused Gaussian distribution of the light is assumed within the junction plane. However, the capacitance C of the device is proportional to $l^{3/2}$ for astigmatic Gaussian light distributions. This dependence contrasts to the diffraction-limited modulator.[4,10] From this it becomes obvious that light confinement in the junction plane is highly desirable because $P_0 \propto l^{-1}$. In addition, b can be chosen as narrow as a few micrometers.

In Table I we have also compiled the characteristic power P_{OP} in order to obtain a phase difference of 1 rad per 1-MHz bandwidth between the orthogonal modes and the corresponding characteristic phase difference per volt and per length, M_{00}. For the case of $E_j \parallel$ [100] and $\lambda = 1.153$ μm, the theoretical predictions have already been verified. It should be emphasized that polarization modulation is readily converted into intensity modulation, by an output polarizer and that 1 rad of polarization modulation corresponds to a linearized intensity modulation of 88%. We have constructed a modulator of $l = 1.04$ mm with a half-wave-voltage of only

10 V and an extinction ratio approaching 20 dB. If the modulator is used in a double-pass fashion, an additional reduction of power by a factor of 2 is possible.

The high-frequency performance of electrical-junction modulators has been discussed in detail in Ref. 7. The situation is expected to be quite similar in DH junction modulators. The series-resistance cutoff frequency f_c and the bandwidth limitation due to the burn-out power level are expected to be higher in DH modulators than in the GaP diodes. Capacitance measurements of DH modulators yielded $f_c \approx 4$ GHz. It is expected that this value can be improved significantly by applying better contacts.

In conclusion, the DH modulator proves to be a phase, polarization, or intensity modulator for the infrared portion of the spectrum of unparalleled efficiency. An even better modulator performance is expected in the red portion of the spectrum by adequately adjusting the Al content of the layers.

We wish to thank R. C. Miller for continuous encouragement, and J. J. Schott, R. Capik, and P. W. Foy for their technical assistance.

[1]F. K. Reinhart, D. F. Nelson, and J. McKenna, Phys. Rev. 177, 1208 (1969).
[2]I. Hayashi, M. B. Panish, and F. K. Reinhart, J. Appl. Phys. 42, 1929 (1971).
[3]F. K. Reinhart, I. Hayashi, and M. B. Panish, J. Appl. Phys. 42, 4466 (1971).
[4]I. P. Kaminow and E. H. Turner, Proc. IEEE 54, 1374 (1966).
[5]M. B. Panish, S. Sumski, and I. Hayashi, Met. Trans. 2, 795 (1971).
[6]B. I. Miller, J. E. Ripper, J. C. Dymant, E. Pinkas, and M. B. Panish, Appl. Phys. Letters 18, 403 (1971).
[7]F. K. Reinhart, J. Appl. Phys. 39, 3426 (1968).
[8]W. G. Oldham and Ali Bahraman, IEEE J. Quantum Electron. QE-3, 278 (1967).
[9]Beams emerging from GaAs lasers are naturally astigmatic. Gaussian beams from gas lasers such as used in the present investigation are readily made astigmatic by introducing a cylindrical lens of suitable focal length at an appropriate place in the incident light beam.
[10]R. T. Denton, F. S. Chen, and A. A. Ballman, J. Appl. Phys. 38, 1511 (1967).

DIGITAL ELECTRO-OPTIC GRATING DEFLECTOR AND MODULATOR

J. M. Hammer
RCA Laboratories, Princeton, New Jersey 08540
(Received 22 October 1970)

A new method of laser light deflection and modulation based on diffraction by electro-optic phase gratings has been demonstrated. The method is applicable to thin-film light guides, requires relatively low power for high-speed operation, and is capable of high diffraction efficiency.

A novel method of producing laser light deflection and modulation has been demonstrated. The method uses simple conducting electrode patterns deposited on thin slabs or films of electro-optic crystals to produce voltage-controlled optical phase or polarization gratings. This approach yields high diffraction efficiency and modulation depth with modest power consumption and is readily adapted to deflect and modulate light traveling in thin-film optical wave guides.[1] In addition, a separate electrode pattern may be provided for each output light position to obtain digital control.

The experimental arrangement is shown in Fig. 1. A z-cut [001] $LiNbO_3$[2] wafer is provided with columns of electrodes on the top surface and a ground plane on the bottom. The elements of a single column are connected by a common bus. Voltage applied between a single column and the ground plane produces the grating characteristic of that column. The y periodic electric field causes periodic variation in refractive index for a sheet of light traveling in the x direction. The z polarization component of the light encounters a simple periodic phase change. Thus, light is diffracted into grating orders along the y direction. The diffracted light is a sheet in the same plane as the undiffracted light.

For the case of rectangular electrons spaced so that $a = d/2 = t$ (see Fig. 1) the index variation in the y direction will be that of a "square wave" near the upper surface. Near the ground plane the index will be uniform. In the mid-plane, the in-

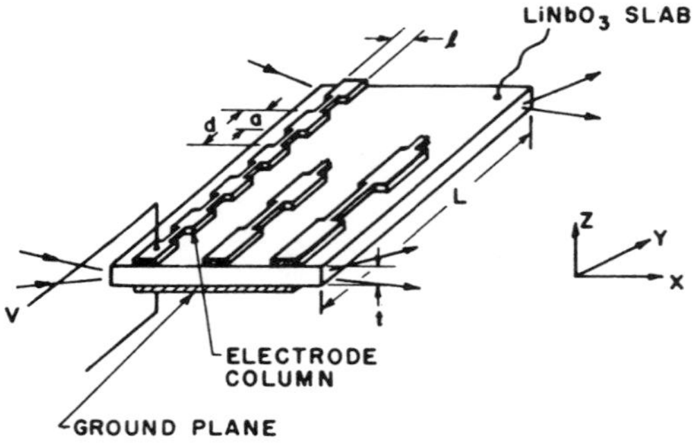

FIG. 1. Digital electro-optic grating. Three electrode columns are shown. Light travels along the x axis. Application of voltage to one of the columns diffracts the light parallel to the y direction. The intensities of the various diffraction orders are characteristic of the periodicity of the particular column and of the magnitude of the voltage.

dex variation will approximate a sinusoidal function. By solving for the fields for each of these regions and taking a linear average of intensities, the relative far-field intensity variation of the grating orders is estimated by Eq. (1),[3]

$$I/I_0 = \begin{cases} \frac{1}{3}\{1 + [J_0(\phi)]^2 + (1+\cos\phi)/2\}; & n = 0 \\ \frac{1}{3}\{[J_n(\phi)]^2 + 2(1-\cos\phi)/n^2\pi^2\}; & n = 1, 3, 5, \cdots, \\ \frac{1}{3}[J_n(\phi)]^2; & n = 2, 4, 6, \cdots \end{cases} \quad (1)$$

where ϕ is the relative phase shift produced by the electro-optic effect. For a transverse linear

Reprinted with permission from *Appl. Phys. Lett.*, vol. 18, pp. 147–149, Feb. 15, 1971.

electro-optic material with half-wave voltage $V_{\lambda/2}$ and applied voltage V,

$$\phi = \pi(l/t)(V/V_{\lambda/2}). \qquad (2)$$

In broadband pulse operation it is difficult to match the load to the generator. Thus, the reactive power P is a realistic measure of the required operating power. Near-optimum conditions are obtained when $\phi = \pi$. Here $V = (t/l)V_{\lambda/2}$ and the reactive power may be shown to be

$$P = \epsilon_0 \epsilon_r (Lt/4l) V_{\lambda/2}^2 f. \qquad (3)$$

ϵ_r is the relative dielectric constant, ϵ_0 is the permittivity of space, L is the length of the optical aperture in the y direction, and f is the average pulse repetition rate. As a comparison, the voltage required to obtain N spots with a conventional isoceles electro-optic prism[4] deflector is $V_{\lambda/2}$ and the power is

$$(3/2)^{1/2} \epsilon_0 \epsilon_r t N^2 V_{\lambda/2}^2 f.$$

This represents a voltage increase by a factor of l/t over the grating. Also, the prism power requirements increase as N^2 while the grating power is independent of N.

Each angular position desired will be produced by a particular grating column. For instance the three columns shown in Fig. 1 each have a different grating space and thus provide three independent deflection angles. Similarly, N columns each with a different grating space would be required to obtain N positions. In order that each deflection position be resolved, the finest grating space d must be equal to or smaller than $d = L/N$ and the other spaces increased in suitable increments.

Measurements were made on an electro-optic grating with dimensions $d = 3\times 10^{-4}$ m, $a = t = d/2$, $L = 5\times 10^{-3}$ m, and $l/t = 3.67$. The gold electrodes were deposited on a z-cut $LiNbO_3$ wafer. The edges parallel to the zy plane were optically polished. 6328-Å light from a He-Ne laser was focused through the wafer with a cylindrical lens. The emergent light was brought to a focus by a second cylindrical lens followed by a 1-m focal length spherical lens. A 60-cycle ac voltage was used to avoid electron trap effects.[5] Recorder plots of the intensity of the far-field pattern were obtained using a moveable 4-μ slit and detector. A typical recording is shown in Fig. 2. As would be expected from $d = L/N$, 16 spots with half-widths equal to those shown can be fitted in the distance between the zero and first orders. The measured and estimated [Eq. (1)] intensities of the zero, first, and second orders are plotted against voltage in Fig. 3. There is fair agreement between the measured and estimated intensity. Approximately 17% of the incident light is placed in each lobe of the first order at 600 V. The zero order is reduced by 60% at 600 V and 74% at 800 V.

Under pulsed operation, light-pulse rise times of less than 10 nsec are observed and seem to be limited by the rise time of the applied voltage. The reactive power is calculated to be 0.73 W for 10^6 Hz average pulse repetition rate.

The digital electro-optic grating thus is useful for both light deflection and modulation. The ob-

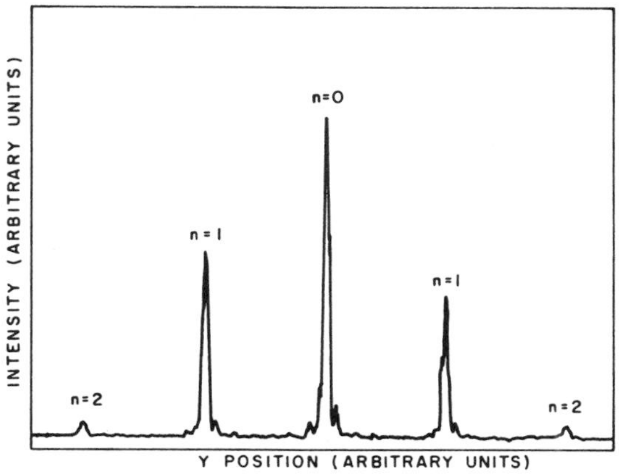

FIG. 2. Recorder tracing of the intensity of the focused far-field pattern. A 4-μ slit is moved parallel to the y direction to map out the pattern.

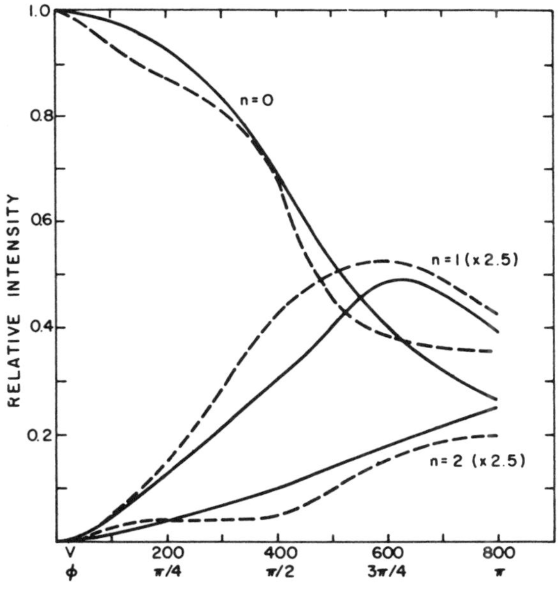

FIG. 3. Relative intensity of the orders $n = 0$, 1, and 2 as a function of voltage and phase shift. The phase shift is calculated from Eq. (2) using $V_{\lambda/2} = 2.94\times 10^3$ V (Ref 2). Solid line—measured value. Dashed line—values estimated with Eq. (1).

served zero-order modulation depth of 64% and first-order intensity are limited by the fringing fields in the particular geometry studied. Reducing the thickness t to that of thin-film wave guides (~ 1 μm) would result in a lessening of the fringe fields and an approximately two orders of magnitude reduction in power. While operation of this device on a waveguide mode has not yet been experimentally shown, there is no theoretical reason to expect it to fail. In addition, the similar effect of diffraction of waveguide modes by acoustic waves has been shown.[6]

If the length l (in the x direction) is increased, a thick grating mode of operation is obtained. Here 100% of the light may be diffracted into first order by operating in a Bragg regime while the power required is reduced by a factor proportional to t/l.

Thus, a new method of laser light deflection and modulation based on diffraction by electro-optic phase gratings has been demonstrated. The method appears to be applicable to both plane-wave and thin-film waveguide transmission, requires relatively little power for high-speed operation, and is capable of high diffraction efficiency.

The author wishes to thank J. Minville for assistance in preparing the gratings.

[1]P.K. Tien, R. Ulrich, and R.J. Martin, Appl. Phys. Letters 14, 291 (1969).

[2]E.H. Turner, Appl. Phys. Letters 8, 303 (1966); P.Y. Lenzo, E.G. Spencer, and K. Nassau, J. Opt. Soc. Am. 56, 633 (1966).

[3]Max Born and Emil Wolf, *Principles of Optics* (MacMillan, London, 1964), 2nd ed., pp. 401-405.

[4]V.J. Fowler and J. Schlafer, Proc. IEEE 54, 1437 (1966).

[5]F.S. Chen, J. Appl. Phys. 38, 3418 (1967).

[6]L. Kuhn, M.L. Dakss, P.F. Heidrich, and B.A. Scott, Appl. Phys. Letters 17, 265 (1970).

Part 8
Parametric Devices

OPTICAL SECOND HARMONIC GENERATION IN FORM OF COHERENT CERENKOV RADIATION FROM A THIN-FILM WAVEGUIDE

P.K. Tien, R. Ulrich, and R.J. Martin
Bell Telephone Laboratories, Holmdel, New Jersey 07733
Received 31 August 1970

We report optical second harmonic generation in form of coherent Cerenkov radiation. The fundamental wave at 1.06 μm propagates in a thin-film optical waveguide which is simply a ZnS film vacuum-deposited on a single-crystal ZnO substrate. The nonlinear polarization excited in the substr te has a phase velocity exceeding that of radiation propagating freely in the substrate material. It thus acts as the source of the observed Cerenkov radiation.

Using a prism-film coupler, [1] it is possible to feed a laser beam into a thin dielectric film which acts as an optical waveguide. Because the film can be made very thin, the light energy can be highly concentrated in the film and its immediate vicinity. The resulting high-light amplitudes make, therefore, the optical waveguide attractive for nonlinear optical experiments at relatively low absolute power levels. Another advantage of such an arrangement is that one can control the phase velocities of the participating waves by adjusting the geometric dimensions of the waveguide. In such experiments a good crystalline perfection and orientation of the nonlinear optical material is important. It can be ensured by electing the substrate, rather than the film, to be the nonlinear material. We report here optical second harmonic (SH) generation utilizing these ideas.

In our experiments, the light beam of a YAG:Nd laser at 1.06 μm is fed into a polycrystalline ZnS film that is vacuum-deposited on a ZnO single-crystal substrate. As shown in Fig. 1, the light propagates as a transverse electric (TE) wave in the $m = 0$ mode[1] along the x direction. The only field components of this wave are E_y, H_x, and H_z. These fields extend as an evanescent wave into the substrate, [2,3] and excite there a wave of SH polarization via the d_{33} nonlinear optical coefficient of the ZnO. For this, the c axis of the ZnO must be oriented in the y direction, and the resulting nonlinear polarization is also along this direction.

By properly choosing the thickness of the ZnS film we can operate the optical guide in such a regime that the forced wave of SH nonlinear polarization travels at a velocity faster than that of the free wave at the same frequency in the substrate medium. Consequently, the nonlinear polarization becomes a source of Cerenkov radiation, which is the second harmonic wave observed in the experiment. A similar experiment has been discussed by Zembrod *et al.*[4] Their fundamental wave was a cylindrical beam in a bulk crystal, and the directions of the SH wave form the familiar Cerenkov cone. In our case the geometry is planar and the nonlinear polarization wave is coherent over the full width of the beam, i.e., over many wavelengths. Therefore, the Cerenkov radiation

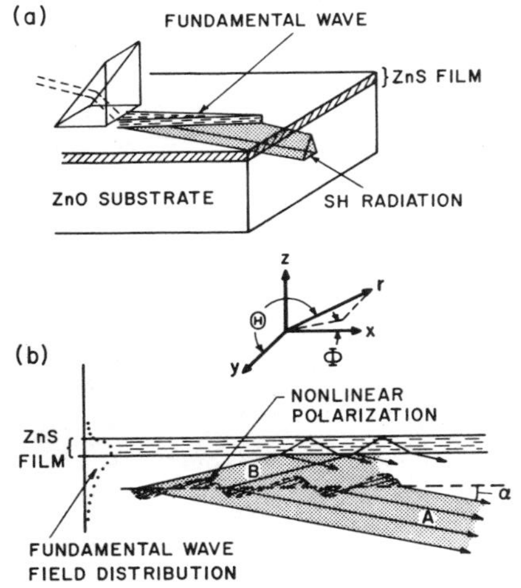

FIG. 1. (a) Experimental arrangement of second harmonic generation from a thin-film light guide, which consists of an evaporated ZnS film on a single-crystal ZnO substrate; (b) each sheet of the nonlinear polarization wave emits two beams of second harmonic radiation, A and B.

Reprinted with permission from *Appl. Phys. Lett.*, vol. 17, pp. 447–450, Nov. 15, 1970.

emerges as a well-collimated beam [Fig. 1(a)] with an angular spread in the order of a few tenths of a degree. This beam forms the Cerenkov angle α with the propagation direction of the nonlinear polarization.

We use the superscripts (1) for the fundamental and (2) for the harmonic. The subscripts 0, 1, and 2 denote the substrate, the film, and the air space above the film, respectively. The vacuum propagation constant of the fundamental is $k=\omega^{(1)}/c$, where $\omega^{(1)}$ is the angular frequency and c is the velocity of light. For the ZnS film, the refractive indices $n_1^{(1)}=2.2782$ and $n_1^{(2)}=2.3902$, and for the ZnO substrate (extraordinary indices), $n_0^{(1)}=1.9565$ and $n_0^{(2)}=2.0454$. The birefringence of the ZnO is irrelevant here since all waves are polarized parallel to the c axis. The fundamental wave propagates as $\exp(-i\omega^{(1)}t+i\beta x)$, and the forced wave of nonlinear SH polarization propagates as $\exp(-i\omega^{(2)}+2i\beta x)$. For thin-film waveguides,[2,3] $n_0^{(1)}<\beta/k<n_1^{(1)}$, and because of the normal dispersion of ZnO, $n_0^{(2)}>n_0^{(1)}$. Therefore, by adjusting the thickness W of the light guide we can achieve

$$n_0^{(1)}<\beta/k<n_0^{(2)}. \qquad (1)$$

Under this condition, the velocity $v^{NL}=\omega^{(2)}/2\beta=\omega^{(1)}/\beta$ of the wave of nonlinear SH polarization exceeds the phase velocity $v^{(2)}=c/n_0^{(2)}$ of the free SH wave in the substrate, and Cerenkov radiation is emitted. We recognize the second part of the inequality (1) as the familiar Cerenkov threshold. The SH radiation is emitted at the Cerenkov angle α where

$$\cos\alpha=v^{(2)}/v^{NL}=\beta/kn_0^{(2)}. \qquad (2)$$

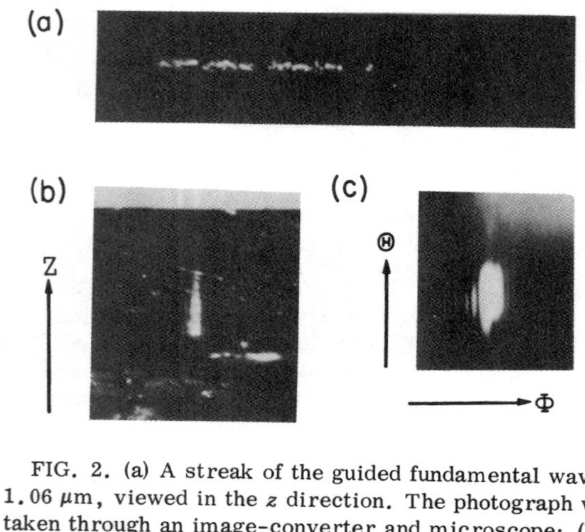

FIG. 2. (a) A streak of the guided fundamental wave at 1.06 μm, viewed in the z direction. The photograph was taken through an image-converter and microscope; (b) the near-field pattern of the second harmonic radiation viewed in the x direction; (c) the far-field pattern of the second harmonic radiation.

TABLE I. SH power, $2P^{(2)}$, computed from (6) using the following data: $P^{(1)}=2W$, $l=0.1$ cm, and $D=50$ μm.

β/k	W(μm)	α	$2P^{(2)}$ (μW)
1.95846	0.1515	16.77°	6.1
1.96144	0.1591	16.47°	13.2
1.96725	0.1698	15.89°	23.4
1.97554	0.1822	15.02°	33.4
1.99067	0.2017	13.28°	44.1
2.00591	0.2202	11.28°	50.2
2.02177	0.2394	8.72°	55.1
2.03353	0.2540	6.17°	59.9
2.03913	0.2612	4.49°	61.9
2.04420	0.2678	1.96°	47.5
2.04501	0.2688	1.12°	31.2
2.04538	0.2693	0.28°	8.4

This dependence of α on β/k is listed, along with the required film thickness, in the first three columns of Table I.

In one experiment the film had a thickness $W=0.204$ μm, as was calculated[1,3] from the measured value $\beta/k=1.992$. For this film, an internal Cerenkov angle $\alpha=13.5°$ was determined from the beam direction observed outside of the ZnO. The agreement with the theoretical α of Table I is within the experimental error of ± 0.3° for this and other films.

The Q-switched YAG:Nd laser produced 1-μsec-long pulses in the fundamental transverse mode. After passing through a polarizer and a system of lenses and apertures, a peak power of typically 3 W was incident on the base of the prism-film coupler. In all experiments, more than 60% of this power could be fed into the film. The distance in x from the coupler to the edge of the film was typically 3 mm. Figure 2(a) is a view of the fundamental beam. This picture was made by photographing through a microscope and image converter the light scattered from the beam as viewed in the z direction. The width of the beam in the y direction was 50 μm; the intensity distribution over this width was Gaussian in the center, the wings being cut off by the apertures, however. Because of imperfections in the film and at the surfaces, the intensity of the fundamental beam in the film decreases along its length. For this and also other films, the entire length of the fundamental beam, as seen from Fig. 2(a), consisted of 2 or 3 separate sections. The exact cause of these discontinuities is not known; it may be due to defects of the crystal, scratches on the surface, or steps in the film thickness. Later, from the SH far field, we can show that most of the SH radiation came from the first section of the fundamental beam which presumably had the largest light intensity. Figure 2(b) is a photograph of the green SH radiation at $\lambda = 0.53$ μm. Here, the camera was focused on the exit surface of the ZnO substrate. This figure may therefore be considered as the near-field pattern. The far-field pattern [Fig. 2(c)] was obtained by exposing, without any lens, a film 45-cm from the ZnO substrate. The angular width of the main radiation lobe in the xz plane (Φ direction) is 6×10^{-3} rad.

For calculation of the SH power, we assume a fundamental beam of uniform intensity, having a l and width D. The wave of nonlinear SH polarization in the substrate has the form (esu units)

$$\mathcal{P}^{NL}(x, y, z) = (16\pi d_{33} P^{(1)} \cos^2 \varphi_{10}^{(1)} / cDW_{eq} n_1^{(1)} \sin\theta_1^{(1)}) \times \exp(-i\omega^{(2)} t + 2i\beta x + 2p_0^{(1)} z) \qquad (3)$$

in the region $0 < x < l$, $|y| < D/2$, and $z < 0$. Elsewhere, $\mathcal{P}^{NL} = 0$. Here, $P^{(1)}$ is fundamental power propagating in the guide. For the other symbols we followed the notation in Ref. 3. Thus

$$\varphi_{10}^{(1)} = \tan^{-1}(p_0^{(1)}/b_1^{(1)}); \quad b_1^{(1)} = [(kn_1^{(1)})^2 - \beta^2]^{1/2};$$

$$p_{0,2}^{(1)} = [\beta^2 - (kn_{0,2}^{(1)})^2]^{1/2}; \quad p_0^{(1)} = [\beta^2 - (kn_0^{(1)})^2]^{1/2};$$

$$W_{eq} = W + (1/p_0^{(1)}) + (1/p_2^{(1)}); \quad \sin\theta_1^{(1)} = \beta/kn_1^{(1)}. \qquad (4)$$

The polarization wave in Eq. (3) is the source of the SH radiation. The far-field pattern resulting from this source can be found from antenna theory[5] or by the theory of Kleinman.[6] In polar coordinates r, Θ, Φ (polar axis in the y direction), the electric field is in the Θ direction, the magnetic field in the Φ direction, and both are proportional to

$$\frac{1}{r}\,\frac{\sin(kDn_0^{(2)}\cos\Theta)}{(kDn_0^{(2)}\cos\Theta)}\,\frac{\sin[kln_0^{(2)}(\cos\Phi - \cos\alpha)]}{[kln_0^{(2)}(\cos\Phi - \cos\alpha)]}. \qquad (5)$$

The main lobe of radiation is directed in $\Theta = \pi/2$ and $\Phi = \alpha$. The observed far field [Fig. 2(c)] did show a fringe pattern in the Φ direction very similar to Eq. (5). This justifies the assumption of a uniform fundamental intensity over the length l. From the fringe spacing in Fig. 2(c), we conclude that all SH radiation came from a section of the fundamental beam of a length $l = 0.9$ mm. In calculating SH power, we consider first the field produced by a sheet of the nonlinear polarization wave given by Eq. (3) at $z = z_1$ and then obtain the total field by an integration over z. As shown in Fig. 1(b), the nonlinear polarization at $z = z_1$ generates two beams of SH radiation, A and B. The beam A is emitted directly into the substrate. The beam B is emitted originally toward the film and is then superimposed on the beam A after being reflected from the film surfaces. Let r_{ij} be the reflection coefficient[3] at interface ij, and $\alpha^2 \gg 2\pi/kln_0^{(2)}$. The total SH power generated is

$$P^{(2)} = \frac{32\pi^3 (d_{33}\omega^{(2)} P^{(1)})^2}{c^3 n_0^{(2)} (kW_{eq} n_1^{(1)})^2}\,\frac{\cos^4\varphi_{10}}{\sin^2\theta_1^{(1)} \sin\alpha}\,\frac{l}{D}\,|S|^2, \qquad (6)$$

where

$$S = \frac{k}{p_0^{(1)} + ikn_0^{(2)}\sin\alpha} \times \left(1 - \frac{r_{10} - r_{12}\exp(2ib_1^{(2)}W)}{1 - r_{10} r_{12}\exp(2ib_1^{(2)}W)}\,\frac{p_0^{(1)} + ikn_0^{(2)}\sin\alpha}{p_0^{(1)} - ikn_0^{(2)}\sin\alpha}\right), \qquad (7)$$

and

$$b_1^{(2)} = 2[(kn_1^{(2)})^2 - \beta^2]^{1/2}.$$

The factor $|S|^2$ in (6) depends on the relative phases of the beams, A and B, at the far field. For small α the beams have opposite phases and they cancel in the far field. Consequently, the quantity $|S|^2$ and so the SH power $P^{(2)}$, decrease rapidly below $\alpha = 2^\circ$.

For $P^{(1)} = 2\times 10^7$ erg/sec (or 2 W), $d_{33} = 17\times 10^{-9}$ esu, $l = 0.1$ cm, and $D = 5\times 10^{-3}$ cm, the values of SH power $P^{(2)}$ have been computed from (6). The theoretical values listed in Table I are twice the values of (6) to account for the sum-frequency mixing between the many simultaneously oscillating longitudinal laser modes. It has been shown[7] that this effect is as efficient as the direct SH generation. We had assumed in the above calculation that the intensity is uniform over the width of the fundamental beam, $|y| < D/2$. If the intensity distribution is Gaussian, $P^{(2)}$ is only $(2)^{-1/2}$ of (6) to account for the difference in beam widths between the fundamental wave and the SH polarization. Under the conditions, $P^{(1)} = 2W$, $D = 50$ μm, and $l = 0.1$ cm with $W = 0.204$ μm and $\alpha = 13.5^\circ$, the measured SH power is 18.2 μW. It is within a factor of 3 of the theoretical value ($2P^{(2)}$ in Table I). We consider the experimental result satisfactory, since the absolute SH power depends critically on the quality and smoothness of the film and of the ZnO substrate. The SH power can easily vary by 50% by simply using a different portion of the same film. When the film thickness is not very uniform, the near field breaks into spots and the far field (5) becomes distorted. We have varied the fundamental power $P^{(1)}$ by a factor of 30 below 2 W. The observed SH power follows the expected square-law dependence on the fundamental power. This shows that power saturation in the film or in the surface region of the ZnO does not occur at this power level. At $P^{(1)} = 2$ W, the average power density in the waveguide is about 5 MW/cm^2.

In conclusion, we have demonstrated that the large power density in a thin-film optical waveguide can be used for nonlinear optical experiments, and that the nonlinear interaction can take place in the substrate.

[1]P.K. Tien, R. Ulrich, and R.J. Martin, Appl. Phys. Letters 14, 391 (1969).

[2]D. F. Nelson and J. McKenna, J. Appl. Phys. 38, 4057 (1967).

[3]P.K. Tien and R. Ulrich, J. Opt. Soc. Am. (to be published).

[4]A. Zembrod, H. Puell, and J.A. Giordmaine, J. Opto-Electron. 1, 64 (1969).

[5]S. Ramo and J.R. Whinnery, *Fields and Waves in Modern Radio* (Wiley, New York, 1953), Chap. 12, 2nd ed

[6]D.A. Kleinman, Phys. Rev. 128, 1761 (1962), see Eq. (139).

[7]A. Ashkin, G.D. Boyd, and J.M. Dziedzic, Phys. Rev. Letters 11, 14 (1963).

Wideband CO_2 Laser Second Harmonic Generation Phase Matched in GaAs Thin-Film Waveguides*

D. B. Anderson and J. T. Boyd

North American Rockwell Science Center, Thousand Oaks, California 91360
(Received 12 July 1971)

Phase-matched CO_2 laser second harmonic generation has been observed in GaAs thin-film waveguides. Phase matching is accomplished through the use of dielectric waveguide dispersion. Large values of the phase-matched bandwidth are observed and attributed partly to the nature of waveguide phase matching and partly to tapering of the waveguide thickness.

We have observed wideband phase-matched second harmonic generation (SHG) of CO_2 laser radiation using GaAs as the nonlinear crystal. Phase matching was accomplished by fabricating GaAs dielectric waveguides and using the waveguide dispersion to compensate for refractive-index dispersion. SHG has been observed for laser wavelengths corresponding to CO_2 rotational lines between 9.2 and 10.8 μ without any tuning of the nonlinear crystal properties. Such large values of the phase-match bandwidth are due partly to the nature of waveguide phase matching[1] and partly due to tapering of the waveguide thickness along the length of the waveguide.

GaAs is a very desirable material for CO_2 laser SHG in that it has both a high nonlinear coefficient and a low absorption coefficient in the wavelength range of interest. Since GaAs is isotropic, it cannot be phase matched by using birefringent phase-matching techniques. Non-phase-matched CO_2 laser SHG in GaAs has been previously observed.[2,3] Phase-matched harmonic generation has also been previously observed by employing total internal reflections iterated at the second-harmonic coherence length to achieve phase matching[4,5] and by utilizing an acoustic wave to induce phase matching.[6] In these previous experiments, harmonic generation has not been particularly efficient and has been implemented for only a single CO_2 laser wavelength. The SHG experiments described in this letter have used dielectric waveguide modal dispersion to achieve phase matching. Harmonic generation in this manner is characterized by a widely variable phase-match bandwidth and potentially high conversion efficiencies. Waveguide phase matching is also compatible with integrated optical concepts.[7]

Waveguide phase matching of various infrared parametric interactions has been discussed previously.[1] For SHG, waveguide phase matching requires that

$$\omega_h = 2\omega_p \tag{1}$$

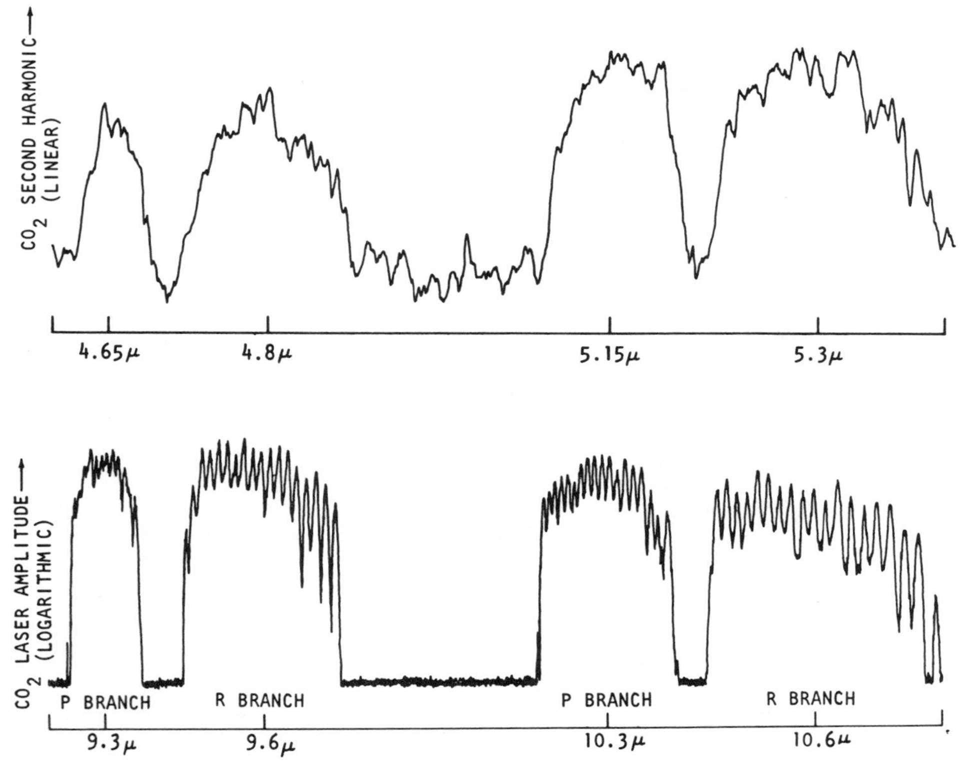

FIG. 1. Second-harmonic power and input laser power as a function of wavelength for a GaAs waveguide having a 3.25×200-μ cross section and $L = 4000$ μ..

Reprinted with permission from *Appl. Phys. Lett.*, vol. 19, pp. 266–268, Oct. 15, 1971.

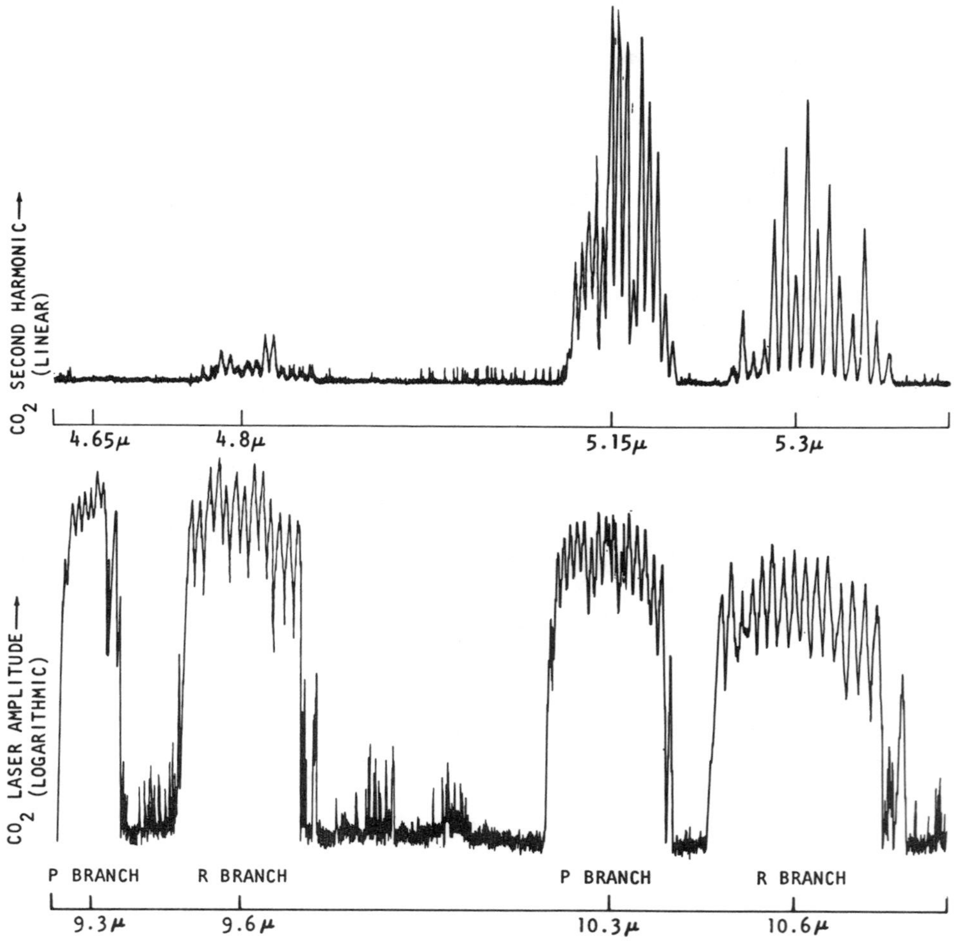

FIG. 2. Second-harmonic power and input laser power as a function of wavelength for a GaAs waveguide having a 110×3.9-μ cross section at one end face tapering to a 110×5.2-μ cross section at the other end face and $L = 3500$ μ.

and

$$k_{zh} = k_{zp} + k_{zp'}, \qquad (2)$$

where ω_p and ω_h are the pump and harmonic frequencies, and k_{zp} and k_{zh} are the pump and harmonic modal propagation constants in the z (longitudinal) direction, respectively, corresponding to the pump mode having index p and the harmonic mode having index h. Various sets of modes denoted by the indices (h, p, p') can satisfy Eq. (2) depending on the waveguide and crystal dispersion properties. The dielectric waveguides used have rectangular cross sections with sufficiently large aspect ratios so that accurate values of k_z can be determined from the dispersion relations for an infinite planar dielectric slab of thickness d, where d corresponds to the smaller transverse dimension of the rectangular dielectric waveguide.[1] Using values of k_z so obtained, we have published elsewhere values of d as a function of λ_p for which sets of modes (h, p, p') are phase matched.[8] The associated phase-match bandwidth is determined by introducing a frequency variation into Eqs. (1) and (2), using the waveguide dispersion relations, and equating the waveguide length to the coherence length arising from the frequency variation introduced. Phase-match bandwidths determined in this way for CO_2 laser SHG have been tabulated elsewhere.[1]

The GaAs rectangular dielectric waveguides were oriented such that the (110) planes correspond to waveguide end faces, $(\bar{1}10)$ planes correspond to the smaller area side, and (001) planes correspond to the larger area side. The waveguides were formed by first chemically abrasive polishing samples of GaAs normal to the ⟨001⟩ direction to thin slices of thickness d with no observable surface damage and then cleaving the end faces and the other two sidewalls. A high degree of surface parallelism can be achieved in this manner. In addition, the cleaved faces are quite smooth with respect to visible radiation. For this crystalline orientation, a pump field excited as a TE mode ($\bar{E}$ parallel to $\langle\bar{1}10\rangle$ direction) will generate a TM second harmonic ($\bar{E}$ parallel to the ⟨001⟩ direction).

The pump source is a low-pressure cw CO_2 laser tunable by an internal diffraction grating over more than 50 rotational lines from 9.2 to 10.8 μ. The laser power is monitored continuously with a Ge : Au photoconductive detector cooled to 77 °K. A con-

trolled portion of the output power from this laser is coupled into the GaAs waveguide by focusing the radiation onto the waveguide end face with a high-magnification Cassegrainian reflective objective. The excited waveguide mode spectrum can be varied by introducing aberrations corresponding to a slight adjustment of the Cassegrainian secondary reflector. The waveguides are supported by a holder similar to that used in previously described up-conversion experiments.[1] Radiation is coupled out of the waveguide by a second reflective objective in conjunction with a cylindrical lens. Discrimination of the harmonic output is accomplished by the use of polarization and sapphire filters prior to detection of the second harmonic by a photovoltaic InSb detector cooled to 77 °K.

Phased-matched harmonic generation was observed to occur in three different GaAs dielectric waveguides. The first sample had a 3.25×200-μ cross section with length $L = 4000\ \mu$. Figure 1 depicts a recording of both the harmonic power output and the CO_2 laser power input as the internal laser cavity grating is mechanically scanned. The peak CO_2 laser power in Fig. 1 is 160 mW. Noise present in the harmonic trace tends to obscure the harmonic response to individual oscillating rotational lines. However, the presence of this response was verified by interrupting similar scans in several places. The phase-match data contained in Ref. 1 imply that the harmonic response most likely arises from the phase-matched mode order combination $(h, p, p') = (2, 1, 1)$. From the scan in Fig. 1 we conclude that the experimentally observed phase-match bandwidth is at least as large as the theoretical value of 12% for the (2, 1, 1) mode combination calculated in Ref. 1. The second sample of GaAs used had a cross section of $3.2 \times 160\ \mu$ at one end face tapering to $3.9 \times 160\ \mu$ at the other end face over an interaction length of 1000 μ. At $\lambda_p = 10.7\ \mu$, we observed an over-all efficiency (product of input and output coupling efficiencies and conversion efficiency) of 10^{-2} for 1.0 W of input CO_2 laser power. Gradual crystalline damage prevented an accurate measure of the bandwidth for this sample. The third sample of GaAs had a cross section of $3.9 \times 110\ \mu$ at one end face tapering to $5.2 \times 110\ \mu$ at the other end face over an interaction length of 3500 μ. A trace of the second-harmonic power and the CO_2 laser power as the cavity grating is mechanically scanned for this waveguide is shown in Fig. 2. The peak input CO_2 laser power in Fig. 2 is 320 mW. Note the absence of any harmonic response at the 9.3-μ band and the peak response in the 10.3-μ band. The broad bandwidth in this case is attributed partly to the waveguide taper and partly to the nature of waveguide phase matching. Note that in contrast to Fig. 1 the harmonic response in Fig. 2 to individual oscillating rotational lines is distinguishable. A more complete interpretation of these experiments along with a description of phase matching in a tapered waveguide will be published elsewhere.

Currently, the input laser power is generally limited to low values due to the damage threshold of the GaAs waveguides. We expect that the waveguide damage threshold could be raised considerably by the use of a substrate transparent at ω_p and ω_h which would act as a heat sink. Multiple layers of $Ga_xAl_{1-x}As$ films deposited on a GaAs substrate by liquid epitaxy may yield a suitable combination. Raising the damage threshold would allow larger input laser powers which would improve the conversion efficiency. The over-all efficiency would then be limited by the input mode coupling efficiency. The input mode coupling efficiency could be improved considerably if one of the existing efficient and mode selective input couplers could be implemented for such a structure.[9–11]

The authors wish to thank Dr. R. B. Thompson for comments on the manuscript, L. Dyal for technical assistance, and Mrs. J. Healy for crystal polishing.

---

*Program supported in part by the U.S. Army Electronics Command, Night Vision Laboratory, Ft. Belvoir, Va.

[1]D. B. Anderson, J. T. Boyd, and J. D. McMullen, *Proceedings of the Symposium on Submillimeter Waves* (Polytechnic Institute of Brooklyn Press, New York, 1971), Vol. 20, MRI Series, p. 191.
[2]C. K. N. Patel, Phys. Rev. Letters **16**, 613 (1966).
[3]J. J. Wynne, Phys. Rev. **188**, 1211 (1969).
[4]G. D. Boyd and C. K. N. Patel, Appl. Phys. Letters **8**, 313 (1966).
[5]J. A. Armstrong, N. Bloembergen, J. Ducuing, and P. S. Pershan, Phys. Rev. **127**, 1918 (1962).
[6]G. D. Boyd, F. R. Nash, and D. F. Nelson, Phys. Rev. Letters **24**, 1298 (1970).
[7]S. E. Miller, Bell System Tech. J. **48**, 2059 (1969).
[8]These data are contained in Fig. 5 of Ref. 1. The mode indices appear in the order *hpp'* rather than *pp'h* as indicated in Fig. 5.
[9]P. K. Tien and R. Ulrich, J. Opt. Soc. Am. **60**, 1325 (1970).
[10]M. L. Dakss, L. Kuhn, P. F. Heidrich, and B. A. Scott, Appl. Phys. Letters **16**, 523 (1970).
[11]P. K. Tien and R. J. Martin, Appl. Phys. Letters **18**, 398 (1971).

LOW-POWER QUASI-cw RAMAN OSCILLATOR

E. P. Ippen
Bell Telephone Laboratories, Holmdel, New Jersey 07733
(Received 23 February 1970)

Quasi-cw stimulated Raman emission in the visible has been obtained from an oscillator cavity with a pump-power input of less than 5 W. This large reduction of oscillator threshold is achieved with the use of a liquid-core optical waveguide structure. Significant conversion of pump to Stokes light has been observed. Extension of the system to continuous operation and to the study of other nonlinear effects is suggested.

This letter reports the first observation of stimulated Raman emission from an oscillator cavity under conditions of low-power quasicontinuous excitation. In contrast to previous Raman lasers which have required pump powers in excess of 10^4 W,[1] the oscillator described here operates with an input power of less than 5 W. The key to this low-power operation is confinement of both the pump and Stokes beams in a liquid-core optical waveguide. It is the maintenance of high optical power densities over relatively long lengths of guide that dramatically increases the gain per pass in this type of oscillator. In our experiments the waveguides are fabricated by filling small-bore glass tubing with carbon disulfide (CS_2). The guide is pumped longitudinally with a repetitively pulsed argon-ion laser operating at $\lambda_p = 5145$ Å. Raman emission occurs at a Stokes wavelength $\lambda_s = 5325$ Å. This observed Stokes frequency shift is characteristic of the 656-cm^{-1} line in CS_2.

Operation of the oscillator is illustrated schematically in Fig. 1. A magnified view of the guide

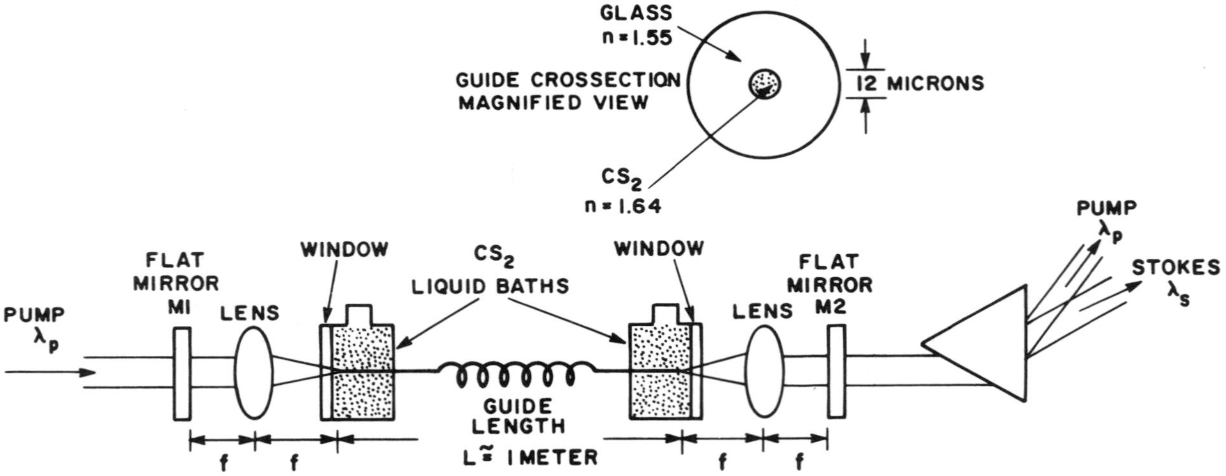

FIG. 1. Experimental arrangement for the observation of Raman oscillation in a liquid-core optical waveguide resonator.

Reprinted with permission from *Appl. Phys. Lett.*, vol. 16, pp. 303–305, Apr. 15, 1970.

cross section is also shown. When filled with CS_2, the 12-μ-diam guide core has an index of refraction of 1.64 as compared to 1.55 for the glass walls. The small dimensions of the guide make it very flexible, and it is easily coiled up to make the oscillator more compact. Both ends of the 1-m-long CS_2-filled guide are immersed in baths of CS_2 to facilitate the coupling of light in and out. External lenses of focal length 2 cm collimate light emerging from the ends and at the same time focus light back into the guide. Mirrors provide the necessary feedback for oscillation. The reflectivities of M1 and M2 at the Stokes wavelength are 99% and 97.5%, respectively. The argon laser pump beam enters the oscillator through M1 which transmits about 80% at $\lambda_p = 5145$ Å. Mirror M2 reflects 97.5% at λ_p, so that the pump beam makes both a forward and backward pass through the oscillator.

A spectrometer has been used at the output of the oscillator to verify the Stokes shift of the observed stimulated emission, but a dispersive prism is sufficient to separate the Raman beam from the transmitted pump light. The color difference between the two beams and the sensitivity of the Stokes intensity to the alignment of either oscillator mirror provide visual verification of operation. Since the output mirror has roughly the same transmissivity for the two wavelengths, visual comparison of the two beams can also give an indication of how much pump light is being converted into Stokes radiation. We have, in fact, observed significant amounts of conversion. The oscilloscope traces shown in Fig. 2 demonstrate this clearly. The top trace (a) shows the transmitted pump pulse when one of the oscillator mirrors is tilted out of alignment. Peak amplitude corresponds to an input power of about 7 W. Traces (b) and (c) illustrate, respectively, a partially depleted pump pulse and the corresponding Stokes signal when the oscillator cavity is aligned. In this instance approximately 40% of the effective peak pump power is converted into Raman light. At higher pump powers considerably more conversion has been observed.

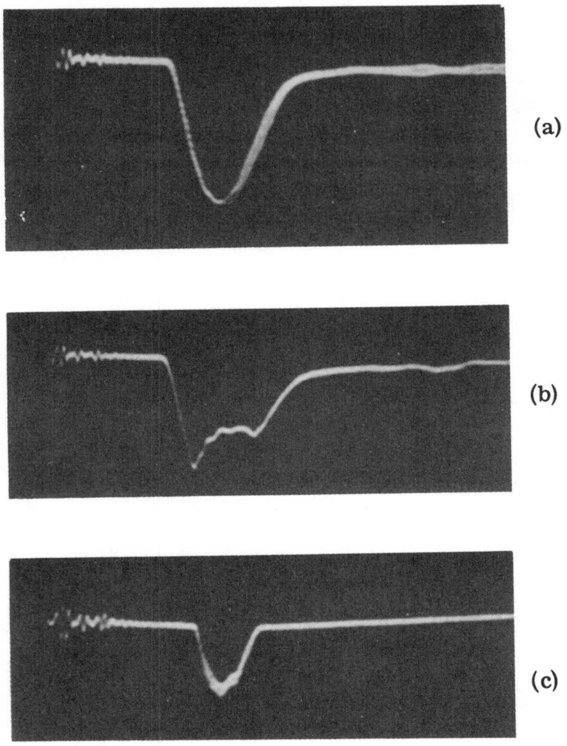

FIG. 2. Oscilloscope traces of (a) transmitted pump pulse with resonator misaligned, (b) transmitted pump pulse resonator aligned, (c) corresponding Stokes output pulse. Same vertical scale in each case. Input pump pulse: 7 W. Time scale: 1 μsec/cm.

The time duration of the pump pulse shown in Fig. 2 indicates that some difficulty is encountered in coupling the pump laser output into the liquid-core guide. Although the argon laser had a pulse duration of about 10 μsec, the pump light in the guide has only a duration of about 2 μsec. This effect is believed due to temporal variations in the transverse mode structure of the argon laser and fast thermal lens effects[2] at the input end of the guide. With careful adjustment of the input coupling and at slightly lower powers we have observed more continuous behavior. It should be noted that it is not necessary to pump an oscillator of this type with either single transverse or single longitudinal laser modes. As long as the linewidth of the pump is less than that of the spontaneous Raman linewidth of the gain medium, all pump frequencies provide gain for a single Stokes frequency. Furthermore, with these guide dimensions and with the indices of refraction involved, propagation of pump light in the guide tends to be multimode, but since stimulated Raman gain is nondirectional and depends only on pump intensity, any particular guided Stokes wave derives gain from all of the pump modes. This suggests that the mode purity of the oscillator output can be considerably better than that of the pump.

Design of an oscillator of this type depends primarily upon two factors: (1) loss per unit length in the guide for both the pump and Stokes beams and (2) end losses due to mirror transmission and coupling light in and out of the guide. With estimates of these two kinds of losses it is possible to calculate optimum guide lengths for either minimum threshold power or maximum power output for a given pump intensity. No concerted effort has yet been made to optimize our present oscillator. Thus, it is informative to estimate just how well one should be able to do. We have made quantitative measurements of the distributed losses in 12-μ-diam liquid (CS_2)-filled guides using light from a He-Ne laser operating at $\lambda = 6328$ Å. These measurements have revealed a guide loss of only 10^{-3} cm^{-1}, which is apparently limited by the expected bulk loss of CS_2 alone. About 70% of this loss can be attributed to Rayleigh scattering by the liquid and about 30% to absorption.

We write the exponential gain factor for stimulated Raman emission in a guide as

$$G = (g_s I_p / \pi r^2)\ \mathrm{cm}^{-1}, \qquad (1)$$

where I_p is the total pump power, r is the guide core radius, and F is some numerical factor to take into account the intensity overlap between modes in the guide. The gain factor g_s for CS_2 at a pump wavelength $\lambda_p = 5145$ Å is 1.3×10^{-8} cm/W, as obtained from the peak spontaneous cross section of Skinner and Nilsen.[3] So that with a 6-μ-diam guide, a distributed loss $\alpha_G = 0.002$ cm^{-1}, a filling factor $F = 0.5$, and with end losses as large as 50% per pass, an oscillator can be constructed with an injected threshold power of less than 0.2 W The optimum threshold length would be 3 m for this example and would vary inversely with α. Threshold power at optimum length varies linearly with α but decreases less rapidly with decreasing end losses. With the configuration of this experiment, a 12-μ-diam guide and a length of only 1 m, calculations based on the same assumptions as above yield a threshold of 2.2 W. This is in good agreement with our observed value of 3.2 W considering present uncertainty in the various parameters.

In light of the above estimates and known Raman gain factors for other liquids as well as CS_2,[3,4] it seems that one should be able to achieve continuous (cw) Raman oscillation with a variety of different liquids and a number of existing laser sources. Oscillation at frequencies corresponding to higher-order Raman effects also appears feasible. The relatively large linewidths of Raman liquids and the high single pass gains achievable in guides make such systems attractive as amplifiers as well as oscillators.

In conclusion, we have demonstrated that low-power quasi-cw Raman oscillation can be achieved in a liquid-core optical waveguide resonator. We have introduced experimental evidence and the results of other calculations based on such a device which indicate that continuous Raman oscillation is within reach using other liquids as well as CS_2. The large reduction of threshold levels obtained in these experiments further encourages the use of guiding techniques in studying other nonlinear effects in both liquids and solids.

The author gratefully acknowledges many helpful discussions with A. Ashkin and the valuable technical assistance of D. Eilenberger.

[1]C. C. Wang, Phys. Rev. Letters 16, 344 (1966).
[2]F. W. Dabby, R. W. Boyko, C. V. Shank, and J. R. Whinnery, IEEE J. Quantum Electron. QE-5, 516 (1969).
[3]J. G. Skinner and W. G. Nilsen, J. Opt. Soc. Am. 58, 113 (1968).
[4]M. J. Colles, Opt. Commun. 1, 169 (1969).

Part 9
Film Deposition Techniques

Optical Waveguides Formed by Proton Irradiation of Fused Silica*†

E. Ronald Schineller, Richard P. Flam, and Donald W. Wilmot‡
Wheeler Laboratories, Inc., Smithtown, New York 11787
(Received 9 November 1967)

Various techniques for fabricating optical waveguides for laser-oriented applications are being studied. A technique utilizing radiation-induced changes of refractive index in optical materials is presented; the particular waveguide considered is a fused-silica slab that has been irradiated with protons to produce a channel with a refractive index slightly higher than the unirradiated silica. This high-index channel serves as the waveguide core and the surrounding unirradiated region as the cladding.

Formulas for determining the size and shape of the high-index channel and the amount of index change have been developed. The primary waveguide parameters, core size, and refractive-index difference, may be adjusted by controlling proton energy and dosage, respectively. The technique is useful for formation of waveguides with core widths of 1 to 50 μ and index differences of 0.01 to 0.0001; it is particularly suitable for forming complex arrays of waveguides and waveguide components.

Preliminary experimental work has used 1.5-MeV protons and dosages of 10^{14} to 10^{17} protons/cm^2. Light propagation has been observed in waveguides formed by this technique.

Index Heading: Refractive index; Fiber optics; Waveguides.

Optical waveguides are currently being investigated for application in many laser systems. Waveguides are being considered for transmission of optical signals and for the construction of optical components.[1–5] The most common type of guide at optical frequencies is the dielectric waveguide, which is a core dielectric surrounded by a cladding dielectric of lower refractive index; light propagates by the process of total internal reflection at the core-cladding interfaces. This is the configuration of conventional fiber-optic light pipes. During the course of a program for development of single-mode optical waveguides and waveguide components, one of the more novel techniques investigated for fabricating such waveguides was the irradiation of optical materials. The waveguide configuration investigated there, and considered herein, is the conventional dielectric waveguide in the slab geometry, as illustrated in Fig. 1.

Various types of radiation are known to cause changes of the refractive index of optical materials. If such changes are to be applied to waveguide formation, they must be predictable in some detail, at least on a macroscopic scale. A model which can be used to make such predictions has been formulated, based on the theoretical and experimental results of various investigations reported in the literature. This model, based on certain simplifying assumptions, is intended to give approximate solutions for the particular cases of interest. Subsequent experimental evaluation has yielded results which support the theory, at least qualitatively, and demonstrate the potential of this technique for waveguide fabrication.

MECHANISM OF WAVEGUIDE FORMATION

The general procedure for forming a waveguide is to irradiate an optical material so as to increase the refractive index in a localized region. This high-index region may then serve as the waveguide core and the surrounding region as the cladding.

The fundamental waveguide parameters of a dielectric waveguide which must be controlled are the size and shape of the core region and the difference of

* An earlier version of this paper was presented at the 1966 Fall Meeting of the Optical Society of America in San Francisco [J. Opt. Soc. Am. 56, 1434A (1966)].

† Work supported by the National Aeronautics and Space Administration Electronics Research Center under contract NAS 12-2.

‡ Present address: Sanders Associates, Nashua, N. H. 03060.

[1] R. A. Kaplan, Proc. IEEE 51, 1144 (1963).
[2] D. W. Wilmot and E. R. Schineller, J. Opt. Soc. Am. 56, 839 (1966).
[3] R. P. Flam and H. W. Redlien, Jr., Proc. IEEE 55, 1750 (1967).
[4] E. R. Schineller, Microwaves 7, 77 (1968).
[5] R. P. Flam and E. R. Schineller, Proc. IEEE 56, 195 (1968).

Reprinted with permission from *J. Opt. Soc. Am.*, vol. 58, pp. 1171–1176, Sept. 1968.

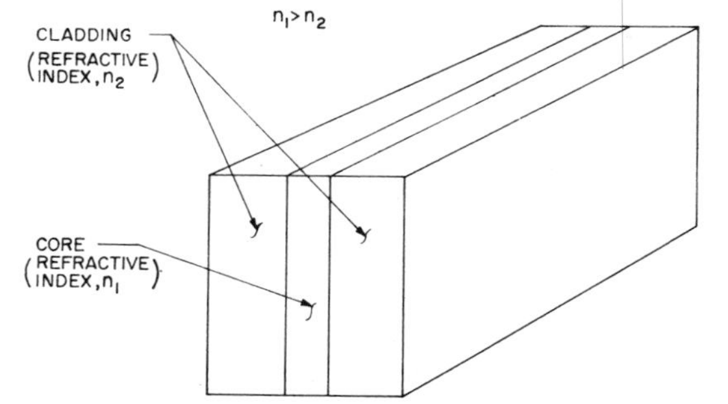

FIG. 1. Dielectric-slab waveguide.

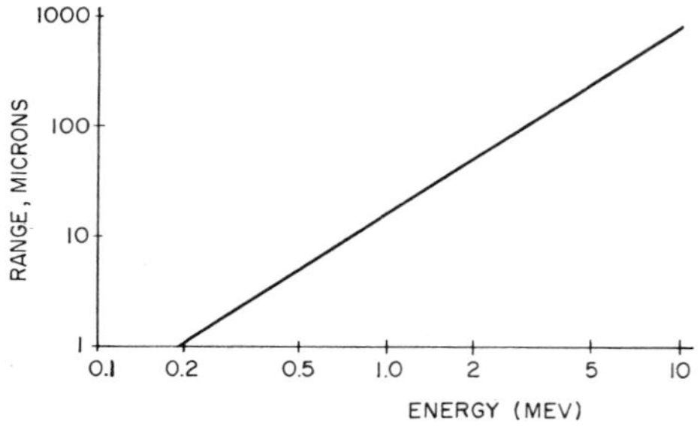

FIG. 2. Range of protons in fused silica.

refractive index between core and cladding. These parameters determine all of the propagation characteristics such as the number of propagating modes, their field patterns, phase velocities, etc. A purpose of the present program has been to develop single-mode waveguides with core sizes of several or many wavelengths (1 to 50 μ). In order to achieve single-mode operation in such guides, the index difference between the core and cladding must be very small[1] (10^{-2} to 10^{-5}). However, the technique described is applicable to multimode guides (light-pipes) as well, where precise control of refractive index and size is not required. To form an optical waveguide by irradiation of a dielectric, it is necessary to determine the amount of index change introduced by the irradiation and the size and shape of the affected region. In the case of uniform irradiation of a flat surface of a material, the index profile in the direction parallel to the incident radiation must be determined.

Radiation-induced changes of refractive index can be attributed to several mechanisms. Of these mechanisms, two which have been considered are ionizing radiation, and displacement creation by atomic collisions. Other possible effects, such as those connected with ion implantation and chemical combination of the incident particles with the target material, are not considered in this paper.

Ionizing radiation such as gamma rays and electrons induce absorption bands (color centers) which have an associated index change.[6,7] It has been found experimentally that this index change is small (in the visible), typically of the order of 10^{-4}. Irradiation with various types of particles such as neutrons, protons, and other ions changes the density and hence the refractive index of a material.[8–11] The index change in this case can be considerably greater, changes from 0.01 to 0.2 having been reported.[8,11] It has generally been assumed that these changes are caused predominately by atomic displacements and that the effect of ionization on the index is negligible. However, recent investigations by Primak[12] indicate that this assumption may not be valid. Nevertheless, based on information available in the literature (including Primak) and the good experimental agreement obtained, the effect of ionization on refractive index has been ignored.

In order to relate the index change to dosage for proton irradiation of fused silica, empirical data obtained by Primak[8] from neutron irradiation have been utilized. These data relate the index change to neutron dosage, from which the change with displacement density can be computed. Up to the saturation dosage, the increase in refractive index is approximately linear, and is given by

$$\Delta n = 10^{-23} N, \tag{1}$$

where N is the total number of displacements per unit volume. From this fundamental relationship, the index change with dosage can be determined for any type of radiation. Above the saturation point, further irradiation causes no additional increase of the index. For fused silica, the saturation dosage results in a positive index change of about 0.01.

The refractive-index profile for proton irradiation of fused silica can now be determined by calculating the density of displacements N as a function of position in the material. When a proton passes through fused silica it gives up energy both by ionization of the atoms and by atomic displacements. In order to determine the distribution of displacements, it is convenient to consider two regions, an ionization region in which the proton loses energy primarily by ionization, and a collision region in which the proton loses energy primarily by the creation of displacements. The transition between these two regions is gradual, but the result is not significantly affected if an abrupt transition is assumed. The exact proton energy E_i at which the transition occurs is difficult to determine. It is certainly less than the threshold energy for ionization

[6] I. H. Malitson, M. J. Dodge, and M. E. Gonshery, J. Opt. Soc. Am. **55**, 1583A (1965).
[7] M. J. Dodge, I. H. Malitson, and M. E. Gonshery, J. Opt. Soc. Am. **56**, 1432A (1966).
[8] W. Primak, Phys. Rev. **110**, 1240 (1958).
[9] W. Primak and M. Bohmann, in *Progress in Ceramic Science*, Vol. 2, J. E. Burke, Ed. (Pergamon Press, Inc., New York, 1962), pp. 103–181.
[10] R. L. Hines and R. Arndt, Phys. Rev. **119**, 623 (1960).
[11] R. L. Hines, J. Appl. Phys. **28**, 587 (1957).
[12] W. Primak (private communication).

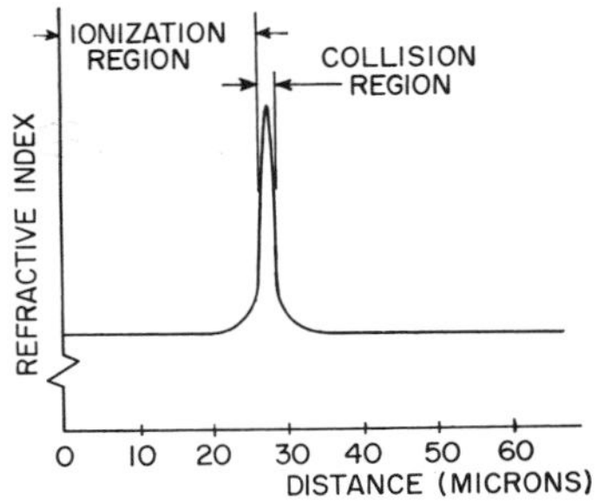

FIG. 3. Calculated refractive-index profile for irradiation of fused silica with 1.5-MeV protons (below saturation).

EXPERIMENTAL IMPLEMENTATION AND EVALUATION

In order to evaluate this technique of waveguide formation and to verify the calculations, several samples of fused silica were irradiated. In each case, slabs of Ultrasil fused silica, 2 cm long, 1 cm wide, and 0.1 cm thick, were uniformly irradiated over one large surface; both protons and deuterons have been used. In the first experiment, an irradiation with 1.5 MeV protons and a dosage of 10^{17} protons/cm^2 (well above saturation) was used. The above-saturation dosage was used to insure that the maximum index change would be obtained, and also to broaden the width of the high-index channel. A sketch of the index profile for this irradiation is shown in Fig. 4; the dosage above saturation results in the flat top. The secondary channel illustrated is caused by the presence of molecular hydrogen in the proton beam used in the experiment. Upon impact with the fused silica, the H_2^+ ions split into two protons, each having half the initial energy. Therefore, they effectively create an additional source of protons of half the energy, having a range of about one-third that of the primary protons. Since the straggling distance is also reduced, the channel created by the molecular hydrogen ions is narrower than the primary channel and located as shown. The asymmetric shape of the channels is caused by displacements produced in the ionization region, which can not be neglected for dosages above saturation.

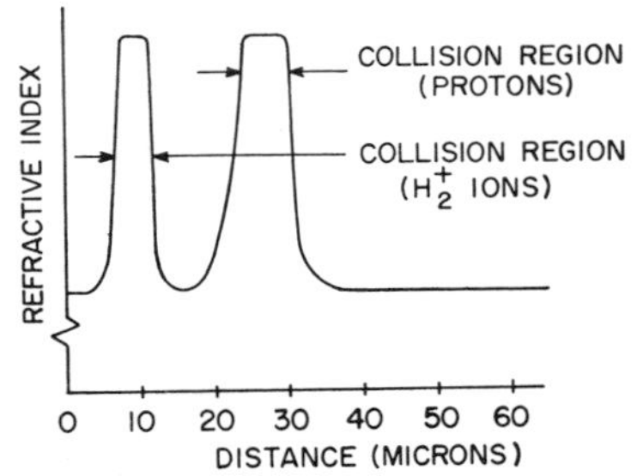

FIG. 4. Calculated refractive-index profile for irradiation of fused silica with 1.5-MeV protons and H_2^+ ions (above saturation).

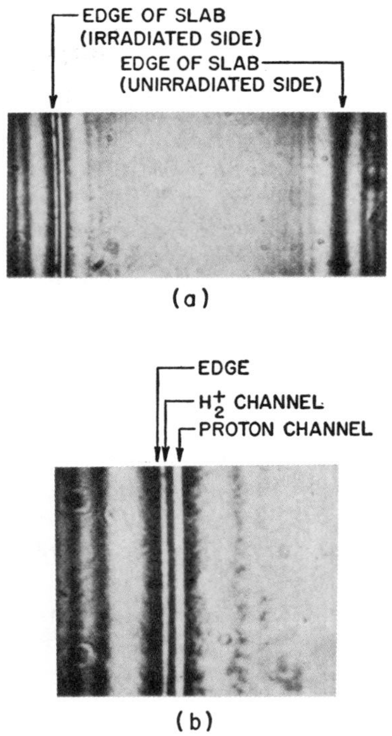

FIG. 5. Light guiding in a fused-silica slab irradiated with 1.5-MeV protons and H_2^+ ions. (a) End view of slab, (b) magnified view of irradiated region.

The irradiated slab was tested for light guiding by illuminating one end with a He–Ne gas laser operating at 0.6328 μ, and viewing the opposite end with a microscope. The slab was immersed in a matching liquid to eliminate reflections from the surface near the irradiated channel.

A photograph of the entire end of the slab is shown in Fig. 5(a). The end is essentially uniformly lighted except for two bright lines on the left-hand side near the irradiated surface. (The edges of the slab are indicated; the fringes outside the slab on each side are caused by a Lloyd's mirror effect.) A magnified view of the above photograph showing the irradiated region in greater detail is given in Fig. 5(b). Again the edge of the slab is indicated; two bright lines occur at the calculated positions of the high-index channels shown in Fig. 4, and so are attributed to light guiding. Comparison with Fig. 4 also indicates that the width of the channels is in good agreement with that calculated.

The actual value of the index change has not been measured directly. However, by off-axis excitation, it was possible to excite higher-order waveguide modes, and the number of modes was approximately equal to that expected for an index difference of 0.01 and the given waveguide size. Several other samples were irradiated with protons and deuterons with dosages nearer to saturation. Deuterons produce essentially the same effects as protons except that the range and straggling are smaller. In most cases, light guiding was observed,

($\sim$20 keV) and is probably greater than the threshold for excitation ($\sim$2 keV). In either case, it is shown that its value does not significantly affect the shape of the calculated profile, only the index change resulting from a given dosage.

In the ionization region most of the proton energy is given up by ionizing the atoms in the fused silica, which does not cause a significant change of the refractive index. However, some displacements are produced, the cross section for displacements being inversely proportional to the proton energy. The density of displacements in the ionization region is given by

$$N = kD/E \quad \text{(displacements per cm}^3\text{)}, \tag{2}$$

where D is the dosage in protons per cm^2 and E is the proton energy in MeV. The constant of proportionality k has been estimated from data obtained with proton irradiation of silicon,[13] and has a value of about 8000.

Since the proton energy is greatest at the surface of a given sample, and decreases as the proton travels through it, the index change is smallest at the surface and increases with distance from the surface. The maximum change in the ionization region occurs at the end and is equal to kD/E_i. The index change is one fifth this value at an energy of $5\,E_i$; since this energy occurs less than $0.3\,\mu$ from the end of the ionization region (for $E_i < 20$ keV), the only significant change of index, in the ionization region, occurs at the end.

The distance to the end of the ionization region is virtually the total penetration for incident energies above 0.2 MeV. This distance, called the range, is determined primarily by the rate of energy loss by ionization, and is a well-known function of the target material and the proton energy.[14] (The data in Ref. 14 actually apply to the range of protons in aluminum, which is a good approximation for fused silica.) The range of protons in fused silica is plotted as a function of the initial energy in Fig. 2. Ranges from one micron to several millimeters can be achieved with readily attainable energies.

At the end of the range, the proton energy is equal to E_i. This remaining energy is given up primarily through collisions with the atoms of the fused silica, which create atomic displacements; this region is called the collision region. The number of displacements created by each proton in this region is given approximately by $E_i/2E_d$, where E_d is the energy required for displacement creation. The length of the collision region for a single proton is less than $0.1\,\mu$. Therefore, the only significant density of displacements, in both the ionization and collision regions, occurs in a region less than one-half micron long. Thus the profile of displacement density for a single proton is simply a narrow channel less than one-half micron long located at the end of the range.

To determine the density of displacements N for a many-particle beam, the effects of range straggling must be considered. Since the ionization process which determines the proton range is random, not all particles of the same incident energy have exactly the same range. Rather, there is a gaussian distribution of ranges, about the mean, where σ (the standard deviation of the gaussian distribution) is a known function of the energy.[14] In general, the standard deviation of the range straggling for a many-particle beam, is greater than the length of the displacement channel for a single proton, and the distribution of displacements within that region is determined by the range straggling.

The consideration of range straggling indicates that a monoenergetic beam of many particles will create a displacement channel having a refractive-index profile of gaussian cross section with standard deviation σ. The density of displacements at the peak of the gaussian channel will depend on the total number of displacements and the width of the gaussian channel (proportional to σ). As an approximation, we assumed that all of the displacements are created in the collision region (total displacements $= DE_i/2E_d$). This assumption is very good for $E_i = 20$ keV and is reasonably good even for $E_i = 2$ keV. For a normalized gaussian distribution, the peak value is given by $1/(2\pi\sigma)^{\frac{1}{2}}$, so the peak value of N is

$$N_{\max} = (E_i D/2E_d)(2\pi\sigma)^{-\frac{1}{2}}, \tag{3}$$

where σ is measured in centimeters. Thus the displacement profile for a many particle beam is a gaussian-shaped channel with peak value given by Eq. (3).

In fused silica, E_d is approximately 25 eV and E_i is estimated to be 10 keV. Using these values in Eq. (3) gives

$$N_{\max} = 80D/\sigma. \tag{4}$$

From Eq. (1) the maximum index change, to order of magnitude, is

$$\Delta n_{\max} = 8 \times 10^{-22} D/\sigma. \tag{5}$$

For a typical case of irradiation of fused silica with 1.5-MeV protons, σ is approximately $\frac{1}{2}\,\mu$, so the maximum index change is approximately $10^{-17}D$.

A sketch of the calculated index profile for a dosage below saturation with 1.5-MeV protons, is given in Fig. 3. It can be seen that a high-index channel is created, which has a width of about $1\,\mu$ centered $27\,\mu$ from the surface. This high-index channel can serve as the core region of a dielectric waveguide. Sketches of index profiles for other specific examples of irradiation are given below.

[13] J. A. Baicker, H. Flicker, and J. Vilms, Appl. Phys. Letters **2**, 104 (1963).

[14] *American Institute of Physics Handbook*, 2nd ed. (McGraw-Hill Book Co., New York, 1963), pp. 8–20 to 8–47.

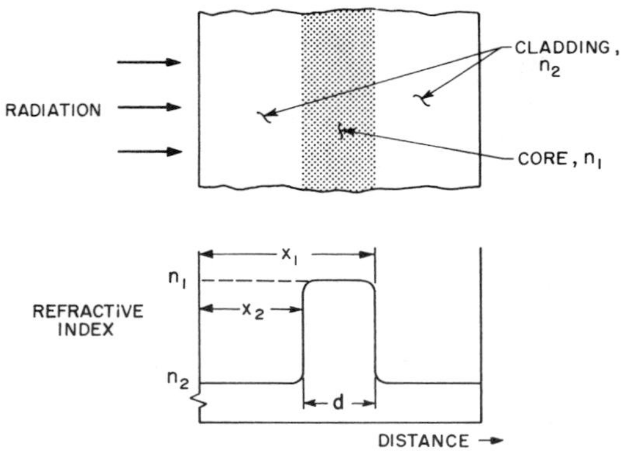

FIG. 6. Formation of single waveguide with variable-energy protons.

in general agreement with the calculations. However, in a few cases, for dosages below saturation, no guiding was observed. Therefore, further experimental evaluation is needed to confirm the variation of index with dosage for dosages below saturation. When very small index differences are desired, it is possible to control the index after irradiation by annealing; this should provide a technique for precise control of the index where required.

Another problem which should be mentioned is the creation of strain in the irradiated material. The strain is usually indicated by crazing (fine cracks) in the irradiated surface. In some of the irradiations, particularly those with large dosages and high dose rates, the strain was sufficient to impair waveguide performance. However, in several cases, the effects of the induced strain have been negligible. Further investigation of ways to minimize the effects of strain is needed.

ADDITIONAL WAVEGUIDE CONFIGURATIONS

The previous sections have outlined the general procedures for formation of simple slab waveguides by proton irradiation of fused silica. This section describes several proposed techniques for formation of multiple

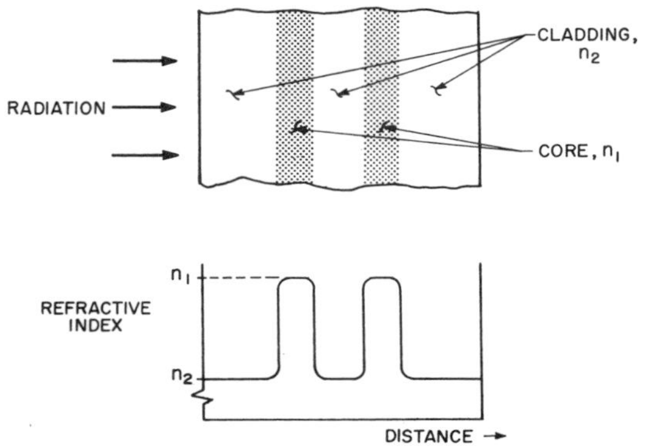

FIG. 7. Formation of multiple waveguides by irradiation over two energy ranges.

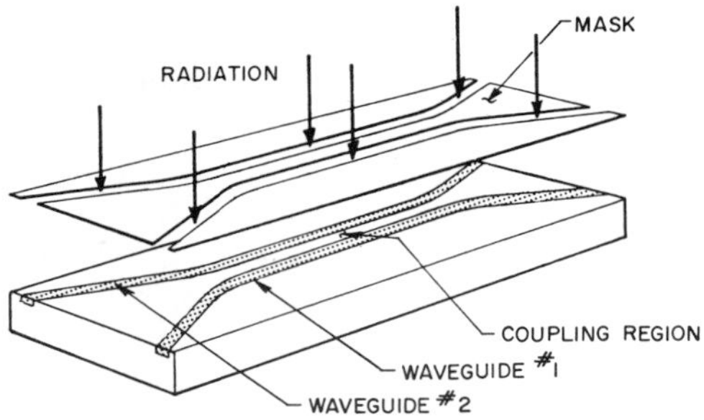

FIG. 8. Formation of printed waveguide coupler by irradiation.

waveguides or waveguide components. Although the discussion applies directly to fused silica, irradiation-induced index changes in other materials have been observed, and so the same or similar techniques could be applied to other materials as well. For example, Hines[11] reports a reduction of refractive index of about 0.2 for heavy-ion irradiation of silica-soda-lime glass.

In order to form a single waveguide of arbitrary width in the interior of a slab, the slab can be irradiated with high-energy protons to obtain great penetration, and the energy can then be reduced to sweep the narrow high-index channel towards the surface of the slab to achieve the desired size. In the illustration in Fig. 6, the initial penetration is x_1 and the final penetration is x_2, resulting in a width $d=x_1-x_2$.

It is possible to form two (or more) waveguides side-by-side with a slight variation of the above technique. In this case the energy is varied over two (or more) discrete energy ranges as illustrated in Fig. 7. It is thus possible to create a number of high-index channels which might form an array of waveguides. Alternatively, two individual waveguides might be formed sufficiently close together to form a directional coupler.

Another approach for forming waveguides with complex shapes is to irradiate through a mask as illustrated in Fig. 8. It is thus possible to form printed waveguides in a manner analogous to the printed-circuit techniques developed in the electronics field. The example in Fig. 8 illustrates how a printed-waveguide component, such as a directional coupler, might be formed.

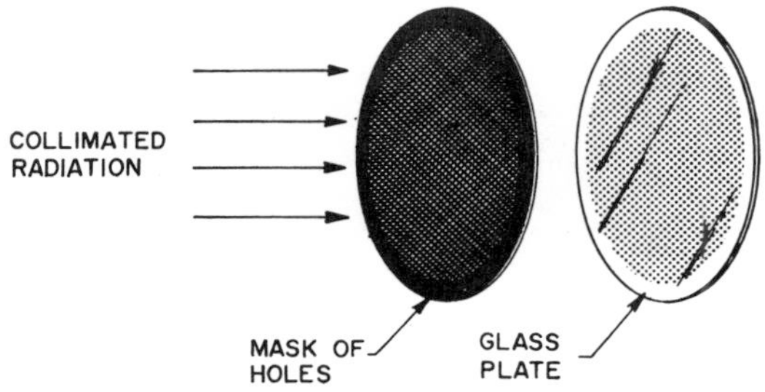

FIG. 9. Formation of a fiber-optic faceplate by irradiation.

A combination of the two techniques described could be used to form a two-dimensional array of waveguides. This array would be similar to conventional fiber-optic faceplates.

An alternative technique for forming a thin fiber-optic faceplate is illustrated in Fig. 9. In this arrangement, the axis of the waveguides is parallel to the direction of the proton beam rather than perpendicular as in all the previous cases. The procedure is to use a mask of small holes to irradiate regions corresponding to the waveguide cross-section with polyenergetic particles, so that the collision region extends through the entire thickness of the plate. This technique could be used to form faceplates up to a few millimeters thick if an appropriate mask can be obtained.

ACKNOWLEDGMENTS

The authors wish to thank J. Hirschfield of NASA–Goddard Space Flight Center and Professor B. Wooten of Worcester Polytechnic Institute who performed the irradiations.

Properties of Ion-Bombarded Fused Quartz for Integrated Optics

R. D. Standley, W. M. Gibson, and J. W. Rodgers

Development of techniques for production of carefully controlled, low-loss optical waveguides in solid dielectric materials is essential to development of integrated optical circuits for signal processing in future optical communications systems. Ion implantation offers an attractive possibility because of the refractive index and film thickness control possible by this technique. To evaluate this possibility we have investigated some of the optical properties of ion-bombarded fused quartz. A variety of ions ranging from helium ions to bismuth ions has been used. We have concentrated on refractive index and optical loss variations (on those implants into which a beam could be launched) as effected by (1) ion species and dose, (2) surface preparation, (3) surface temperature during bombardment, and (4) postbombardment annealing. This paper does not attempt to give an inclusive account of all the results obtained but principally discusses the best results so far, which are those using lithium ions. For lithium ion bombardment we have observed approximately linear variation of refractive index at 6328 Å with dose $n = n_0 + 2.1 \times 10^{-21}C$, where n_0 is the prebombardment value ($=1.458$ for fused quartz), and C is the ion concentration in ions/cm^3 ($C < 2.2 \times 10^{19}$). The optical absorption decreases significantly with increase in substrate temperature during implantation, and losses less than 0.2 dB/cm have been achieved. The refractive index change appears to be primarily due to disorder produced by the incident particles rather than a chemical doping effect as evidenced by postbombardment annealing studies.

I. Introduction

Several techniques exist for fabricating low-loss thin films required for dielectric waveguide integrated optical circuits.[1,2] In this paper, promising results achieved by ion bombardment of fused quartz are presented. Ion bombardment techniques, now well established for doping semiconductor materials, have the advantage of being easily controlled with the additional potential possibility of directly writing a circuit pattern using a well-focused beam.

Workers at Wheeler Laboratories were the first to create optical waveguides by proton bombardment of fused quartz.[3] However, no information was given on the losses in such guides. In our work, we have utilized other ions concentrating on refractive index and optical loss variations as affected by (1) ion species and dose, (2) surface preparation, (3) surface temperature during bombardment, and (4) postbombardment annealing.

For room temperature bombardments the refractive index varies as

$$n = 1.458 + 2.1 \times 10^{-21}C,$$

where C is the impurity ion concentration in particles/cm^3. The range of n corresponding to total ion dose variations of 10^{13}/cm^2 to 2.2×10^{15}/cm^2 spread uniformly over a 1-μ thickness is $1.46 \leq n \leq 1.51$. We have not attempted to reach a saturation limit by further increasing the dose.

II. Refractive Index of Ion-Bombarded Fused Quartz

A. Ion Distribution

It is well known that ion bombardment of *amorphous* materials at a specified ion energy yields a concentration of implanted ions whose depth profile is Gaussian in the region where the incident particle comes to rest.[4] The width of the Gaussian (the straggling) is determined by the variations in particle range which is, in turn, caused by random collisions in the material prior to ion stoppage. It is also energy dependent. By multiple bombardment at prescribed energies and appropriately chosen doses at each energy, it is possible to sum the resulting Gaussian profiles to derive a nearly uniform implanted ion concentration from the surface to a chosen maximum depth.[5] This, of course, assumes that range and straggling information are available for the particular bombarding ion and the material to be bombarded.

Assuming that a uniform ion concentration will result in a nearly uniform refractive index profile, such multiple energy bombardments were used in the lithium ion implantations of the present experiments.

The authors are with Bell Telephone Laboratories, Inc.; R. D. Standley is at Holmdel, New Jersey 07733; the other authors are at Murray Hill, New Jersey 07974.

Received 1 November 1971.

Reprinted with permission from *Appl. Opt.*, vol. 11, pp. 1313–1316, June 1972.

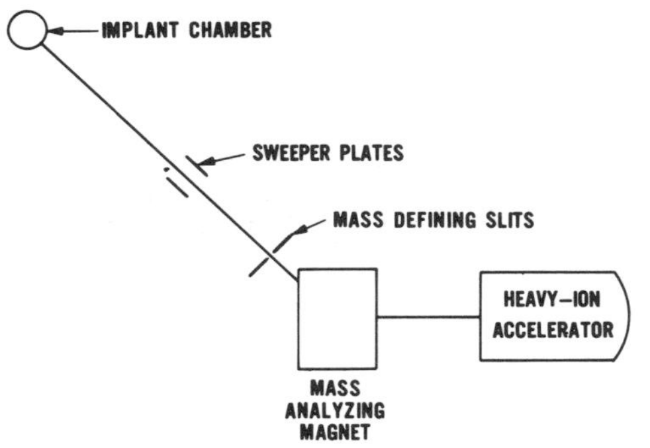

Fig. 1. Block diagram of ion implantation equipment.

However, annealing studies to be discussed indicate that at least the initial refractive index change is primarily due to atomic displacement, and thus these energies and doses should presumably be chosen to yield uniform concentration of atomic displacements. The necessary information is not yet available for a modification of the implantation schedule in this way.

B. Refractive Index Measurement

The method of refractive index measurement used in this work is described by Abeles[6]; in essence, reflectivity measurements vs angle of incidence are made on both the film-on-substrate sample and the substrate alone. Assuming a uniform, lossless film, these two reflectivities are equal when the angle of incidence, φ_i, equals Brewster's angle of the film. Difficulties encountered with this method due to lossy films are discussed by Heavens,[7] while the errors anticipated due to some particular index gradient profiles in a film have been briefly discussed by Goell and Standley.[2] The uncertainty of the index of refraction depth profile of the bombarded films introduces a potential error into the nidex measurements.

C. Choice of Ions

It appears that bombardment of fused quartz with virtually any energetic ion produces an initial increase in the refractive index, each ion species having its own range and straggling properties. Our studies began by implanting the glass formers but has been extended to include He, Li, C, P, Xe, Tl, and Bi. However, the most complete studies have been made for Li implants because of the solubility of Li in SiO_2, and the relatively long ion range is a convenience in forming layers of the desired thickness. All species suffer from surface damage due to charging and discharging of the dielectric during the implant or, in the case of He, what may be bubbling or He pocket formation within the film. Both effects increase scattering losses beyond tolerable limits. For reasons that are not completely understood, the lithium-bombarded samples show less surface damage. It may be that the lithium induces sufficient conductivity in the quartz to prevent excessive charge buildup.

In our early work, bombardment through thin gold films was attempted to reduce surface charging. Although the surface charging effect was avoided in this way, the resulting films were found to be extremely lossy after removal of the gold. It is not known if this is a result of the chemical treatment necessary to remove the gold or is related to gold mixed into the surface of the sample during irradiation as reported by Thompson.[8]

For the lithium bombardment, an aluminum strip coated with a conductive adhesive was applied to the sample edges to help bleed off any charge. As will be shown in Sec. V, substrate heating during bombardment also increases the conductivity of fused quartz and was found effective in reducing optical losses.

III. Losses in Ion-Bombarded Fused Quartz

A. Sources

Optical loss in thin film dielectrics is attributable to (1) bulk scattering and absorption and (2) scattering due to (a) imperfections internal to the film and (b) surface imperfections. At 6328 Å, Rayleigh scattering of fused quartz is on the order of a few dB to 10 dB/km as determined by Tynes.[9] The bulk absorption in the type fused quartz used in this work is about 1 dB/m. This can be increased by impurity doping due to the bombardment process. As noted previously, such effects were observed when attempting to bombard through a thin gold film. Scattering losses (1) result primarily from surface imperfections caused by poor substrate preparation or (2) can be induced by surface discharge during bombardment. Both effects appear significant for the samples used here. No attempt was made in this study to separate the two contributions; rather, the total loss in the high index region was determined.

B. Measurement

The loss of the thin films was determined by measurement of the intensity of light scattered from a beam in the film as described in Ref. 2. This method yields the total film loss, i.e., scattering plus absorption. The required beam was injected via a prism coupler.

IV. Ion Bombardment Apparatus

Figure 1 shows a block diagram of the ion implantation equipment used. The accelerator used is capable of a maximum potential of 300 kV and a maximum ion energy that may be proportionately higher if multiply charged ions are available in sufficient quantities. The analyzing magnets and defining slits serve to select the appropriate ion, while the deflection plates allow sweeping the beam over the desired area. The substrate to be implanted was located in the implant chamber. The beam current obtained for a typical Li^{+7} implant is a few μA so that about 5 min are required to complete a 10^{15}-particle/cm^2 implant over a 2-cm^2 area.

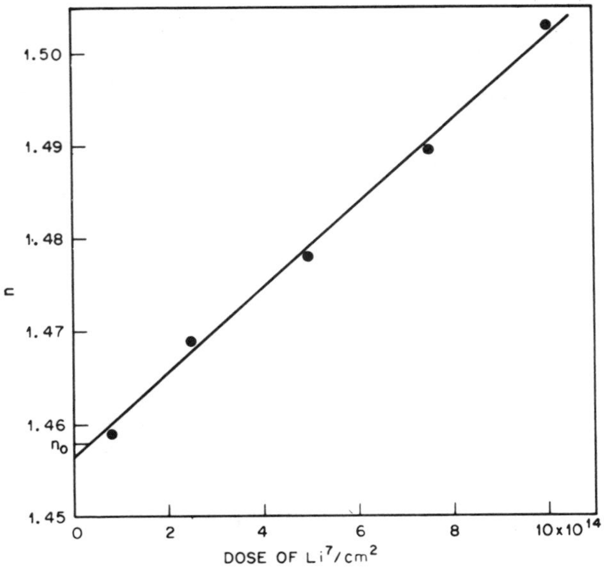

Fig. 2. Index of refraction of fused silica as a function of lithium ion implantation dose.

V. Experimental Results

A. Bombardment of Fused Quartz with Li^{+7} Ions

1. *Bombardment Parameters*

The object was to create a uniformly implanted region about 1 μm thick extending from the surface. To accomplish this, data on range and straggling vs energy of lithium into silicon were used.[5] The set of four energies with appropriate dose at each energy to accomplish uniform Li^{+7} density was as follows:

Energy (keV)	Dose (particles/cm²)	
200	10^{15}	
108	5.9×10^{14}	(Doping of 2.2×10^{19}
58	3.6×10^{14}	Li/cc over a region 1
32	2.4×10^{14}	μm in depth)

2. *Refractive Index*

For these bombardment conditions we have determined the refractive index of the bombarded region vs total dose as shown in Fig. 2. The refractive index varies approximately as $n = 1.458 + 2.1 \times 10^{-21}C$, where C is the implanted ion concentration in particle/cm³. This change is attributed to atomic displacement rather than chemical change since we have not approached the dosage required to see significant lithium silicate formation. Further, most of the effects of bombardment can be removed by annealing the samples. This was demonstrated by using 1-h anneal cycles at successively higher temperatures. A typical example of refractive index change vs annealing temperature is shown in Fig. 3. This information is, of course, very significant when subsequent processing involves operating at elevated temperatures. Room temperature stability appears to be good, but a more thorough study is required.

3. *Loss*

As stated in Sec. III-A, numerous factors are involved in determining the losses of these thin films. Our best results have been achieved using carefully polished and cleaned substrates. (Vitreosil was the only substrate material used.) Further, we have found that bombardment at elevated temperatures contributes to reducing loss. To cite a typical result, we implanted 10^{15} Li^7 particles/cm² with the substrate at 220°C. The initial refractive index was 1.493 with a loss of 1.8 dB/cm. After annealing for 1 h at 300°C, the refractive index was 1.468 and the loss below 0.2 dB/cm.

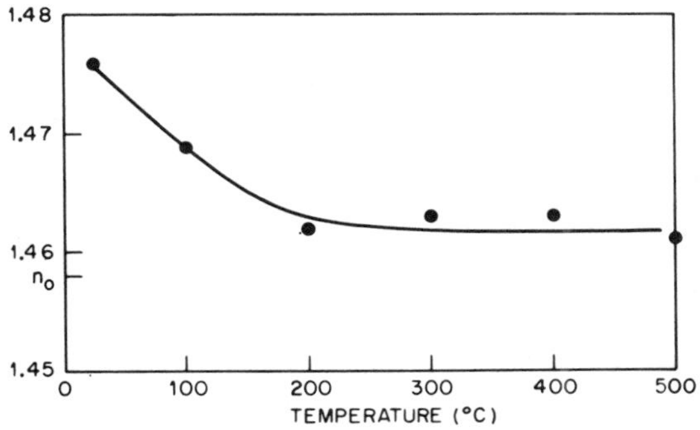

Fig. 3. Partial annealing of the index of refraction change produced by 5×10^{14} lithium/cm².

Table I. Data on Ion-Bombarded Fused Quartz

Ion	Energy (keV)	Implanted dose at the maximum energy indicated in each case (atoms/cm²)	Initial n	Remarks
	Sample	Sample		
Bismuth	A. 120	A.1. 10^{15}	1.58	
	80			
	50			
	25			
	B. 200	B.1. 3×10^{13}	1.465	
	100			
	50	0.2. 10^{15}	1.508	
Carbon	A. 250			
	200	A.1. 10^{15}	1.574	
	100	0.2 6×10^{14}	1.475	
	50			
	B. 300	B.1. 10^{15}	1.616	Beam launched
	162			
	87			
	47			
	25			
Thallium	A. 150	A.1. 10^{15}	1.489	
	100	0.2. 10^{14}	1.480	
Helium	A. 150	A.1. $\sim 10^{15}$	1.595	Beam launched
	90			
	48			
	23			

B. Bombardment with Other Ions

Table I shows refractive index data for samples bombarded by a variety of ions. The fact that beams have been launched into samples implanted with helium and carbon further indicates that refractive index changes due to atomic displacement and not chemical doping are the key factor in producing these initial changes.

C. Buried Layers

In an attempt to avoid the scattering losses resulting from imperfect surfaces, efforts are being made to produce a buried layer. If the region of high ion concentration can be placed well below the surface (a few micrometers) and this region tapered back to the surface at one end to provide a launching pad, most of the surface scattering should be avoided.

VI. Conclusions

Initial results on ion bombardment of fused quartz show that after annealing, the region of increased refractive index produced is sufficiently low loss to be of interest for integrated optics. We are currently attempting to create guiding regions a few micrometers below the substrate surface. The purpose of this effort is to determine whether scattering losses due to surface imperfections can be reduced.

Additional problems to be resolved are associated with the production of rectangular dielectric waveguides by ion implantation as contrasted with dielectric sheets as in studies discussed. One possibility under experimental investigation is the suitability of electron-resist materials as implant masks. The potentially attractive possibility of direct writing of a waveguide using an ion beam of a few micrometers in diameter is also under consideration, although it presents formidable ion optics problems at the implantation doses of interest.

References

1. S. E. Miller, Bell Syst. Tech. J. **48,** 2059 (1969).
2. J. E. Goell and R. D. Standley, Proc. IEEE **58,** 1504 (1970).
3. E. R. Schineller, R. Flam, and D. Wilmot, J. Opt. Soc. Am. **58,** 1171 (1968).
4. D. Grey, Ed., *American Institute of Physics Handbook* (McGraw-Hill, New York, 1963, pp. 8–20.
5. R. M. Allen, Electron. Lett. **5,** 111 (1969).
6. F. A. Abeles, in *Advanced Optical Techniques* (Wiley, New York, 1967), p. 143.
7. O. S. Heavens, *Optical Properties of Thin Solid Films* (Academic, New York, 1955).
8. M. W. Thompson, in *Proc. European Conference on Ion Implantation* (P. Peregrinus Ltd., Stevenage, Herts, England, 1970), p. 109.
9. A. R. Tynes, A. D. Pearson, and D. L. Bisbee, J. Opt. Soc. Am. **61,** 143 (1971).

Sputtered Glass Waveguide for Integrated Optical Circuits

By J. E. GOELL and R. D. STANDLEY

(Manuscript received September 16, 1969)

A series of papers which appeared in the September 1969 issue of the Bell System Technical Journal treated the theory of dielectric waveguides and stressed the potential use of such media for optical communication circuits.[1–4] Here we report on the realization of low-loss, thin glass films which can be used for circuit fabrication. Methods of preparing planar films and waveguides having rectangular cross section are described along with the techniques used in evaluating their optical characteristics.

The films we used for waveguide fabrication have been prepared by RF Sputtering of suitable glasses. The sputtering system used was oil-diffusion pumped and had five-inch diameter electrodes. Oxygen was used as the sputtering gas. The best films obtained to date were made by sputtering Corning 7059 glass. For convenience, in the early stages of this work, laboratory slides have been used as substrates. Necessary steps were taken to ensure that the substrates were clean.

The index of refraction of the films was measured to be 1.62 by determining Brewster's Angle for the films as described by Abeles.[5] From the color of the film and by interferometer methods the film thickness was found to be about 0.3 μm.

The transmission loss of the films was measured by two methods. Both use prisms to launch a light beam into the film.[6,7] In method 1 it is assumed that the scattering centers in the films are uniformly distributed. A fiber optic probe is then used to measure the intensity of the light scattered at right angles to the film. In method 2, the intensity of the output beam is measured as a function of launcher position along the film. Method 2 appears least accurate due to variations in launching efficiency as a function of prism movement. Method 1 works well to losses of the order of 1 db per cm. Below this level, the variability in the strength of the scattering centers makes reliable measurements difficult. An increase in film length would partially overcome the difficulty of measuring low level scattering from random centers.

Reprinted with permission from *Bell Syst. Tech. J.*, vol. 48. pp. 3445–3448, Dec. 1969.

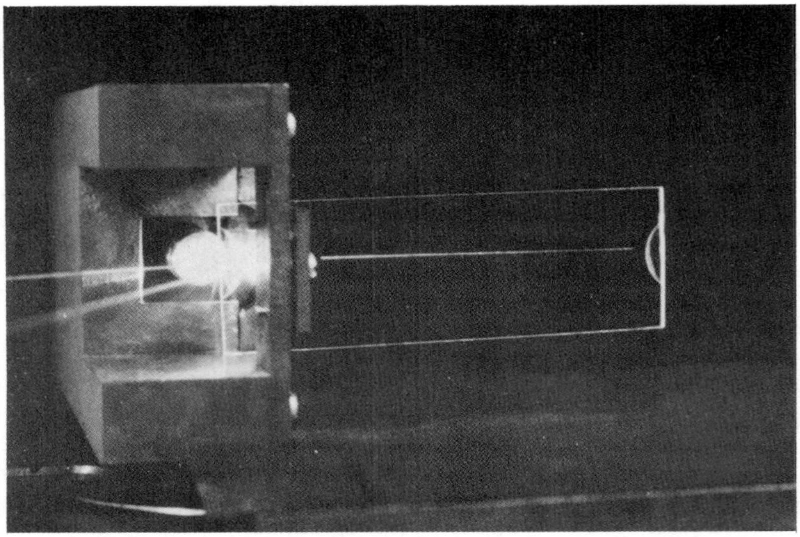

Fig. 1 — Light scattered from a beam propagating in a Corning 7059 glass film.

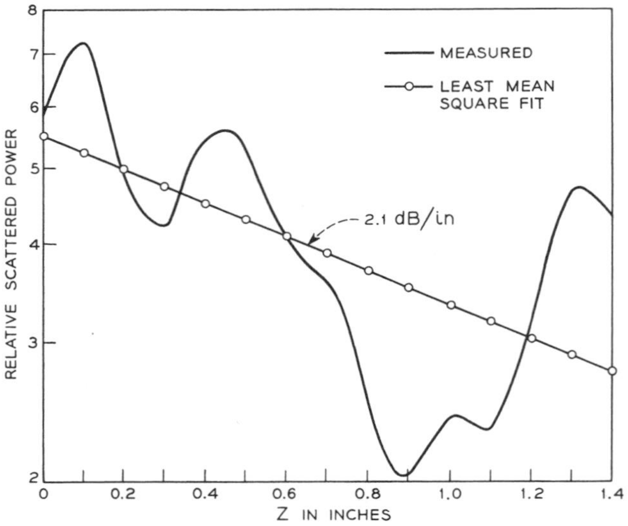

Fig. 2 — Relative scattered power versus length (7059 glass film).

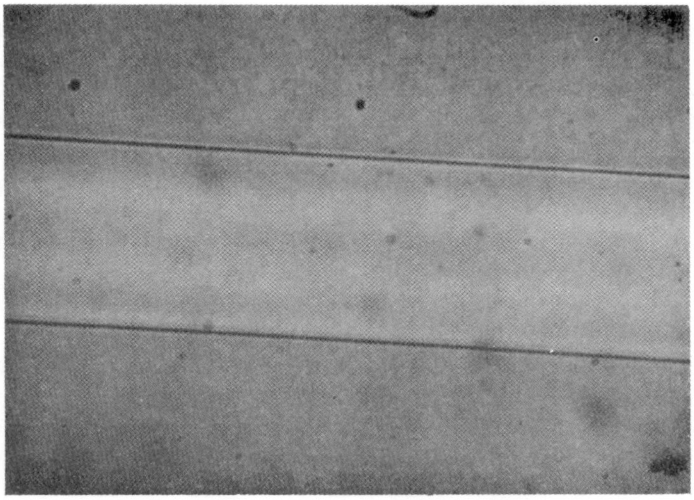

Fig. 3 — Section of a rectangular waveguide (×1000).

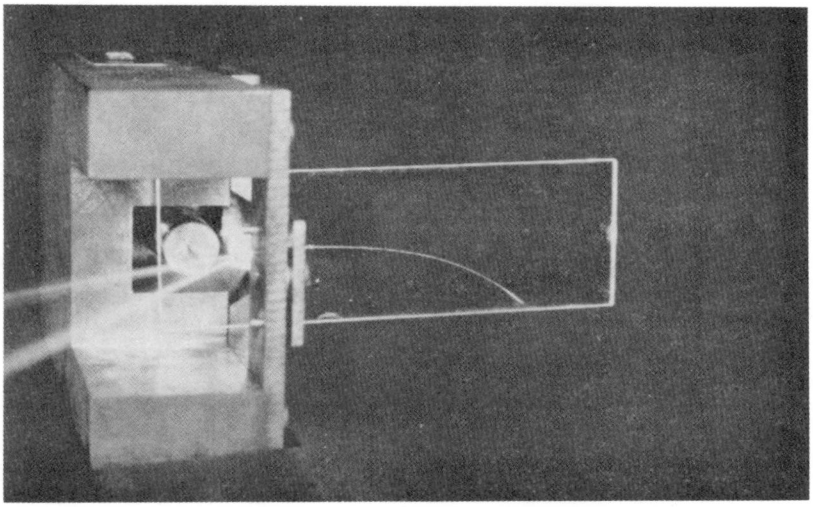

Fig. 4 — Light propagating in a curved section of rectangular waveguide.

Figure 1 is a picture of the light scattered from a beam propagating in the film. The intensity of scattered light as measured by the fiber optic probe is plotted in Figure 2. The average slope is less than -1 dB/cm. This result is in agreement with measurements made by the second method. The lack of uniformity of the scattered light intensity is due, at least in part, to inhomogeneities in the substrate. By using a higher quality substrate this source of scatter can be eliminated.

Curved sections of rectangular waveguides have been constructed from 7059 glass films by back-sputtering using quartz fibers as shadow masks. The waveguides were about 0.3 μm thick, 20 μm wide, and had a radius of curvature of about $\frac{1}{2}$ inch. A photograph of a typical section is shown in Figure 3. Figure 4 shows prism-launched light propagating in such a waveguide. Due to the small size of the waveguide our instrumentation will have to be improved before loss measurements can be made.

Our initial efforts have demonstrated the feasibility of using sputtered glass films and sputter etching in the fabrication of optical waveguides. This approach shows promise as a method of producing low-loss optical integrated circuits.

The authors are indebted to W. R. Sinclair for his valuable comments regarding the sputtering of glass films and the preparation of substrates, and to R. R. Murray who assisted in the preparation of the films and waveguides.

REFERENCES

1. Miller, S. E., "Integrated Optics: An Introduction," B.S.T.J., *48,* No. 7 (September 1969), pp. 2059–2069.
2. Marcatili, E. A. J., "Dielectric Rectangular Waveguide and Directional Coupler for Integrated Optics," B.S.T.J., *48,* No. 7 (September 1969), pp. 2071–2102.
3. Marcatili, E. A. J., "Bends in Optical Dielectric Guides," B.S.T.J., *48,* No. 7 (September 1969), pp. 2103–2132.
4. Goell, J. E., "A Circular-Harmonic Computer Analysis of Rectangular Dielectric Waveguides," B.S.T.J., *48,* No. 7 (September 1969), pp. 2133–2160.
5. Abeles, F., "Determination of the Refractive Index and the Thickness of Transparent Thin Films," J. Phys. Radium, *11,* No. 7 (July 1950), pp. 310–314.
6. Osterberg, H., and Smith, L. W., "Transmission of Optical Energy Along Surfaces: Part II, Inhomogeneous Media," J. Opt. Soc. Amer., *54,* No. 9 (September 1964), pp. 1078–1084.
7. Tien, P. K., Ulrich, R., and Martin, R. J., "Modes of Propagating Light Waves in Thin Deposited Semiconductor Films," Applied Physics Letters, *14,* No. 9 (May 1969), pp. 291–294.

Thin Organosilicon Films for Integrated Optics

P. K. Tien, G. Smolinsky, and R. J. Martin

The continued development of integrated optics is heavily dependent upon the availability of materials that are suitable for the construction of thin-film optical circuitry and devices. We report here an investigation of new films made by an rf discharge polymerization process of organic chemical monomers. We concentrate our discussion on films prepared from vinyltrimethylsilane and hexamethyldisiloxane. These films are smooth, tough, pinhole-free, transparent from 0.4 μm to 0.75 μm, and exhibit very low loss (<0.04 dB/cm) for light-wave propagation. More importantly, experiments demonstrate the possibility of controlling the refractive index of the films either by the mixing of the two monomers before deposition or by chemical treatment after the film is deposited. The use of the prism–film coupler for studying the refractive index of each material is discussed in detail.

I. Introduction

In this paper, we draw attention to the excellent properties of organosilicon films prepared by a plasma polymerization process as new media for guiding light waves in integrated optical circuitry. Several recent papers[1–4] have stressed the importance and wide variety of applications of thin-film optical circuits and devices that can be made a part of an integrated electronic circuit. It is likely that such integrated optoelectronic devices will be useful in new laser communication systems.

Clear, smooth, pinhole-free films suitable for light guides have been fabricated on glass substrates in an argon-monomer discharge from each of the following compounds: cyclohexane, acetone, hexene-1, isopropyl alcohol, perfluorocyclohexene, methyl methacrylate, diethyl ether, vinyltrimethylsilane, and hexamethyldisiloxane. Only the film prepared from methyl methacrylate proved to be unsatisfactory, since after several days' exposure to daylight, it turned brown. In this paper, we shall focus our attention exclusively on films prepared from vinyltrimethylsilane (VTMS) and hexamethyldisiloxane (HMDS) monomers.

At the 0.6328-μm line of the helium–neon laser, a typical VTMS film has a refractive index of 1.531, which is about 1% larger than that of ordinary glass (1.512). Since for light-wave propagation experiments we often prefer a film having a refractive index only slightly larger than that of the glass substrate, VTMS films are excellent for this purpose. A typical HMDS film has an index of 1.488, which is smaller than ordinary glass but larger than that of Corning 744 Pyrex glass (1.4704). A superposition of a VTMS on an HMDS film thus permits construction of optical circuitry in multilayers.

We will describe our measurements of the refractive indices and losses of VTMS and HMDS films, the preparation of films from mixtures of these two monomers with predetermined refractive indices, and some interesting phenomena observed on heat treatment of these films with oxygen, nitrogen, and chlorine and exposure to nitric oxide. Finally, we will explore important features of the prism–film coupler used as a tool in the study of these films.

II. Film Fabrication

The films were prepared in a vacuum chamber 20 cm in diameter and 21.2 cm in height. The substrate rested on the active electrode, a brass plate 15 cm in diameter fitted with a copper cooling coil. A congruent steel plate serving as a ground electrode was located 4 cm above and parallel to the active electrode. Monomers and argon were introduced into the system through separate variable leak valves.

In a typical run, the chamber was evacuated initially to a pressure of less than 2×10^{-6} Torr and the active electrode was maintained at 25°C with circulating water from a constant-temperature bath. Monomer and argon were metered into the chamber to a pressure of 0.3 Torr and 0.1 Torr, respectively, while the system was continuously pumped with a mechanical pump. A discharge was established by coupling 200 W of rf energy from a 13.56-MHz generator to the electrodes. Polymer formation began immedi-

The authors are with Bell Telephone Laboratories, Inc., Holmdel, New Jersey 07733.

Received 29 July 1971.

Reprinted with permission from *Appl. Opt.*, vol. 11, pp. 637–642, Mar. 1972.

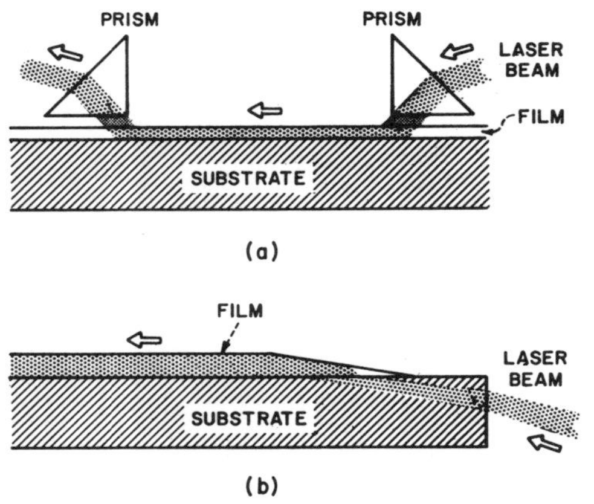

Fig. 1. (a) A method of coupling a laser beam into and out of a thin VTMS film. The prism at the right excites a light wave into the film, and a few centimeters away, the same light wave is coupled out of the film by the prism at the left. (b) Another method of coupling a laser beam into a VTMS film. The film forms a smooth tapered edge on the substrate. The laser beam enters into the film through this tapered edge.

ately, and VTMS film grew at the rate of about 2000 Å/min while HMDS film grew at about half that rate.

In another paper, Vasile and Smolinsky[5] discuss in detail the effect of pressure, rf power, argon concentration, and substrate temperature on film formation. In addition, they describe the physical and chemical properties of these films as well as elemental analysis, infrared spectra, and electron spin resonance experiments. Briefly, VTMS (1) and HMDS (2) undergo

$$CH_2 = CH{-}Si(CH_3)_3 \quad \textbf{(1)} \qquad (CH_3)_3Si{-}O{-}Si(CH_3)_3 \quad \textbf{(2)}$$

ionization and fragmentation in the discharge, producing a highly cross-linked and therefore insoluble polymer, which has a carbon–silicon ratio appreciably less than that of the starting monomer. Obviously polymerization does not occur by a conventional process such as addition of a radical or ionic species to a double band, producing a new reactive species which adds to another unsaturated molecule in a sequential process. Indeed, HMDS (2) has no unsaturated group in the molecule and thus polymerizes by a more complex mechanism. The study of chemistry in a discharge is presently an active field of research.

III. Propagation of Light in Thin Films

Typical light-wave propagation experiments performed with organosilicon compounds deposited on microscope glass slides 2.5 cm × 7.6 cm are shown in Fig. 1. In Fig. 1(a), a light wave was fed into the right side of a coated glass slide by a prism–film coupler,[6] propagated inside the film, and taken out of the left side of the slide by a second coupler. To serve as a light guide, the film must have a refractive index larger than that of the substrate. The film has several waveguide modes, but only one at a time was used for light-wave propagation. It is possible to excite any of the waveguide modes by varying the direction of the laser beam incident on the prism. Directions of the beam for proper excitation of waveguide modes are called the synchronous directions. Accurate knowledge of these synchronous directions allows an independent determination of the refractive index and thickness of the film.[6] It is not difficult to process a large number of films in a short time and still determine the refractive index to better than one part in a thousand. Thus we were able to detect minute changes in refractive indices, ranging from a few tenths of a percent to several percent, caused by suitable treatment of the films or by impurities in the materials.

Figure 1(b) illustrates another method[7] used for feeding a laser beam into a film. In this case, a mechanical mask is used, and the film was deposited with a tapered edge on the glass slide. A light wave entered first the glass substrate and then the film at the tapered edge. The VTMS film used here formed an excellent tapered edge.

The refractive indices of more than fifty VTMS and HMDS films deposited on glass substrates under different reaction conditions (varying combinations of rf power, total and partial pressure of monomer and/or argon) have been determined by the prism–coupler method.[6] These films ranged in thickness between 1 μm and 4 μm. The refractive indices of chemically or thermally untreated VTMS films (see Sec. VI) prepared under different conditions vary by only about 5 parts in 1000: $n_{0.6328\ \mu m}$ (red) = 1.5279–1.5356; $n_{0.5145\ \mu m}$ (green) = 1.5370–1.5440; $n_{0.4880\ \mu m}$ (blue) = 1.5398–1.5469. We believe that with good control, films can be produced within a very small tolerance in refractive index. A typical HMDS film has a refractive index of 1.4880 at 0.6328 μm, 1.4960 at 0.5145 μm, and 1.4996 at 0.4880 μm. Remeasurement of the refractive indices of some of the above films after several months' time indicated no noticeable change. The films can be washed in water without damage, but they are frequently scratched by the corners of the glass prism used in the experiments.

IV. Light Scattering, Film Defects, and Electron Micrographs

A light wave propagating inside a film appears as a bright streak to the eye. What we see is largely the light scattered from the defects in the film and from the roughness of the surfaces. Dirt particles on the film surface also scatter light. The light streak in a high quality film is barely visible and can only be photographed by long exposure. Such a photograph of a VTMS film is shown in Fig. 2(a). Except for a few specks of dirt, the helium–neon laser red streak shown is uniform in brightness with a clean sharp beam boundary over the entire length of the film indicating that the film is smooth, homogeneous, and free from defects. Photographs of light streaks excited in a VTMS film by the green (0.5145 μm) and blue (0.4880 μm) lines of the argon laser are shown in